4. wissenschaftliche Konferenz
der Gesellschaft Deutscher Naturforscher und Ärzte
Berlin 1967

# *Molecular Genetics*

Edited by H. G. Wittmann and H. Schuster

Springer-Verlag  Berlin · Heidelberg · New York  1968

ISBN-13: 978-3-540-04276-1     e-ISBN-13: 978-3-642-87534-2
DOI: 10.1007/ 978-3-642-87534-2

# Preface

A small informal symposium on "Molecular Genetics" was organized by us on behalf of the ,,Gesellschaft Deutscher Naturforscher und Ärzte" and took place in Berlin in October 1967. There were about 40 participants from Europe and the United States.

Molecular Genetics represents today an extraordinarily comprehensive research field. Therefore the organizers of the symposium had the choice either of limiting the meeting to a particular topic or of covering a wider selection of current problems. The latter alternative was chosen. The fields of research of the participants covered the broad range of scientific problems in which molecular genetics is nowadays involved: Genetic code; chemistry and biosynthesis of proteins; mutation, modification and reactivation of nucleic acids; biochemistry of regulation; complementation; structure; replication and function of viruses, etc.

The meeting took place in the Hotel Schweizerhof where the participants were also accommodated. This permitted close contact between the participants outside of the official program and allowed informal discussions, which started during the meetings, to be continued afterwards. Owing to the informal nature of these discussions, only a part of them could be included in this book.

Berlin, September 1968

H. G. WITTMANN

H. SCHUSTER

# Contents

# Participants

Dr. *W. Arber*, Institut de Biologie Moléculaire, Université de Genève, Genève, Switzerland.

Dr. *J. R. Beckwith*, Dept. of Bacteriology and Immunology, Harvard Medical School, Boston, Mass., USA.

Dr. *G. Bertani*, Institution för Mikrobiologisk Genetik, Karolinska Institutet, Stockholm, Sweden.

Dr. *K. Beyreuther*, Max-Planck-Institut für Biochemie, München, Germany

Dr. *L. Bockstahler*, Max-Planck-Institut für Biologie, Tübingen, Germany.

Dr. *F. Bonhoeffer*, Max-Planck-Institut für Virusforschung, Tübingen, Germany.

Dr. *H. Delius*, Institut de Biologie Moléculaire, Université de Genève, Genève, Switzerland.

Dr. *J. R. S. Fincham*, The University of Leeds, Leeds, Great Britain.

Dr. *A. Garen*, Yale University, Dept. of Molecular Biophysics, New Haven, Conn., USA.

Prof. Dr. *A. Gierer*, Max-Planck-Institut für Virusforschung, Tübingen, Germany.

Dr. *D. Grünberger*, Institute of Organic Chemistry and Biochemistry, Czechoslovak Academy of Sciences, Prague, Czechoslovakia.

Prof. Dr. *U. Henning*, Max-Planck-Institut für Biologie, Tübingen, Germany

Prof. Dr. *H. Hoffmann-Berling*, Max-Planck-Institut für medizinische Forschung, Heidelberg, Germany.

Prof. Dr. *P. H. Hofschneider*, Max-Planck-Institut für Biochemie, München, Germany.

Dr. *P. Howard-Flanders*, Yale University, School of Medicine, New Haven, Conn., USA.

Dr. *N. O. Kjeldgaard*, University Institute of Microbiology, University of Copenhagen, Copenhagen, Denmark

Dr. *D. Marvin*, Yale University, Dept. of Biophysics, New Haven, Conn., USA.

Priv.-Doz. Dr. *J. H. Matthaei*, Max-Planck-Institut für experimentelle Medizin, Göttingen, Germany.

Prof. Dr. *G. Melchers*, Max-Planck-Institut für Biologie, Tübingen, Germany.

Dr. *G. Nass*, Max-Planck-Institut für molekulare Genetik, Berlin, Germany.

Dr. *F. C. Neidhardt*, Dept. of Biological Sciences, Purdue University, Lafayette, Indiana, USA.

Dr. *M. Nomura*, Laboratory of Genetics, University of Wisconsin, Madison, Wisconsin, USA.

Prof. Dr. *A. Novick*, Institute Pasteur, Paris, France.

Dr. *W. Rombauts*, Departement Biochemie, Universität Leuven, Leuven, Belgium.

Dr. *I. Rychlik*, Institute of Organic Chemistry and Biochemistry, Czechoslovak Academy of Sciences, Prague, Czechoslovakia.

Dr. *H. L. Sänger*, Universität Gießen, Institut für Pflanzenpathologie, Gießen, Germany.

Priv.-Doz. Dr. *C. Scholtissek*, Universität Frankfurt, Institut für Virologie, Gießen, Germany.

Dr. *H. Schuster*, Max-Planck-Institut für molekulare Genetik, Berlin, Germany.

Dr. *J. Seehafer*, Max-Planck-Institut für molekulare Genetik, Berlin, Germany.

Prof. *P. Starlinger*, Universität Köln, Institut für Genetik, Köln, Germany.

Dr. *G. Stöffler*, Max-Planck-Institut für molekulare Genetik, Berlin, Germany.

Dr. *I. Tessman*, Dept. of Biological Sciences, Purdue University, Lafayette, Indiana, USA.

Dr. *H. Tichy*, Max-Planck-Institut für Biologie, Tübingen, Germany.

Priv.-Doz. Dr. *T. A. Trautner*, Max-Planck-Institut für molekulare Genetik, Berlin, Germany.

Prof. *W. Vielmetter*, Universität Köln, Institut für Genetik, Köln, Germany.

Prof. Dr. *U. Winkler*, Ruhr-Universität Bochum, Arbeitsgruppe Genetik der Mikroorganismen, Bochum, Germany.

Prof. Dr. *H. G. Wittmann*, Max-Planck-Institut für molekulare Genetik, Berlin, Germany.

Dr. *W. Zillig*, Max-Planck-Institut für Biochemie, München, Germany.

# Complementation and Dominance Relationships between Protein Subunits

By

J. R. S. FINCHAM

The University, Leeds

When it was thought that proteins were the primary products of gene translation it was difficult to see how alleles could interact to produce something other than the products of their individual action. Now that it is known that most enzymes are oligomeric in structure, consisting often of a number of normally identical polypeptide chains, it is apparent that the joint action of two alleles may produce hybrid protein. Many examples of hybrid enzyme formation in heterokaryons or heterozygous diploids are known, and such hybridization is thought to be the general explanation for allelic complementation (for review see FINCHAM, 1966). It is the purpose of this paper to show how this sort of allelic interaction can act negatively as well as positively, so as to bring about so-called "negative complementation" of mutant alleles or even dominance of a defective mutant over a functional wild type allele.

The formation of mutant-wild protein hybrids has been demonstrated by GAREN and GAREN (1963) and analyzed very completely by FAN, SCHLESINGER, TORRIANI, BARRETT, and LEVINTHAL (1966) in the alkaline phosphatase system of *Escherichia coli*. Alkaline phosphatase is a dimer, and so only one kind of hybrid protein is possible with any two *p* alleles. FAN et al. isolated several kinds of hybrid, including some formed by interaction of mutants and wild type, and showed that, where the mutant by itself produced a thermolabile protein, the hybrid was generally less stable than the wild type enzyme and, to this extent, the product of the normal allele was "spoiled" by the presence of the mutant. So far as enzyme activity was concerned, however, the wild type monomers remained fully functional when hybridized with mutant monomers. Where the mutant monomer was one capable of complementation it could be complemented by the wild monomer in a hybrid but the opposite phenomenon — the depression of the activity of wild type by mutant monomers — was not recorded.

In the case of the NADP-linked glutamate dehydrogenase (GDH) of *Neurospora crassa*, controlled by the gene *am*, I and my colleagues,

Drs. A. Coddington, and T. K. Sundaram, have encountered a rather more complex situation. The complexity is in part due to the fact that this enzyme is a higher oligomer. Preliminary electron microscope pictures (obtained by Dr. Arthur Rowe of Leicester University) appear to show four subunits arranged in a tetrahedron, but on the basis of amino acid analysis and peptide mapping studies we believe that there are in reality eight identical chemical subunits which may, therefore, be arranged in four dimers. Even if we suppose that there are only four subunits capable of idependent reassortment to form hybrids there are still three possible hybrids (of composition 3:1, 2:2 and 1:3) which can be formed from any two different monomers. We have, in fact, obtained evidence that at least this many kinds of hybrid can be formed from the products of two complementary *am* alleles (Coddington, Sundaram, and Fincham, 1966) either *in vivo* (in heterokaryons) or *in vitro* (following freezing and thawing of a mixture of the purified proteins in the presence of sodium chloride).

Our analysis of the effects of mutations on GDH activity depends on the use of several different systems for assaying the enzyme. Neurospora GDH is an allosteric enzyme (West, Tuveson, Barratt, and Fincham, 1967) which, at pH 7.2, is inactivated by NADPH alone but activated by NADPH plus α-oxoglutarate, or by various dicarboxylic acids including glutamate especially in the presence of NADP. Activation occurs rapidly at pH values above 7.5 even in the absence of substrates or activators. Several of the mutant forms of GDH are ineffective *in vivo* because their inactive conformation is unduly stable. These mutant enzymes can, however, be activated *in vitro* by more extreme versions of the procedures which activate the wild type enzyme. Our assay system A, in which the enzyme is added as final addition to an α-oxoglutarate − NADPH − $NH_4Cl$ mixture, gives full activity practically from zero time with the normal enzyme but essentially no activity for several minutes with the mutant proteins formed by alleles $am^2$, $am^3$ and $am^{19}$. Assay system C involves preincubation of the enzyme at pH 8.5 with NADP and a high concentration of glutamate; this procedure activates $am^3$ enzyme completely and $am^2$ enzyme partially but gives very little activity with the $am^{19}$ protein. System S involves preincubation at pH 8.5 with NADP and a high concentration of succinate for an hour, and this treatment succeeds in activating the $am^{19}$ enzyme to a very large extent. In the case of $am^{19}$ enzyme we have strong evidence that the activation is due to a conformational change in that it is fully reversible and is not accompanied by any detectable change in sedimentation coefficient. The activation is accompanied by an alteration in net surface charge as shown by a change from an abnormal to a normal electrophoretic mobility (Sundaram, and Fincham, 1964). We take the degree of electrophoretic abnormality as an indication of how close the $am^{19}$ protein is to the conformation characteristic of the active wild type enzyme.

ALAN CODDINGTON (1966) was the first to look at *in vitro* mutant-wild interactions in the GDH system. His results showed an interesting contrast between the behaviour of $am^1$ mutant protein, on the one hand, and $am^2$ and $am^3$ mutant proteins on the other. Mutant $am^1$ produces a protein which never has any activity in any of our assay systems. The trouble with $am^1$ protein seems likely to be not an upset in its allosteric balance — which may be essentially normal — but an irreparable loss of some component of the active centre. This guess is borne out by the fact that it seems quite neutral in hybrids with wild type — it neither contributes activity of its own nor does it detract from that of wild type. It is relevant here to recall that $am^1$ complements $am^2$, $am^3$ and $am^{19}$, presumably by correcting the conformation of the potentially active monomers produced by these mutants (FINCHAM and STADLER, 1965). This too suggests that $am^1$ tends to promote a relatively "correct" conformation in hybrids in which it occurs. In contrast, the $am^2$ and $am^3$ proteins at different concentrations with a constant lower concentration of wild type enzyme, actually depress the activity in system A, the depression becoming greater the greater the excess of mutant protein. The "lost" activity could however still be realized in assay system C. It seems reasonable to conclude that mutant-wild hybrids with a preponderance of $am^2$ or $am^3$ subunits tends to resemble these mutant proteins in their conformational properties and require the activation treatment of system C; in other words, that the wild type monomers, when outnumbered, are forced to conform to the anomalous behaviour of the mutant monomers.

SUNDARAM and I (J. molec. Biol., in the press, 1967) have followed up these observations by studying the interaction of wild type with $am^{19}$ protein *in vitro*. A similar experiment to those which CODDINGTON performed with $am^2$ and $am^3$ proteins showed that $am^{19}$ protein, too, has a negative effect on GDH activity when hybridized in excess with wild type (Fig. 1). Analysis of a $2:1$ $am^{19} - am^+$ mixture by electrophoresis showed that, as a result of the hybridization procedure, the slower (originally wild type) protein component became fainter and was somewhat modified in its electrophoretic mobility so as to make it run closer to the $am^{19}$ position. The faster component, on the other hand, appeared to gain material (Fig. 1).

This impression was confirmed by an experiment in which the $am^+$ protein was labelled with [35]S, and the electrophoresis gels were sliced, dried and subjected to radioautography by the procedure of FAIRBANKS, LEVINTHAL, and REEDER (1965). The result (SUNDARAM, and FINCHAM, 1967) makes it quite clear that as a result of the hybridization a substantial part of the $am^+$ material is incorporated into hybrids whose

1*

electrophoretic properties, and, by implication, conformation, approximate to those of *am*[19] protein. These hybrids are presumably electrophoretically abnormal because they contain a sufficient proportion of *am*[19] monomers to override the conformation which the *am*[+] monomers would have if aggregated with themselves. The *am*[19] component may gain in ease of activation (note the increase in system C activity in Fig. 1) but the *am*[+] monomers decidely lose.

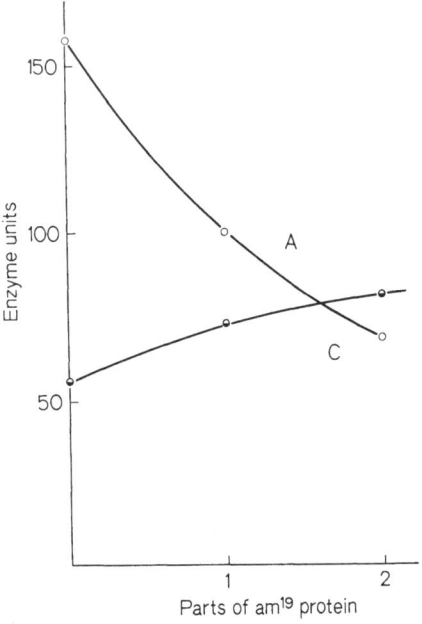

Fig. 1. The effect on the activity of wild type GDH of hybridizing with different proportions of *am*[19] mutant protein. Assay system A measures enzyme which is already active at the start of the assay; system C measures, in addition, enzyme capable of being activated by incubation with glutamate and NADP

We have, as yet, no evidence that similar hybrids are made *in vivo* in *am*[+] + *am*[19] heterokaryons. We have no doubt that they must be made since, so far as complementary interaction between mutant proteins is concerned, the *in vitro* reaction mixtures seem to mirror the *in vivo* situation (CODDINGTON et al., 1966). However, since such a small proportion of the normal level of GDH will suffice to sustain a normal rate of amination, it is doubtful whether the formation of such hybrids would have a noticeable effect on growth. Thus the case under discussion is not one in which the mutant is visibly dominant. It is easy to see, however, that an otherwise similar mechanism operating in a case in which the enzyme level was more critical could result in an effective dominance of the mutant over the wild type. This is one way in which the type of dominant defective mutant which Muller (1932) called an *antimorph* might work. In the *am* series of mutants in Neurospora however, it is impossible to categorize mutants according to whether they act positively or negatively. Depending on the nature of the other allele and the proportion in which it is present a given mutant can either act positively, as in complementation, or negatively, as in the interaction with wild type discussed here. This ambiguous behaviour may be expected of mutations which have their effects through shifting the equilibrium between the active and inactive forms of an allosteric enzyme.

## References

GODDINGTON, A.: Biochem. J. 99, 9c (1966).
—, J. R. S. FINCHAM, and T. K. SUNDARAM: J. molec. Biol. 17, 503 (1966).
FAIRBANKS, G. JR., C. LEVINTHAL, and R. H. REEDER: Biochem. biophys. Res. Commun. 20, 393 (1965).
FAN, D. P., M. J. SCHLESINGER, A. TORRIANI, K. J. BARRETT, and C. LEVINTHAL: J. molec. Biol. 15, 32 (1966).
FINCHAM, J. R. S.: Genetic complementation. New York: W. A. Benjamin 1966.
—, and D. R. STADLER: Genet. Res. 6, 121 (1965).
GAREN, A., and S. GAREN: J. molec. Biol. 7, 13 (1963).
MULLER, H. J.: Proc. 6th. Int. Cong. Genetics 1, 213 (1932).
SUNDARAM, T. K., and J. R. S. FINCHAM: J. molec. Biol. 10, 423 (1964).
— — J. molec. Biol. 29, 433 (1967).
WEST, D. J., R. W. TUVESON, R. W. BARRATT, and J. R. S. FINCHAM: J. biol. Chem. 9, 2134 (1967).

## Discussion

*Starlinger:* Your thermal inactivation curves showed different slopes: Do you think that your hybrid molecules are heterogeneous with respect to thermolability?
*Fincham:* Yes.

# Ribosomal Proteins of E. coli and Yeast

By

E. Kaltschmidt, V. Rudloff, G. Stöffler, A. Chersi, M. Dzionara,
D. Donner, and H. G. Wittmann

Max-Planck-Institut für Molekulare Genetik Berlin, Germany

Initial steps characterizing the nature of proteins present in E. coli ribosomes dealt with the dissociation of the ribonucleoprotein particles, solubilization of the proteins and end-group analyses of bulk-proteins of the two ribosomal subunits. The starch-gel electrophoreses of these early experiments revealed a rather complex pattern (Waller and Harris, 1961). These results were confirmed by later experiments applying the technique of polyacrylamide gel-electrophoresis (Leboy, Cox, and Flaks, 1964; Traut, 1966). A first chromatographic fractionation procedure using carboxymethylcellulose, also reported by Waller and Harris, showed only poor resolution. Changing the original conditions Waller was able to achieve a separation of the acid-soluble ribosomal proteins into twelf subgroups (Waller, 1964). Further modifications of this procedure resulted finally in profiles with very high resolution when proteins of native and altered ribosomal subunits were subjected to carboxymethylcellulose chromatography (Traut, Moore, Delius, Noller, and Tissieres, 1967; Otaka, Itoh, and Osawa, 1967; Delius, Traut, Moore, Noller, and Pearson, this volume).

On a quite different approach fractional dissociation of ribosomes, and ribosomal subunits, respectively, by differential extraction, salt precipitation, and chromatography yielded split-proteins and functionally inactive ribosomal fragments (Lerman, Spirin, Gavrilova, and Golov, 1966; Traub, Nomura, and Tu, 1966; Marcot-Queiroz and Monier, 1966; Salas, Hille, Last, Wahba, and Ochoa, 1967). A re-assembly of ribosome-like particles was reported on mixing the split-proteins and subparticles, the reconstituted particles being functionally active in an in-vitro assay (Spirin, Lerman, Gavrilova, and Belitsina, 1966; Hosokawa, Fujimura, and Nomura, 1966; Staehelin and Meselson, 1966). Only recently, the restoration of biological activity could be related to specific components of the split-proteins (Nomura, this volume).

Although these investigations have centered on ribosomal proteins of E. coli, some preliminary examples of studies with ribosomes from other organism are available, so with proteins of yeast ribosomes (HORSTMANN, 1967) and rat liver ribosomal proteins (HAMILTON and RUTH, 1967).

Most recently ribosomal proteins of Neurospora crassa were characterized (ALBERGHINA and SUSKIND, 1967) mainly with immunochemical methods.

Our studies reported here are concerned with the physical and chemical characterization of 19 purified proteins of E. coli ribosomes and of 17 proteins of yeast ribosomes.

E. coli B was purchased from Fallek Inc., N. Y. The cells were broken by grinding with alumina and the ribosomes were isolated by differential centrifugation. Ribosomal protein was prepared by the acetic acid method WALLER and HARRIS, 1961). Freeze dried ribosomal protein was dissolved in potassium acetate buffer containing 8 molar urea, and the various proteins were separated by preparative polyacrylamide electrophoresis, according to a method developed by KALTSCHMIDT and WITTMANN (manuscript in preparation).

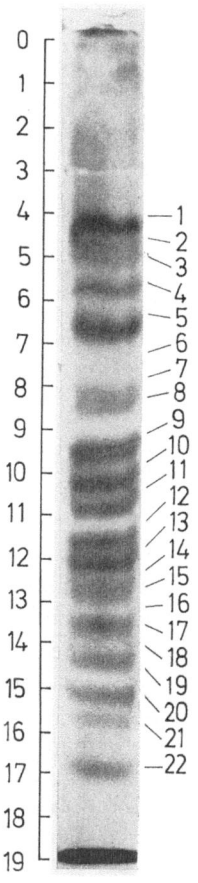

Fig. 1. Separation pattern of ribosomal proteins of E. coli by preparative polyacrylamide gel electrophoresis. A cross section through a polyacrylamide block ($17 \times 120 \times 200$ mm) is shown; the gel piece was stained with amido black. For more details see KALTSCHMIDT and WITTMANN (manuscript in preparation) and KALTSCHMIDT et al. (1967). The numbers of proteins refer to these used in the text

The separation pattern of ribosomal protein on the acrylamide gel is shown in Fig. 1. 22 main protein bands were cut out of the gel and the proteins were extracted by homogenizing the gel pieces with 0.5 % acetic acid and freed of polyacrylamide by centrifugation. The purity of these isolated proteins was checked by analytical disc-electrophoresis; impure proteins contaminated with more than 5 % of a neighbouring protein-band were rerun on the preparative acrylamide gel electrophoresis for further purification. Remaining traces of contaminating acrylamide and salts were removed from the pure proteins by gelfiltration on Sephadex G-25 or Bio-Gel P-10, equilibrated with 10 % acetic acid.

Fig. 2 shows the purity of the isolated proteins which were used for further experiments. The proteins were named according to their migration RF in the acrylamide gel with numbers 1 to 22.

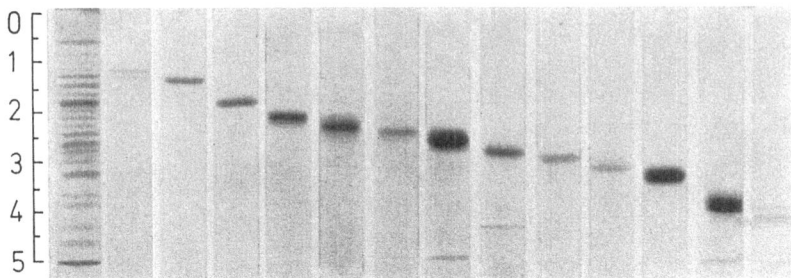

Fig. 2. Acrylamide gel disc electrophoresis (LEBOY et al., 1964; REISFELD, LEWIS, and WILLIAMS, 1962) of isolated and purified protein bands of E. coli ribosomal proteins, used for amino acid analysis, peptide mapping and determination of molecular weights

## Amino Acid Composition

The amino acid composition of 16 purified ribosomal proteins of E. coli is shown in Table 1. Each figure represents the mean value of at least two amino acid analyses.

In Fig. 3 each graph concerns one protein and the amount of the various amino acids present in this protein is symbolized by the height of the columns expressed in mole %.

The amount of each amino acid present in the different proteins of E. coli ribosomes is illustrated in Fig. 4. From the values given in Table 1 and the graphs in Fig. 3 and Fig. 4 it is evident that there is a remarkable over all agreement between the different proteins, however, also distinct differences are noticable and it is obvious that closely migrating proteins can have quite different amino acid compositions as for instance proteins No. 7 and No. 8. However, there are also examples for a similar amino acid composition of very different migrating proteins as number 9 and number 16.

The amino acid composition of 8 proteins, isolated from the 30S subunit, analyzed by TRAUT et al. (1967) revealed a rather pronounced difference between individual proteins. At the moment the reason for this discrepancy is not explainable.

Comparing the amino acid composition of one of the E. coli ribosomal proteins (No. 12) with some completely unrelated proteins (Fig. 5; expressed in the same way as in Fig. 4 for the different E. coli ribosomal proteins) it can be demonstrated that here, with randomly chosen pro-

Table 1. *Amino acid compositions of 16 ribosomal proteins from E. coli* (in mole %)

| Amino acid | Protein number | | | | | | | | | | | | | | | |
|---|---|---|---|---|---|---|---|---|---|---|---|---|---|---|---|---|
| | 1 | 3 | 4 | 5 | 7 | 8 | 9 | 10 | 11 | 12 | 13 | 14 | 16 | 17 | 19 | 20 |
| Lys | 9.14 | 10.15 | 9.30 | 9.03 | 8.85 | 11.36 | 10.43 | 8.07 | 9.06 | 11.04 | 12.79 | 13.35 | 9.15 | 13.68 | 12.35 | 10.51 |
| His | 2.00 | 1.90 | 2.09 | 2.03 | 2.70 | 3.41 | 1.93 | 1.59 | 2.40 | 2.66 | 2.03 | 2.07 | 3.49 | 4.15 | 3.58 | 2.80 |
| Arg | 6.87 | 8.25 | 7.60 | 6.58 | 8.98 | 9.85 | 9.84 | 11.28 | 10.57 | 8.37 | 8.07 | 6.41 | 11.10 | 8.78 | 9.29 | 8.90 |
| Met | 1.62 | 1.28 | 1.56 | 2.09 | 1.86 | 1.20 | 1.07 | 1.09 | 0.83 | 1.34 | 1.00 | 0.98 | 0.91 | 1.40 | 0.80 | 0.85 |
| Asp | 9.29 | 8.28 | 8.37 | 8.14 | 8.05 | 5.66 | 6.92 | 7.51 | 6.79 | 9.17 | 8.88 | 9.58 | 8.33 | 7.20 | 7.62 | 7.13 |
| Thr | 5.70 | 6.13 | 5.41 | 5.56 | 5.67 | 6.37 | 6.00 | 4.95 | 4.69 | 5.10 | 4.74 | 3.48 | 5.65 | 6.27 | 4.19 | 5.94 |
| Ser | 4.59 | 5.00 | 4.77 | 4.75 | 4.91 | 2.87 | 4.44 | 5.96 | 5.83 | 5.55 | 5.09 | 4.53 | 5.60 | 5.04 | 6.66 | 5.02 |
| Glu | 9.48 | 10.25 | 10.68 | 10.09 | 10.52 | 9.83 | 9.82 | 11.21 | 11.97 | 7.11 | 10.10 | 10.78 | 9.65 | 7.59 | 8.20 | 15.08 |
| Pro | 4.38 | 3.58 | 4.36 | 3.91 | 4.14 | 3.80 | 3.71 | 3.32 | 2.90 | 2.69 | 3.03 | 3.43 | 3.77 | 4.66 | 2.23 | 2.09 |
| Gly | 10.39 | 10.34 | 8.93 | 10.81 | 7.99 | 8.77 | 10.06 | 8.90 | 7.00 | 12.56 | 8.60 | 8.52 | 7.33 | 8.18 | 13.07 | 6.63 |
| Ala | 12.05 | 10.10 | 11.73 | 12.43 | 11.26 | 10.97 | 10.74 | 10.72 | 12.55 | 10.09 | 10.46 | 10.32 | 9.88 | 8.40 | 11.27 | 9.33 |
| Val | 9.66 | 9.91 | 8.00 | 9.84 | 8.87 | 8.97 | 7.66 | 7.31 | 8.50 | 8.62 | 9.05 | 9.60 | 7.33 | 7.69 | 6.56 | 6.75 |
| Ile | 4.47 | 4.40 | 4.93 | 5.15 | 4.47 | 5.51 | 5.39 | 4.90 | 5.31 | 6.51 | 6.15 | 6.09 | 5.14 | 5.01 | 4.26 | 3.80 |
| Leu | 6.52 | 6.31 | 7.21 | 6.05 | 7.25 | 7.44 | 7.22 | 7.30 | 7.00 | 6.54 | 6.27 | 5.28 | 7.94 | 7.07 | 4.92 | 12.68 |
| Tyr | 1.37 | 1.26 | 1.95 | 0.96 | 1.84 | 1.32 | 1.62 | 2.61 | 1.67 | 1.09 | 0.84 | 1.36 | 1.45 | 1.30 | 0.55 | 0.30 |
| Phe | 2.46 | 2.91 | 3.15 | 2.61 | 2.65 | 2.68 | 3.20 | 3.29 | 2.94 | 1.59 | 2.88 | 4.25 | 3.30 | 3.60 | 4.47 | 2.28 |

For amino acid analysis the lyophylized proteins, desalted by gel filtration an Sephadex G-25 or Bio-Gel P-10, were hydrolyzed with 2.0 ml of 6 N HCl in sealed evacuated tubes for 18 h at 110° C, without any further chemical modification. Analyses were run with a Beckman-Unichrom automatic amino acid analyzer. Methionine was calculated by summarizing the values for methionine, methionine sulfone and methionine sulfoxide. Cystine, tryptophane and amides were not determined. Calculation of amino acid residues was expressed in mole-%. Each value represents the mean of duplicate runs.

10    E. Kaltschmidt et al.:

teins, the differences in the amino acid composition are very obvious, and
ribosomal proteins are rather similar.

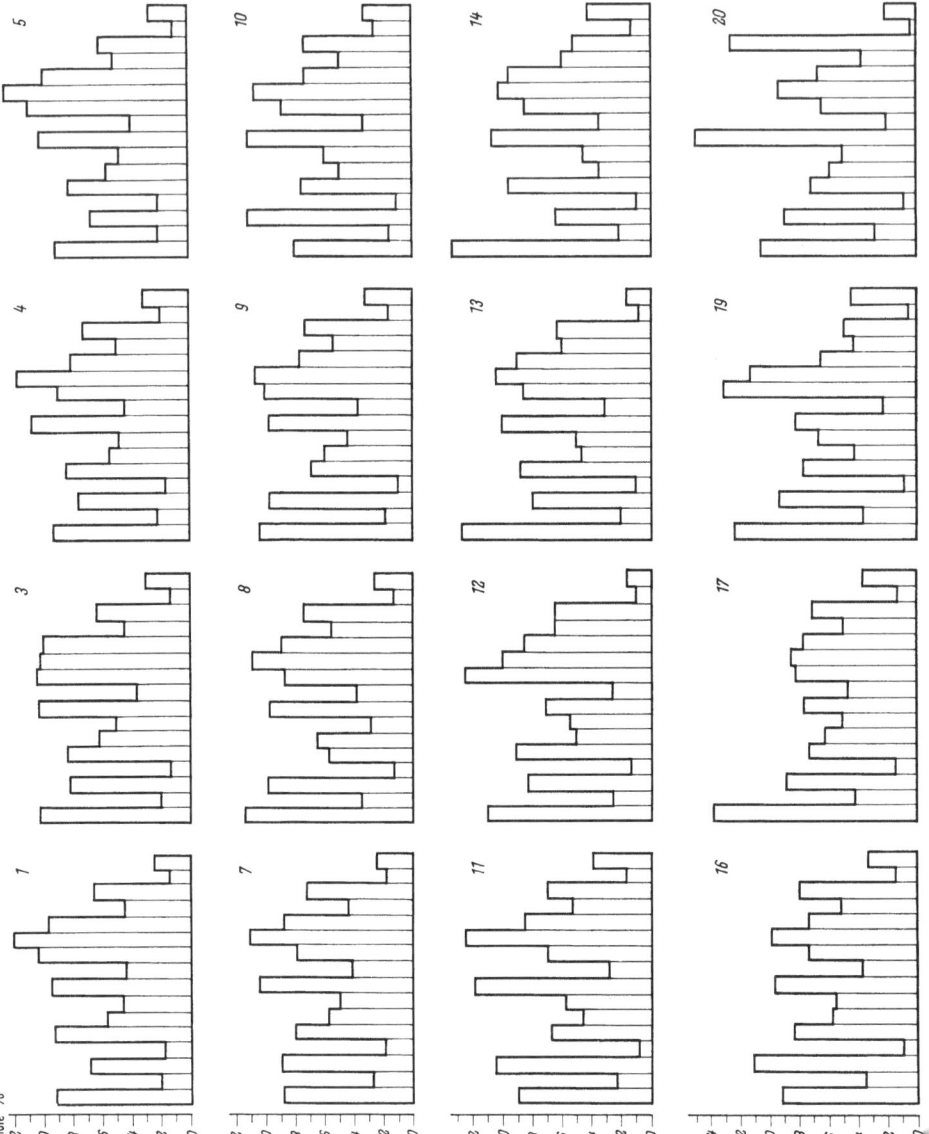

Fig. 3. Comparison of amino acid compositions in several isolated ribosomal proteins of E. coli. The experimental conditions for amino acid analysis are the same
as described in legend of Table 1. The height of the columns represents the amount
of each amino acid determined, expressed in mole %. The columns in each graph
represent from left to right: Lys, His, Arg, Met, Asp, Thr, Ser, Glu, Pro, Gly, Ala,
Val, Ile, Leu, Tyr, Phe

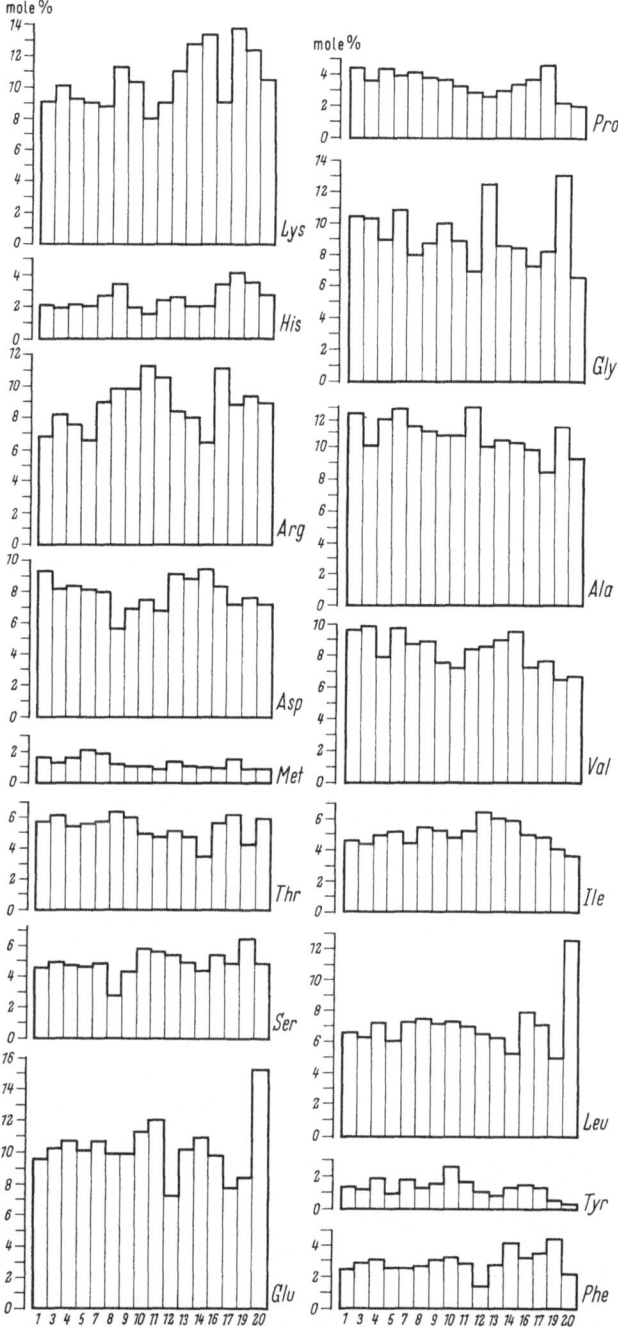

Fig. 4. Comparison of the amount of one amino acid (in mole-%) present in various purified E. coli ribosomal proteins. For experimental details of amino acid analysis see legend of Table 1

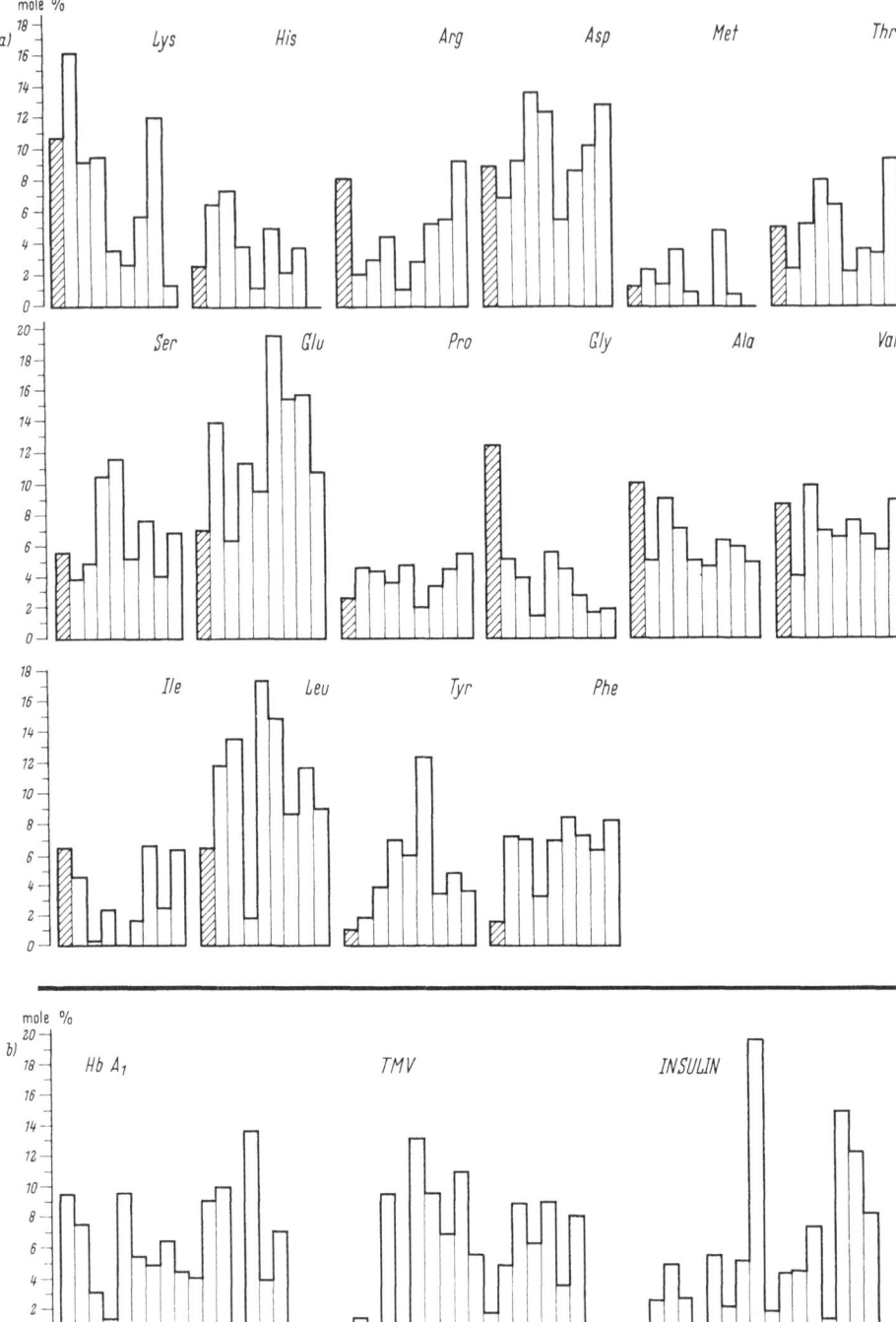

Fig. 5. Comparison of amino acid composition of E. coli ribosomal protein No. 12 with 8 completely unrelated proteins. In part a) the same graphical expression as in Fig. 4 was chosen. The columns in each graph represent the amino acid content (mole-%) for one given amino acid (from left to right) *1* E. coli ribosomal protein, *2* human myoglobin, *3* human hemoglobin A₁, *4* bovine pancreatic ribonuclease A, *5* porcine pepsinogen, *6* bovine insulin, *7* ovalbumin, *8* bovine serum albumin, *9* wild type of tobacco mosaic virus. In part b) the amino acid compositions of HbA₁, TMV (wild type) and bovine insulin are demonstrated in the same way as in Fig. 3

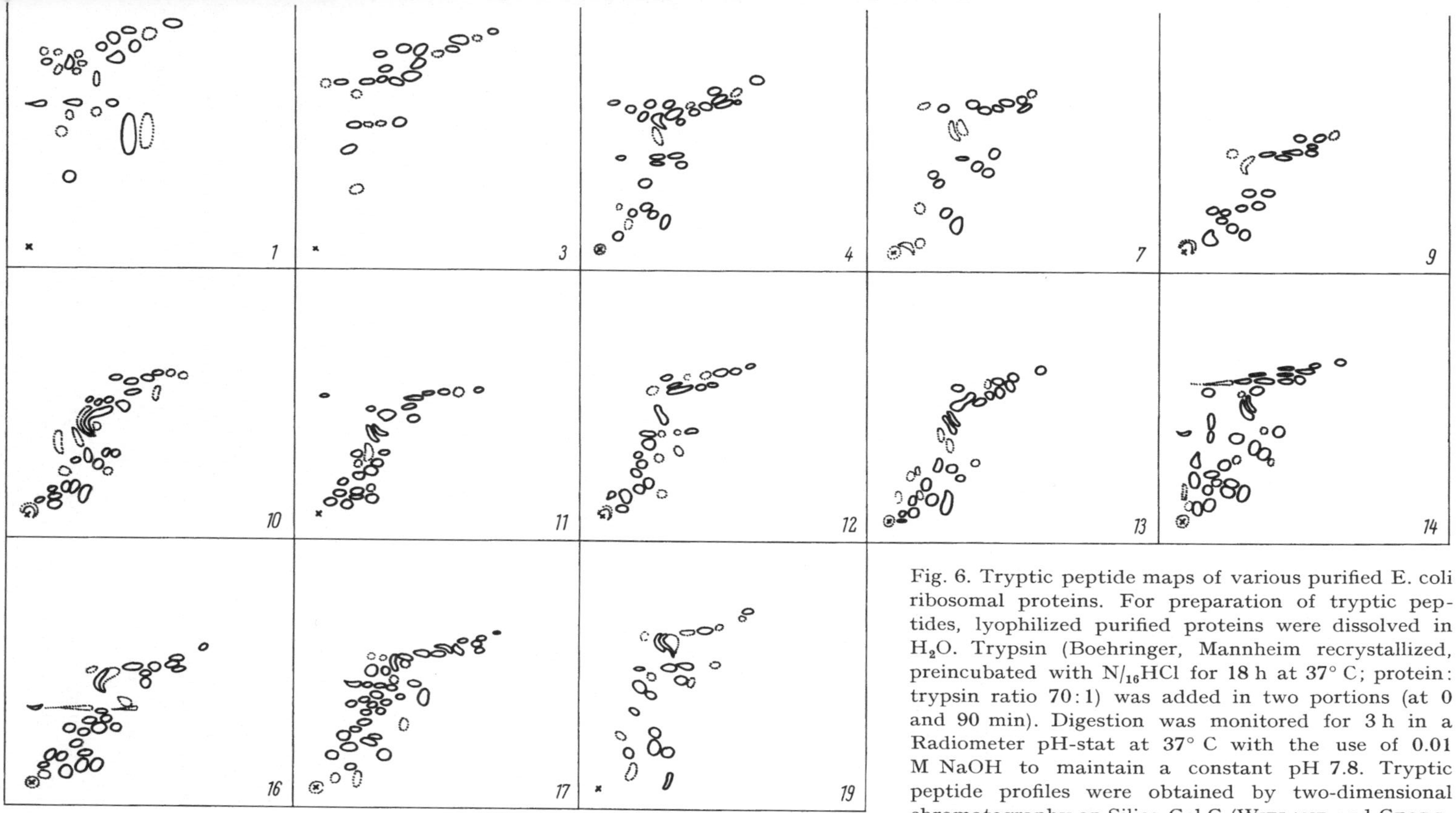

Fig. 6. Tryptic peptide maps of various purified E. coli ribosomal proteins. For preparation of tryptic peptides, lyophilized purified proteins were dissolved in $H_2O$. Trypsin (Boehringer, Mannheim recrystallized, preincubated with $N/16$ HCl for 18 h at 37° C; protein: trypsin ratio 70:1) was added in two portions (at 0 and 90 min). Digestion was monitored for 3 h in a Radiometer pH-stat at 37° C with the use of 0.01 M NaOH to maintain a constant pH 7.8. Tryptic peptide profiles were obtained by two-dimensional chromatography on Silica-Gel G (WIELAND and GEORGOPOULOS, 1964). The solvents used were methanol-chloroform-conc. ammonia (2:2:1) and in the second direction pyridine-acetic acid-n-butanol-water (40:14:68:25) both in solvent saturated chambers at 30° C. After drying the plates at 110° C for 30 min, peptides were located by use of a ninhydrine-collidine reagent (v. ARX and NEHER, 1963)

## Peptide Mapping

Tryptic peptide profiles of all investigated proteins were obtained by two-dimensional thin-layer chromatography on silica-gel G (WIELAND and GEORGOPOULOS, 1964). Their peptide maps are shown in Fig. 6. Again, a certain similarity of the fingerprints of the different proteins is remarkable, however the number and distribution of the peptide spots in the different chromatograms show many clear differences. There is, for example, an increasing number of tryptic peptides in the more basic proteins, as would have been expected from their electrophoretic mobility and amino acid composition.

## Molecular Weight Determinations

In contrast to the relatively similar amino acid composition the ribosomal proteins differ strikingly in their molecular weights: values between 9000 and 41,000 were obtained (Table 2).

Table 2. *Molecular weights of 15 ribosomal proteins of E. coli*

| Proteins | Molecular weights | Proteins | Molecular weights |
|----------|-------------------|----------|-------------------|
| 1 | 41,000 ± 4,000 | 12 | 13,500 ± 1,500 |
| 2 | not determined | 13 | 35,000 ± 2,500 |
| 3 | not determined | 14 | 22,000 ± 2,500 |
| 4 | 28,000 ± 3,500 | 15 | not determined |
| 5 | 36,000 ± 4,000 | 16 | 12,500 ± 1,000 |
| 6 | not determined | 17 | 19,500 ± 1,500 |
| 7 | 20,000 ± 1,000 | 18 | not determined |
| 8 | 26,500 ± 3,000 | 19 | 11,000 ± 1,000 |
| 9 | 15,000 ± 2,000 | 20 | 9,000 ± 500 |
| 10 | 18,000 ± 1,000 | 21 | not determined |
| 11 | 41,000 ± 4,000 | 22 | not determined |

For molecular weight determinations, the low speed method of YPHANTIS (1960) using short columns of solution was employed. The analytical ultracentrifuge Spinco E, equipped with a Schlieren optic was used throughout all experiments. All runs were carried out at $20 \pm 0.1°$ C; protein concentration 0.8 and 0.4%.
Two buffers were used:

a) 10% acetic acid, 0.2 M NaCl, 6 M urea, $\varrho = 1.107$ g/cm³ (pycnometrically determined)

b) 0.075 M sodium acetate, 0.2 M NaCl, 3.5 M urea, pH 4.6; $\varrho = 1.059$ g/cm³. For all ribosomal proteins a partial specific volume of 0.743, as calculated by MÖLLER and CHRAMBACH (1967), was used.
As a reference lysozyme was used in the same buffers. Mean values and their mean derivations (s) are listed in Table 2.

Molecular weight determinations were performed with the low-speed method of YPHANTIS (1960); in addition, sedimentation and diffusion experiments of some proteins were performed. The analytical ultra-centrifuge Spinco E was used throughout all these experiments.

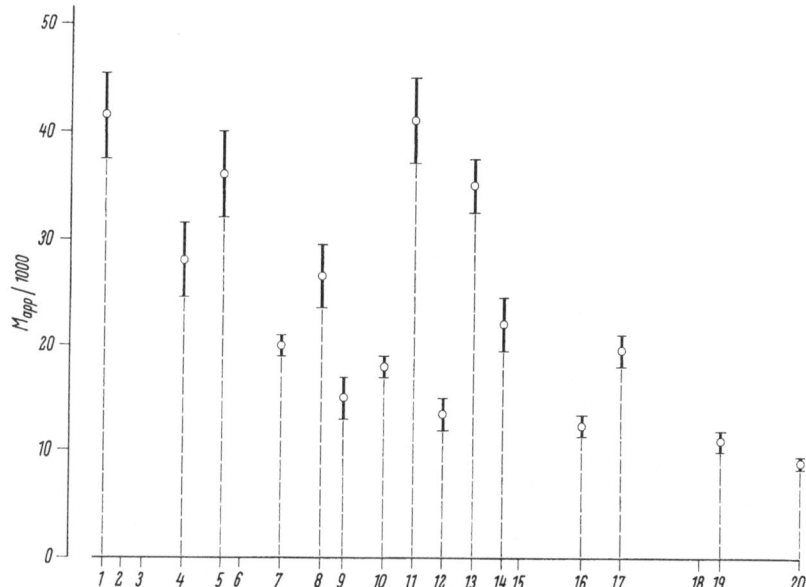

Fig. 7. Molecular weights are plotted versus electrophoretic mobility under standard conditions (refer to legend of Fig. 1) of each protein. Column heights represent the molecular weights, with the maximal deviation from the mean indicated. For experimental details concerning determinations of molecular weights, see legend to Table 2

Fig. 7 shows the correlation between the molecular weights and the migration of the proteins in the preparative acrylamide gel. In general, the molecular weight values decrease with increasing values according to mobility of the proteins toward the cathode. This is in good agreement with results obtained by gel filtration of ribosomal protein (STÖFFLER, RUDLOFF, and WITTMANN, 1968). However, protein No. 11 (molecular weight 41,000) and protein No. 7 (molecular weight 20,000), respectively, are exception of this rule. The still missing molecular weights of a few proteins are presently under investigation in the analytical ultracentrifuge. Studies with gel-filtration technique on Sephadex G-100 show molecular weights for proteins No. 2 and No. 3 higher than 30,000, for proteins No. 21 and No. 22 around 9000 (STÖFFLER, unpublished).

All available evidence seems to indicate that the cytoplasmatic ribosomes of eucaryotic organisms as well as the ribosomes of the procaryotic microorganisms serve the same purpose, i. e. protein biosynthesis. On the other hand it is well established that the two kinds of ribosomes are characterized by different sedimentation properties. In order to gain some information on possible relationships between ribosomal proteins of

the 70S ribosomes of bacteria and the 80S ribosomes of yeast we extended
our investigation on yeast ribosomal proteins.

Yeast cells (Saccharomyces cerevisiae) strain 5 (kindly provided by Dr.
LASKOWSKI, Biophysikalisches Institut der Freien Universität Berlin-Dahlem) were grown aerobically in a synthetic medium, and the harvested cells
were broken by shaking with glass-beads. The ribosomes were isolated and

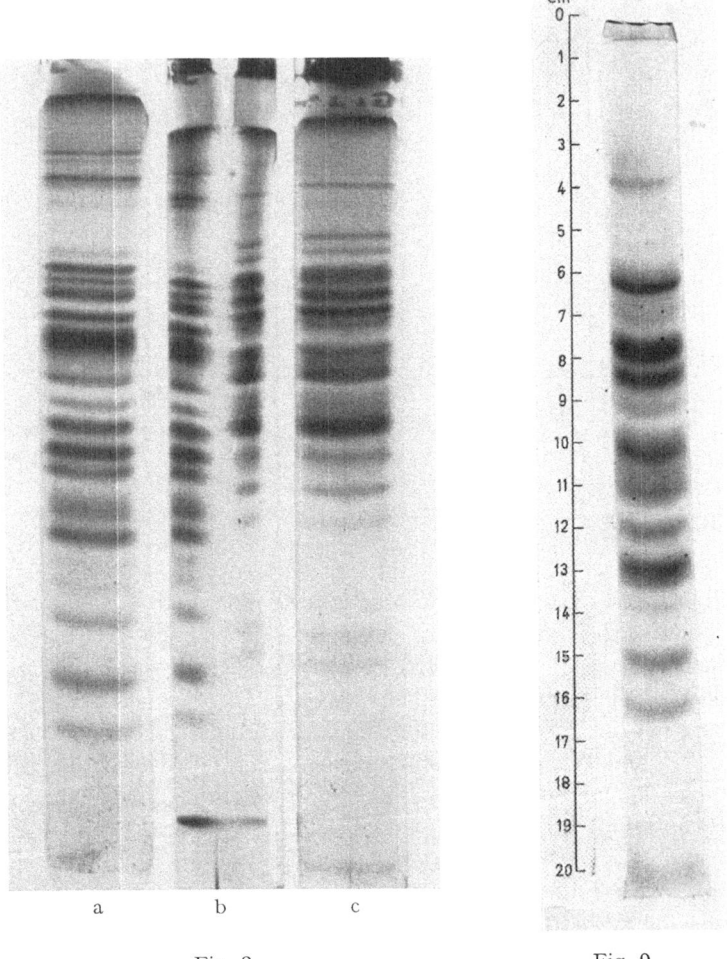

Fig. 8                                          Fig. 9

Fig. 8. Acrylamide gel disc electrophoresis, comparing ribosomal proteins from
E. coli and yeast. a) E. coli ribosomal protein, 250 μg. b) Split gel (left: E. coli
ribosomal protein, right: yeast ribosomal protein) c) Yeast ribosomal protein, 250μg.

Fig. 9. Separation pattern of yeast ribosomal protein by preparative polyacrylamide gel electrophoresis. Experimental conditions refer to Fig. 1

their protein prepared by the acetic acid method. All other methods used for the purification of the proteins and their characterization were the same as already described for the proteins of E. coli ribosomes.

The disc electrophoretic pattern of yeast ribosomal protein exhibits a degree of heterogeneity comparable to that observed in E. coli; however, ribosomal protein from E. coli and yeast differ remarkably in their disc electrophoresis profiles. This can be demonstrated by comparing two polyacrylamide columns loaded with these proteins. Using the split gel technique (LEBOY et al., 1964) these differences between the two proteins become much more convincing (Fig. 8).

Preparative separation of yeast ribosomal proteins was accomplished by preparative polyacrylamide gel electrophoresis, except that the gel contained 12.5 % acrylamide (for E. coli 15 %). The separation pattern of yeast ribosomal proteins in acrylamide gel allowed the preparation of about 20 well separated protein-bands (Fig. 9).

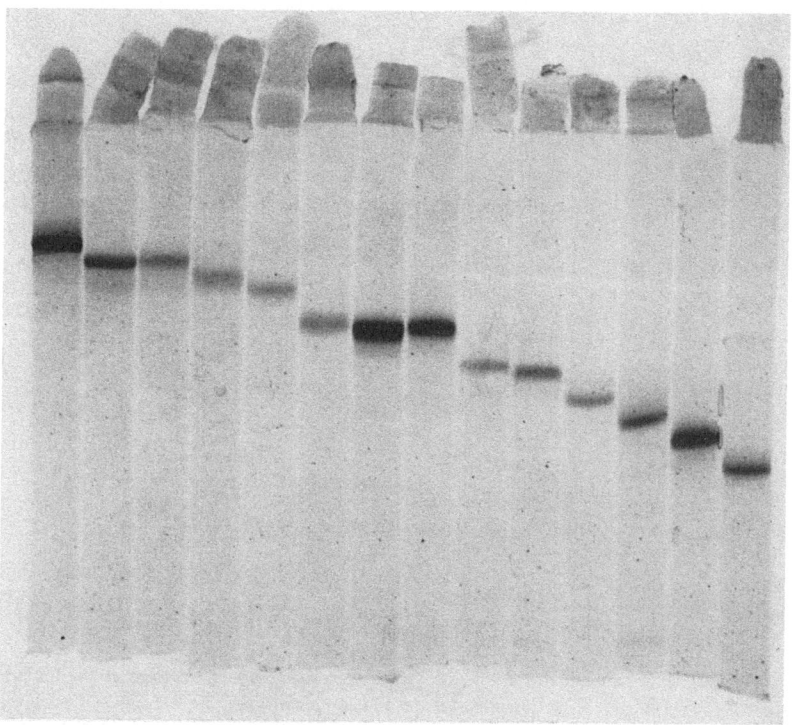

Fig. 10. Acrylamide gel disc electrophoresis at pH 4.2 of isolated and purified protein bands of yeast ribosomal proteins, used for amino acid analysis, peptide mapping and determination of molecular weights

Determinations of amino acid compositions, peptide mapping and molecular weight determinations were performed only with proteins which gave a single band in the analytical disc electrophoresis. Fig. 10 shows the purity of the isolated proteins from yeast ribosomes.

The amino acid composition of 17 yeast proteins is presented in Table 3. Comparing the yeast proteins with each other, it appears, inspite of a certain general similarity, that they seem to show a more expressed heterogeneity than in the case of E. coli ribosomal proteins.

The same is true for the tryptic peptide profiles of the different isolated yeast proteins (Fig. 11). However, beside very different peptide maps, e. g. 59-6 and 59-8, similarities between fingerprints, e. g. 58-5 and 59-10, are demonstrable, too.

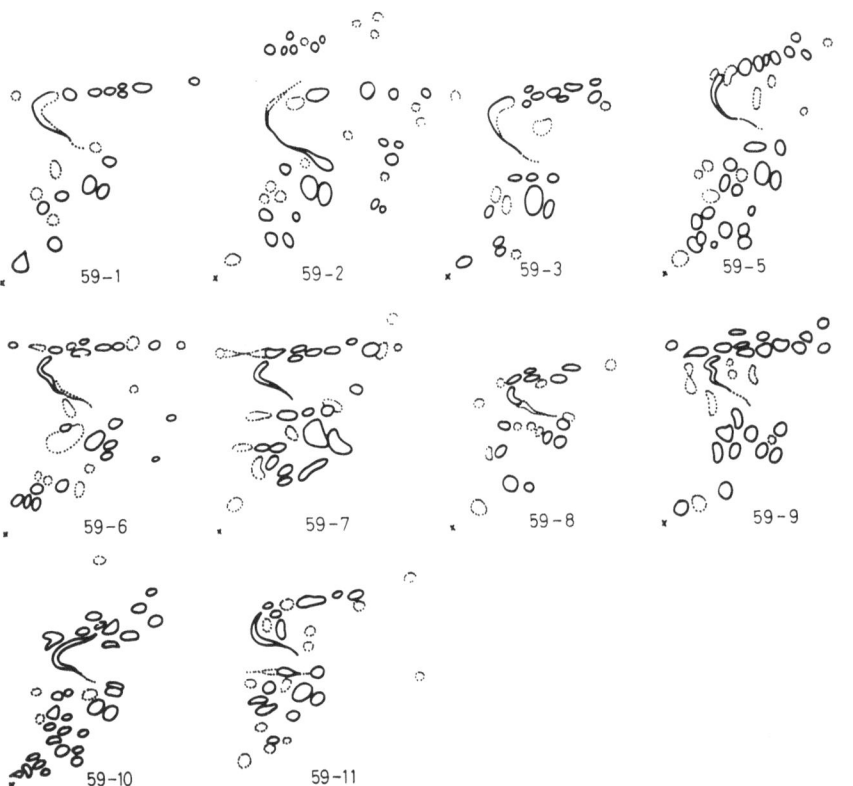

Fig. 11. Tryptic peptide maps of various purified yeast ribosomal proteins. For experimental details see legend to Fig. 6

So far, molecular weight determinations have been carried out for 12 different isolated yeast proteins (Table 4). As with E. coli proteins a

Table 3. Amino acid compositions of 17 ribosomal proteins from yeast (in mole-%)

| Amino acid | 57-7 | 57-8 | 57-11 | 57-13 | 57-15 | 57-18 | 59-1 | 59-2 | 59-3 | 59-5 | 59-6 | 59-7 | 59-8 | 59-9 | 59-10 | 59-11 | 59-12 |
|---|---|---|---|---|---|---|---|---|---|---|---|---|---|---|---|---|---|
| Lys | 8.85 | 8.81 | 11.84 | 11.86 | 10.31 | 13.53 | 6.55 | 11.84 | 12.81 | 11.94 | 10.82 | 11.18 | 10.43 | 10.50 | 10.22 | 13.04 | 13.92 |
| His | 2.21 | 2.56 | 1.96 | 2.11 | 1.00 | 0.75 | 3.63 | 1.93 | 2.54 | 2.25 | 2.84 | 2.43 | 2.07 | 2.95 | 2.96 | 2.74 | 3.22 |
| Arg | 7.24 | 7.16 | 6.55 | 8.19 | 5.78 | 8.27 | 5.60 | 5.85 | 7.50 | 7.30 | 7.02 | 7.17 | 7.14 | 6.49 | 7.39 | 7.26 | 6.18 |
| Asp | 10.23 | 8.57 | 7.28 | 7.53 | 7.58 | 8.83 | 9.12 | 11.37 | 9.38 | 8.73 | 8.44 | 7.95 | 8.37 | 8.08 | 8.08 | 7.97 | 7.40 |
| Thr | 5.05 | 5.06 | 5.87 | 3.65 | 6.93 | 9.10 | 6.59 | 5.88 | 5.14 | 5.34 | 5.38 | 5.09 | 5.39 | 6.16 | 5.68 | 4.12 | 5.10 |
| Ser | 6.53 | 6.73 | 8.34 | 9.41 | 10.15 | 6.51 | 10.53 | 10.36 | 7.26 | 7.02 | 7.68 | 8.67 | 8.01 | 8.20 | 9.30 | 10.66 | 11.68 |
| Glu | 10.38 | 10.66 | 6.89 | 8.95 | 8.67 | 7.63 | 9.66 | 9.66 | 9.12 | 11.25 | 9.73 | 10.85 | 10.52 | 8.36 | 9.30 | 8.75 | 8.30 |
| Pro | 8.36 | 4.77 | 5.70 | 3.45 | 1.70 | 5.99 | 4.41 | 4.07 | 4.10 | 4.96 | 4.02 | 3.22 | 4.26 | 3.22 | 3.27 | 3.84 | 3.13 |
| Gly | 8.00 | 9.63 | 9.04 | 7.76 | 11.68 | 6.90 | 9.77 | 9.70 | 9.35 | 7.00 | 9.90 | 9.50 | 10.41 | 9.14 | 10.33 | 9.96 | 10.40 |
| Ala | 8.81 | 8.34 | 10.60 | 8.11 | 8.95 | 7.33 | 10.49 | 9.98 | 9.73 | 8.13 | 10.47 | 10.67 | 9.64 | 10.40 | 8.86 | 8.28 | 8.71 |
| Val | 6.35 | 8.19 | 6.46 | 9.60 | 10.53 | 6.15 | 7.44 | 6.42 | 6.87 | 8.24 | 6.64 | 7.91 | 7.59 | 9.10 | 7.29 | 7.84 | 7.70 |
| Ile | 5.56 | 6.52 | 5.69 | 5.36 | 6.06 | 6.08 | 4.35 | 3.70 | 5.00 | 5.18 | 5.12 | 4.79 | 5.23 | 4.34 | 4.32 | 3.78 | 4.32 |
| Leu | 6.76 | 7.51 | 7.68 | 10.32 | 8.51 | 11.29 | 6.78 | 5.70 | 7.67 | 8.16 | 6.73 | 6.23 | 6.07 | 7.14 | 7.21 | 7.72 | 5.41 |
| Tyr | 1.55 | 1.10 | 2.88 | 1.33 | — | — | 2.26 | 1.25 | 0.95 | 1.38 | 1.92 | 1.41 | 2.01 | 2.68 | 1.84 | 1.98 | 2.26 |
| Phe | 2.89 | 3.32 | 3.50 | 2.39 | 3.11 | 5.23 | 2.82 | 2.29 | 2.56 | 3.38 | 3.28 | 2.92 | 2.44 | 3.25 | 3.09 | 2.09 | 2.26 |

Experimental details are the same as in legend to Table I, with the exception that also methionine values were not calculated.

2*

relatively wide distribution of sizes was observed. The range of molecular
weights varies from 12,000 to 30,000. These values are comparable to the
range of isolated proteins from E. coli, obtained by two independent
investigators (DELIUS et al., 1967; KALTSCHMIDT, DZIONARA, DONNER,
and WITTMANN, 1967).

Table 4. *Molecular weights of 12 ribosomal proteins from yeast*

| Proteins | Molecular weights | Proteins | Molecular weights |
|----------|-------------------|----------|-------------------|
| 57- 1    | 31,000 ± 4,000    | 59- 7    | 17,500 ± 4,500    |
| 57-11    | 18,500 ± 2,500    | 59- 8    | 25,500 ± 5,000    |
| 57-13    | 12,000 ± 2,000    | 59- 9    | 15,500 ± 2,000    |
| 59- 1    | 29,500 ± 4,000    | 59-10    | 19,500 ± 2,500    |
| 59- 2    | 25,500 ± 4,000    | 59-11    | 20,000 ± 2,500    |
| 59- 6    | 21,000 ± 4,000    | 59-12    | 13,500 ± 1,500    |

For experimental details refer to legend of Table 2.

Considering the specific function of ribosomes in protein biosynthesis
and their apparent morphological universality with respect to their RNA
and protein moieties, one is tempted to expect certain similarities in
primary structure between ribosomal proteins of different species, at
least if one were to believe that conservation of genetic information is
phylogenetically adventageous. The comparison of differences in amino
acid composition between different E. coli ribosomal proteins and yeast
ribosomal proteins, respectively, on the one side and between comparable
proteins of the two species on the other side, leads to the recongnition of
much more similarity than is observed with any unrelated proteins
(Table 5). Also the results of ROMBAUTS (1967) hint into the same direction. In the disc electrophoretic patterns of CNBr treated total E. coli
ribosomal proteins he found a significant decrease of the number of bands
compared with the pattern of untreated proteins. Ribosomal proteins
from Neurospora crassa can be resolved into approximately 30 bands in
disc electrophoresis. In tryptic digests of total Neurospora crassa ribosomal protein one could theoretically expect more than a hundred fingerprint spots (assuming no homologous regions in primary structure), but
only some 45 to 55 spots were actually found by ALBERGHINA and SUS-
KIND (1967). These authors also reported partial immunological cross
reactions between E. coli — and N. crassa — ribosomes and antisera
against these ribosomal species. This is at least compatible with the
assumption of similar antigenic determinants in the different ribosomal
species.

Table 5. *Variation of amino acid compositions within four groups of protein*

| | TMV strains | Ribosomes E. coli | yeast | Unrelated proteins |
|---|---|---|---|---|
| Lys | 34.3 | 16.6 | 16.7 | 67.4 |
| His | — | 29.4 | 40.0 | 57.3 |
| Arg | 12.0 | 17.3 | 11.1 | 60.9 |
| Asp | 11.2 | 13.2 | 12.4 | 29.0 |
| Thr | 15.2 | 14.7 | 21.3 | 50.1 |
| Ser | 25.3 | 16.5 | 18.9 | 42.8 |
| Glu | 19.9 | 18.1 | 12.6 | 22.3 |
| Pro | 10.3 | 21.7 | 31.4 | 26.2 |
| Gly | 16.3 | 20.0 | 13.5 | 48.3 |
| Ala | 21.4 | 10.4 | 10.7 | 24.7 |
| Val | 24.3 | 13.1 | 12.3 | 25.9 |
| Ile | 10.2 | 14.5 | 16.1 | 74.9 |
| Leu | 13.5 | 24.1 | 19.9 | 42.4 |
| Tyr | 22.9 | 41.8 | 50.6 | 59.5 |
| Phe | 11.4 | 23.5 | 23.5 | 23.3 |

The variations in the amino acid compositions of the ribosomal proteins of yeast and E. coli are compared with those of related proteins (strains of tobacco mosaic virus) and various unrelated proteins:

a) TMV strains: five strains of tobacco mosaic virus;

b) E. coli: isolated ribosomal proteins of E. coli (KALTSCHMIDT et al., 1967);

c) yeast: isolated ribosomal proteins of yeast (Table 2);

d) unrelated proteins: human hemoglobin $A_1$, ovalbumin, wild type of tobacco mosaic virus, bovine serumalbumin, bovine pancreatic ribonuclease A, porcine pepsinogen, bovine insulin, human myoglobin. The figures give the variations $\pm \sqrt{\frac{\sum (x_i - \bar{x})^2}{n-1}}$ in rel.-% of the mean value for each amino acid.

Of course, all these results and their interpretation do neither prove structural homologies between different ribosomal proteins nor between ribosomal proteins from different species. Such homologies can solidly be proven only by sequence analyses of peptides isolated from the proteins in question.

In summary, the proteins of E. coli ribosomes differ significantly *in their molecular weights* from each other. The same is true for yeast proteins. There are also differences in the amino acid compositions and the fingerprint-patterns of the ribosomal proteins. The relatively similar amino acid composition of the various E. coli proteins on the one hand and their pronounced difference in molecular weights on the other hand could be explained by the hypothesis that the evolution of the ribosomal proteins took place by gene duplication. The demonstration of identical tryptic peptides and the amino acid sequence of the proteins will finally prove or disprove this hypothesis. First attempts in this direction have already been done by separating the tryptic peptides of three E. coli proteins on Dowex-1 X 2 columns. The amino acid composition of these separated peptides is presently under investigation.

## References

ARX, E. V., u. R. NEHER: Eine multidimensionale Technik zur chromatographischen Identifizierung von Aminosäuren. J. Chromatogr. **12**, 329—341 (1963).

ALBERGHINA, F. A. M., and S. R. SUSKIND: Ribosomes and ribosomal proteins from Neurospora crassa. J. Bact. **94**, 630—649 (1967).

DELIUS, H., R. R. TRAUT, P. B. MOORE, H. NOLLER, and P. PEARSON: this volume.

HAMILTON, M. G., and M. E. RUTH: Characterization of some of the proteins of the large subunit of rat liver ribosomes. Biochemistry **6**, 2585 (1967).

HORSTMANN, H. J.: Zur Frage analoger und homologer Primärstrukturen bei Ribosomenproteinen. Herbsttagg. der Ges. für Biologische Chemie, Frankfurt a. M., Oct. 1967. In: Hoppe-Seyler's Z. physiol. Chem. **348**, 1234 (1967).

HOSOKAWA, K., R. K. FUJIMURA, and M. NOMURA: Reconstitution of functionally active ribosomes from inactive subparticles and proteins. Proc. nat. Acad. Sci. (Wash.) **55**, 198 (1966).

KALTSCHMIDT, E., M. DZIONARA, D. DONNER, and H. G. WITTMANN: Ribosomal proteins: I. Isolation, amino acid compositions, molecular weights and peptide mapping of proteins from E. coli ribosomes. Molec. Gen. Genet. **100**, 364—373 (1967).

LEBOY, P. S., E. C. COX, and J. G. FLAKS: The chromosomal site specifying a ribosomal protein in E. coli. Proc. nat. Acad. Sci. (Wash.) **52**, 1367 (1964).

LERMAN, M. I., A. S. SPIRIN, L. P. GAVRILOVA, and V. F. GOLOV: Studies on the structure of ribosomes. II. Stepwise dissociation of protein from ribosomes by caesium cloride and the re-assembly of ribosome-like particles. J. molec. Biol. **15**, 268 (1966).

MARCOT-QUEIROZ, J., et R. MONIER: Préparation de particules 18S et 25S à parir des ribosomes d'Escherichia coli. Bull. Soc. Chim. Biol. **48**, 446 (1966).

MÖLLER, W., and A. CHRAMBACH: Physical heterogeneity of the ribosomal proteins from Escherichia coli. J. molec. Biol. **23**, 377 (1967).

NOMURA, M.: this volume.

OTAKA, E., T. ITOH, and S. OSAWA: Protein components in the 40S ribonucleoprotein particles in E. coli. Science **157**, 1452 (1967).

REISFELD, R. A., U. J. LEWIS, and D. WILLIAMS: Disc electrophoresis of basic proteins and peptides on polyacrylamide gels. Nature (Lond.) **195**, 281 (1962).

ROMBAUTS, W.: Structural studies on ribosomal proteins: peptides from cyanogen bromide cleavage and tryptic digestion. This volume.

SALAS, M., M. B. HILLE, J. A. LAST, A. J. WAHBA, and S. OCHOA: Translation of the genetic message, II. Effect of initiation factors on the binding of formyl-methionyl-tRNA to ribosomes. Proc. nat. Acad. Sci. (Wash.) **57**, 387 (1967).

SPIRIN, A. S., M. I. LERMAN, L. P. GAVRILOVA, and N. V. BELITSINA: Reconstruction of biologically active ribosomes from proteinpoor ribonucleoprotein particles and ribosomal protein. Biokhimiya **31**, 424 (1966).

STAEHELIN, M., and M. MESELSON: In vitro Recovery of ribosomes and of synthetic activity from synthetically ribosomal subunits. J. molec. Biol. **16**, 245 (1966).

STÖFFLER, G., V. RUDLOFF, and H. G. WITTMANN: Ribosomal proteins. III. Preparative separation of escherichia coli and yeast ribosomal proteins by means of gel filtration. Molec. Gen. Genetics 101, 70—81 (1968).

TRAUB, P., M. NOMURA, and L. TU: Physical and functional heterogeneity of ribosomal proteins. J. molec. Biol. 19, 215 (1966).

TRAUT, R. R.: Acrylamide gel electrophoresis of radioactive ribosomal protein. J. molec. Biol. 21, 571 (1966).

--, P. B. MOORE, H. DELIUS, H. NOLLER, and A. TISSIERES: Ribosomal proteins of E. coli, I. Demonstration of different primary structures. Proc. nat. Acad. Sci. (Wash.) 57, 1294 (1967).

WALLER, J. P.: Fractionation of the ribosomal protein from E. coli. J. molec. Biol. 10, 319 (1964).

—, and J. I. HARRIS: Studies on the composition of the protein from E. coli ribosomes. Proc. nat. Acad. Sci. (Wash.) 47, 18 (1961).

WIELAND, TH., u. D. GEORGOPOULOS: Zweidimensionale Auftrennung von Peptiden (Fingerprinttechnik) auf Dünnschichtplatten. Biochem. Z. 340, 476—482 (1964).

YPHANTIS, D. A.: Rapid determination of molecular weights of peptides and proteins. Ann. N. Y. Acad. Sci. 88, 586—601 (1960).

## Discussion

*Matthaei:* Have you analyzed the protein from 30- and 50S-subunits, and how much material is needed to do the amino acid analysis for the proteins of these subunits?

*Stöffler:* We have separated the 30S and 50S subunits of E. coli ribosomes by sucrose density gradient centrifugation. Ribosomal proteins of both subunits were subjected to preparative polyacrylamide gel electrophoresis and the results (NASS, STÖFFLER, and WITTMANN unpublished) are shown in the figure on page 24. For amino acid analyses of the different proteins of both subunits, about 100—200 mg protein of each subunit is necessary.

*Gierer:* Is it possible to calculate from the sum of the molecular weights of the isolated proteins, given in your table 2 the overall molecular weight of the reconstituted ribosome. And have you any estimates how many molecules of each individual protein are present in one ribosome?

*Stöffler:* So far molecular weight determination have been done only for seventeen isolated proteins, and there are still some proteins not yet investigated. Therefore, we cannot make a decission on this part of your question.

Actually, we did not make any investigation as to the number of molecules of particular polypeptide chains present in one ribosome. However the yields of certain protein components from preparative polyacrylamide gel electrophoresis as well as from column chromatography indicate that at least a few proteins seem to be present in higher amounts than others, on a molar basis.

*Tiedemann:* Is there any evidence that ribosomes from different organs of higher organisms differ in their protein composition?

*Stöffler:* In our investigations we are not concerned with ribosomes of higher organism. Low, R. V., and I. G. WOOL presented recently a paper [Science 155, 330 (1967)] in which they compared disc electrophoretic profiles from heart muscle-, skeletal muscle-, reticulocyte-, liver- and kidney ribosomal proteins from rat, with E. coli ribosomal protein. They found no differences between ribosomal preparations derived from the different rat organs, but these electrophoretic patterns were very different from that of E. coli ribosomes.

H. Bielka (Internationales Symposium Reinhardtsbrunn) compared ribosomal proteins prepared from different organs of many species of different phylogenetic level. He always found differences between ribosomal proteins of the various species but never between ribosomal proteins of selected organs of one species. This is consistent with the hypothesis that the same cistrons direct synthesis of ribosomal protein in the different mammalian tissues.

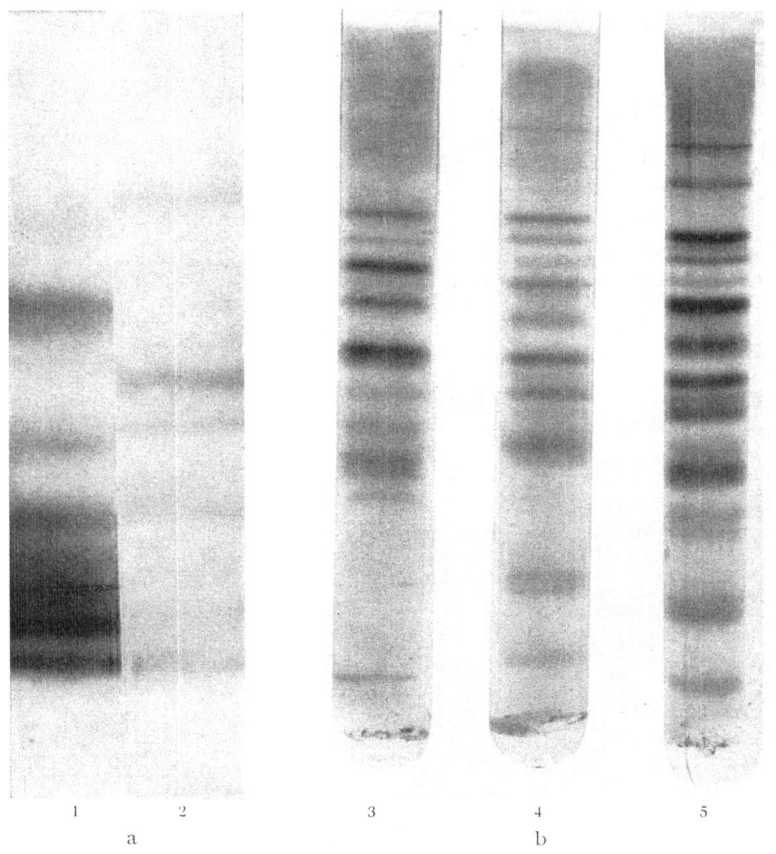

Comparison of electrophoretic patterns of the proteins of the 30S und 50 subunit of E. coli ribosomes in preparative and analytical polyacrylamide gel electrophoresis. a) preparative polyacrylamide gel electrophoresis; *1* E. coli ribosomal proteins of the 50S subunit; *2* E. coli ribosomal proteins of the 30S subunit. For more details see legend to Fig. 1. b) analytical disc electrophoresis. *3* E. coli ribosomal proteins of the 30S subunit; *4* E. coli ribosomal proteins of the 50S subunit; *5* E. coli ribosomal proteins from 70S ribosomes

*Winkler:* In a paper by Apirion it was pointed out that is not easy to distinguish between true ribosomal proteins and those which are only attached to the ribosomes during preparation. How did you exclude the last possibility?

*Stöffler:* All our ribosomal preparations (E. coli as well as yeast) were washed twice with 0.05 M Tris-buffer, pH 7.8, $10^{-2}$M Mg$^{++}$ and gave a single peak of 70S or 80S, respectively, in the analytical ultracentrifuge. Polyacrylamide disc-electrophoresis showed in all the different preparations of ribosomal proteins a unique and always reproducible pattern. The same was true for ribosomal protein derived from ribosomes passed through sucrose gradients or washed up to 7 times with the Tris buffer.

# Studies on Purified E. coli Ribosomal Proteins

By

H. DELIUS, R. R. TRAUT, P. B. MOORE, H. NOLLER, and P. PEARSON

Institut de Biologie Moléculaire, Université de Genève (Suisse)

## Introduction

The role of ribosomes in protein synthesis has appeared increasingly complex in recent years. It is known that they specifically bind messenger RNA, aminoacyl- and peptidyl-transfer RNA, formylmethionyl-transfer RNA, GTP, and probably at least six different protein factors. They are the sites for the reactions of chain initiation, chain termination and amino acid polymerization itself.

The view that a large number of different protein molecules are bound to ribosomal RNA to form specific sites with catalytic and binding activities is consistent with the complexity of ribosome function. Further elucidation of the mechanism of protein synthesis seems to be intimately related to investigation of ribosomal protein.

We have found the weight average molecular weight of both the total 30 S and 50 S protein to be about 15,000 (MOORE, TRAUT, NOLLER, PEARSON, and DELIUS, in press). With the protein part of the 70 S ribosome representing approximately $10^6$ daltons (TISSIÈRES, WATSON, SCHLESSINGER, and HOLLINGWORTH, 1959) one should therefore expect to find at least 65 polypeptide chains per 70 S ribosome. We have investigated the following questions: Are these polypeptides all the same, do they fall into a few classes or are they all different?

It was found several years ago that the ribosomal protein can be separated by electrophoresis and ion exchange chromatography into a large number of components (WALLER and HARRIS, 1961; WALLER, 1964; LEBOY, COX, and FLAKS, 1964) and experiments of various kinds have suggested that these components do not result from aggregation among a small number of protein species (WALLER, 1964; TRAUT, 1966).

It was shown that the fingerprint of 70 S ribosomal protein labelled with [35]S and digested with trypsin yielded a number of spots far greater than could be explained by the existence of only a few ribosomal protein species (TRAUT, MOORE, DELIUS, NOLLER, and TISSIÈRES, 1967). Several pure proteins were isolated and shown by amino acid composition, N-terminal groups, and tryptic fingerprints to have different primary structures.

We shall describe here our work on the isolation and characterization of 13 proteins from the 30 S ribosome subunit and 14 proteins from the 50 S subunit. Moreover we shall further describe experiments which led to the conclusion that the ribosome is a unique structure composed of one copy of each of the component protein species.

### Identification of Proteins

As a reference for the identification of the proteins and a simple method to check their purity we used the electrophoresis of the protein fractions on polyacrylamide gels at pH 4.5. Fig. 1a and b shows the gel

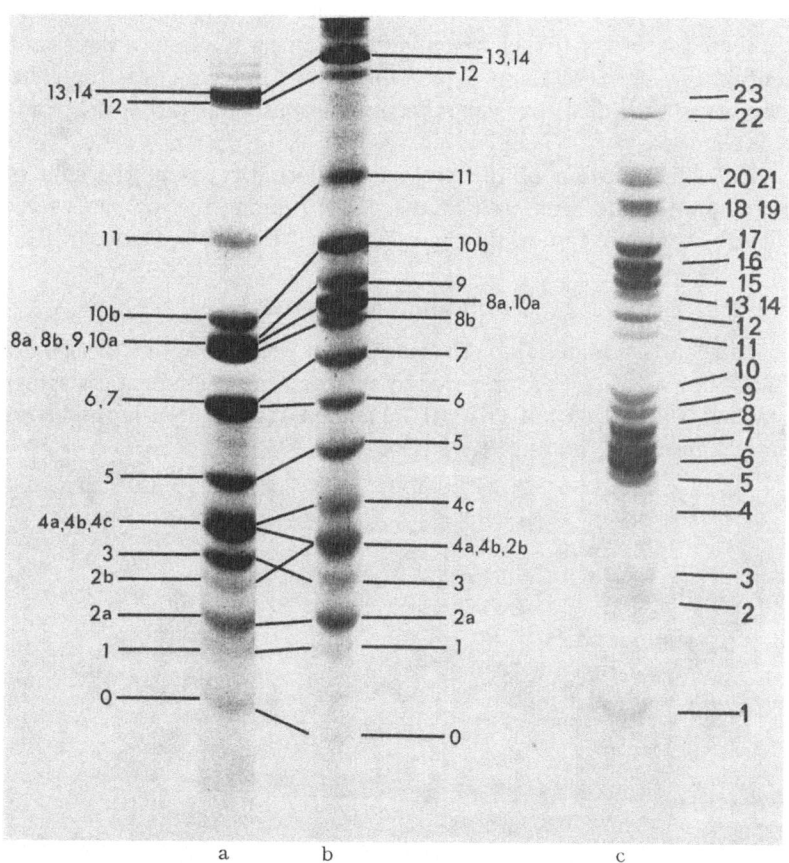

Fig. 1. Polyacrylamide Gel Electrophoresis of Total 30S (a, b) and 50S (c) Protein. The 30S pattern is shown for two gel conditions: (a) 0.2% and (b) 0.8% bisacrylamide (bis). The 50S pattern is shown only for a 0.2% bis gel (TRAUT, 1966; TRAUT et al., 1967). The proteins are numbered for identification. The a's and b's designate proteins, separable by other means, which coelectrophorese under the conditions employed

pattern produced by the total protein of the 30 S subunit of *E. coli* strain MRE 600 (Cammack and Wade, 1965) after staining with amido schwarz. About 13 protein bands are resolved. The proteins are numbered according to their electrophoretic mobility, starting with the fastest moving band. The second pattern is produced in a gel with a higher degree of crosslinking. The resulting shifts in position of the different proteins are revealed by comparison of the gel patterns with bands produced by purified proteins. The 0.2 % bis gel is included only in order to reveal band 2b which coincides with 4a + 4b in the 0.8 % gel. By comparing these patterns with the one obtained by CMC chromatography (see below) 20 different components were identified. In the pattern of the 50 S protein (Fig. 1c) about 21 electrophoretic bands can be discerned and these are numbered according to their mobility in the same way. Many of these bands contain two or more components which have the same electrophoretic mobility, but which are separable by various preparative procedures (see below).

The identification of the proteins is done by use of the split gel technique (Leboy, Cox, and Flaks, 1964) running the sample together with the total mixture in two halves of the same gel.

## Purification of Proteins

Our starting material for the isolation of proteins were either 30 S or 50 S subunits. They were prepared by sucrose density gradient centrifugation in 0.1 M KCl, 0.01 M Tris, pH 7.4 and 0.001 M $MgCl_2$ using an inverse sample gradient (Brakke, 1964; Bolton, 1966).

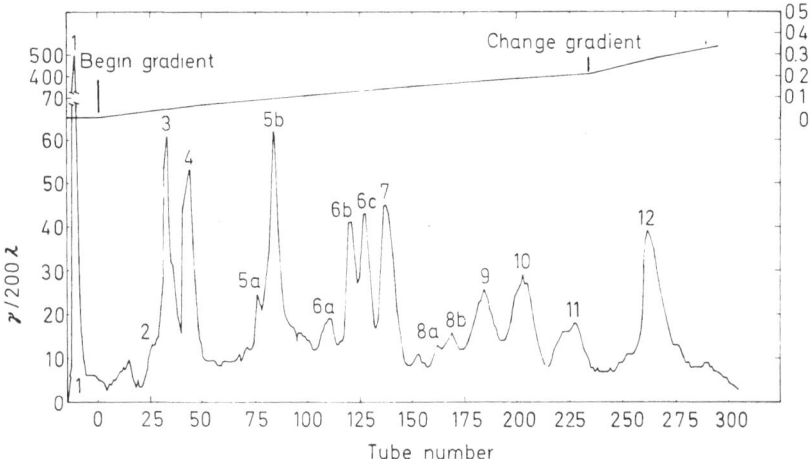

Fig. 2. CMC Column Elution Profile of 30S Protein. The protein was prepared by acetic acid extraction and eluted with a salt gradient from 0.01 to 0.4 M sodium acetate, pH 5.6 in 6 M urea. The amount of protein was determined by Lowry assay

The protein was extracted either by 66% acetic acid (FRAENKEL-CONRAT, 1957; WALLER and HARRIS, 1961) or by treatment with RNase in 4 M urea.

The first step in the fractionation of the 30 S proteins was chromatography on carboxymethyl cellulose (CMC) columns employing a linear salt gradient in a buffer containing 6 M urea and sodium acetate, pH 5.6. This method had already been employed for fractionation of ribosomal protein using stepwise elution (WALLER, 1964). As is shown in Fig. 2 the 30 S protein is fractionated into at least 12 peaks.

The 50 S protein mixture is more complex than that of the 30 S proteins and in order to simplify the CMC column elution pattern it was divided into two complementary fractions by treatment of the 50 S subunits with 2 M LiCl, $10^{-5}$ M MgCl and $10^{-2}$ M Tris, pH 7.4 (ATSMON,

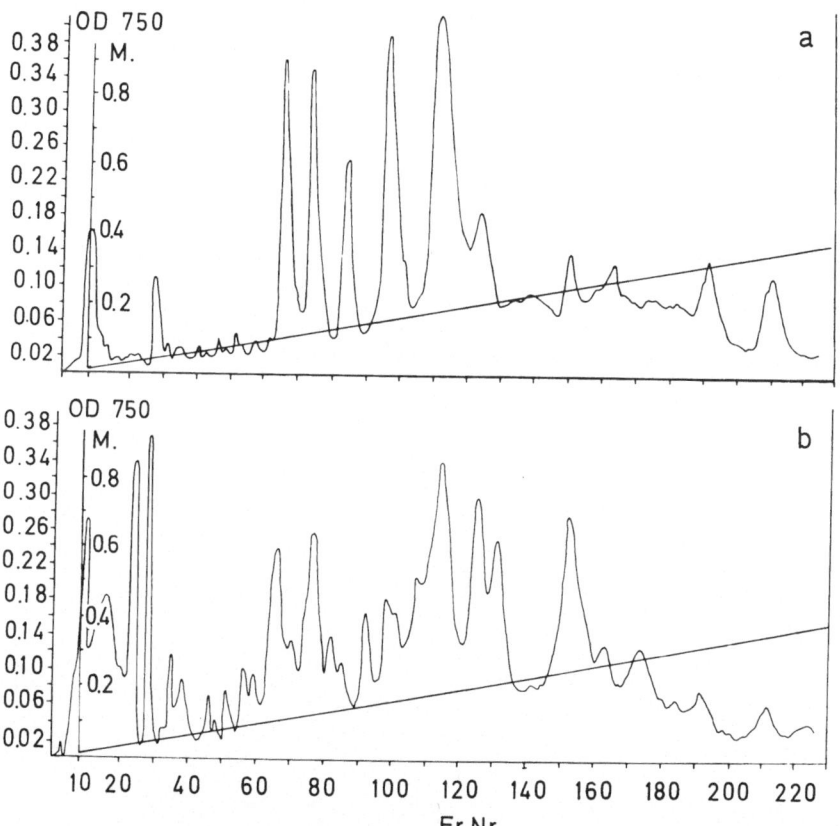

Fig. 3. CMC Column Elution Profile of 50 S Protein. (a) Protein released from the particle fraction of LiCl treated 50 S ribosomes by RNase digestion (LiCl) particle protein). (b) Protein released from the 50 S ribosomes by the LiCl treatment (LiCl supernatant protein). The conditions of the elution are the same as in Fig. 2

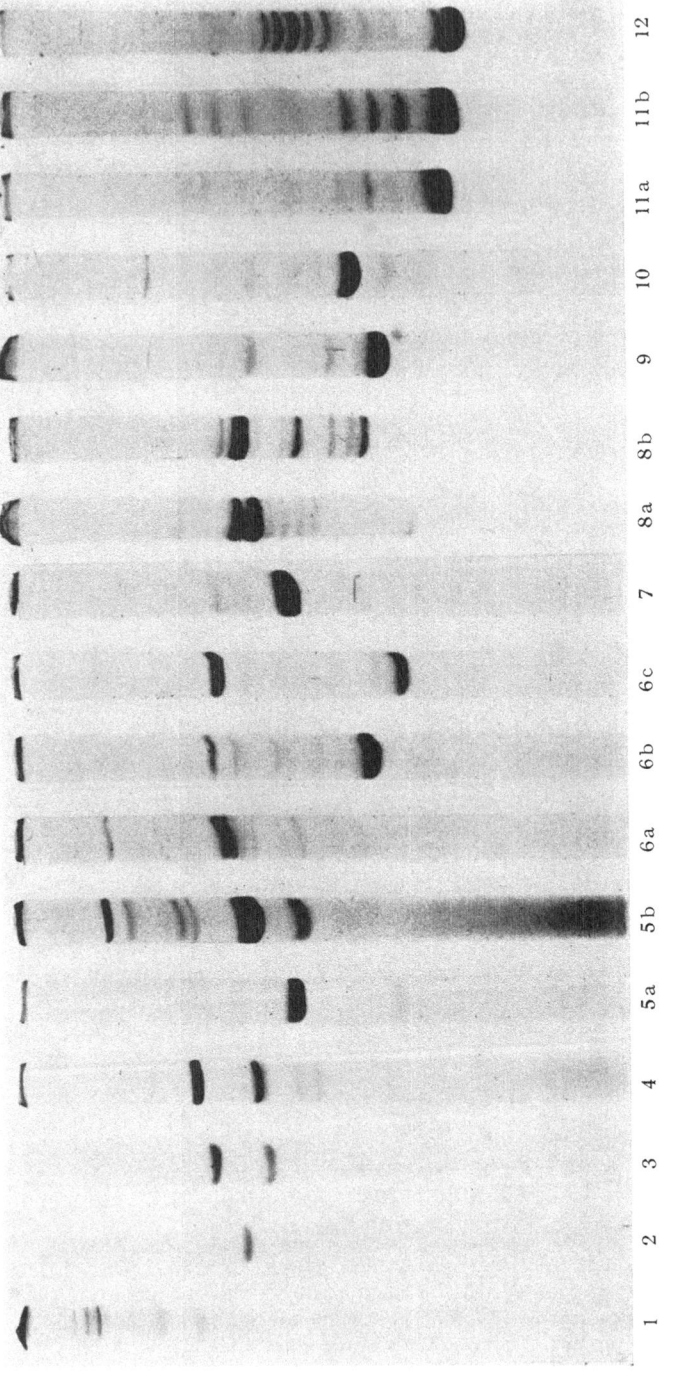

Fig. 4. Polyacrylamide Gel Analysis of the 30S Protein Peaks Eluted from the CMC Column

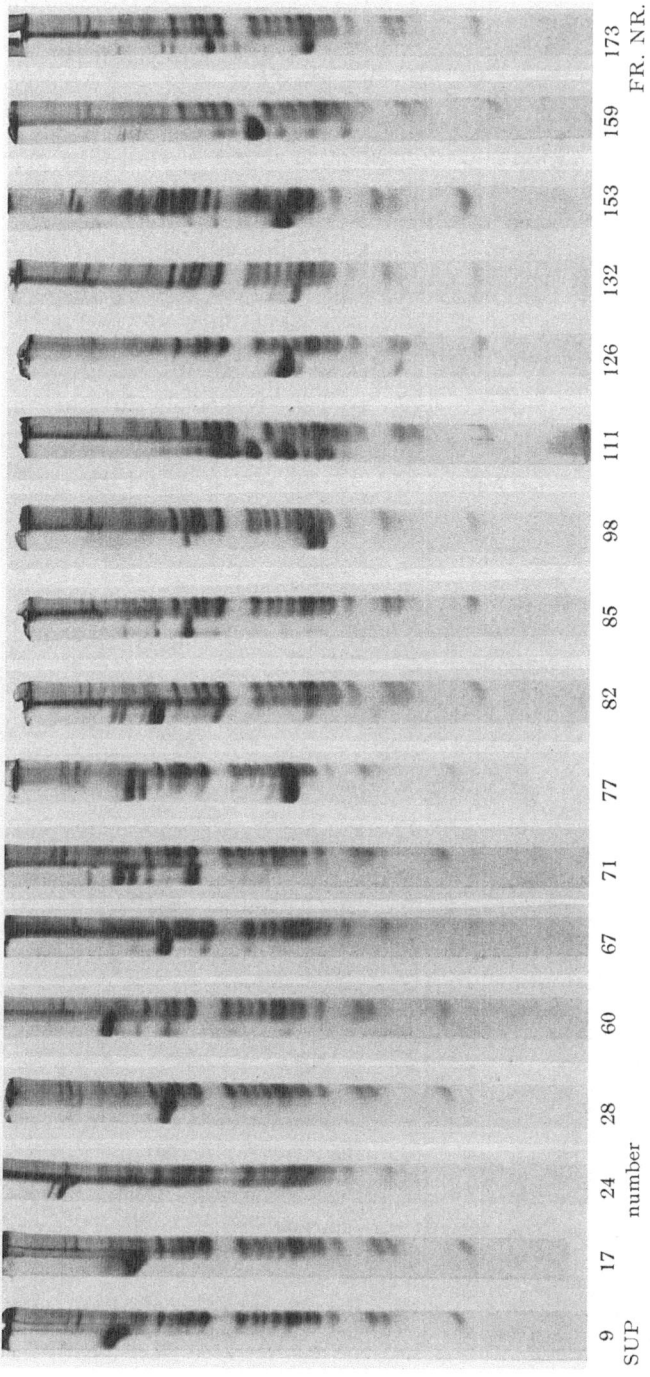

Fig. 5. Polyacrylamide Split Gel Analysis of CMC Fractions of the 50S LiCl Supernatant Protein. The fractions are from the experiment of Fig. 3 and are compared with 70S protein as standard

Spitnik-Elson, and Elson, 1966). Fig. 3 shows the elution profiles of the two fractions which were chromatographed on CMC columns in parallel. The upper graph shows the protein of the LiCl particles. Well separated protein peaks are eluted. The LiCl supernatant protein in the lower curve is more complex. In general a given protein is found only in one of the two LiCl fractions. For example almost none of the proteins contained in the first three big CMC peaks of the particle fraction is found in the corresponding region of the supernatant CMC profile.

Figure 4 shows analytical gels loaded with the pooled protein from the different peaks of the 30 S CMC chromatography. Most of the peaks contain only one or two electrophoretically discernible proteins present as major components. In general the proteins eluted at low ionic strength have a low mobility in gel electrophoresis. The last peaks eluted at high ionic strength consist of faster migrating proteins.

Figure 5 shows split gels of CMC fractions of the 50 S LiCl supernatant protein compared to 70 S ribosomal protein as a standard. Many of the

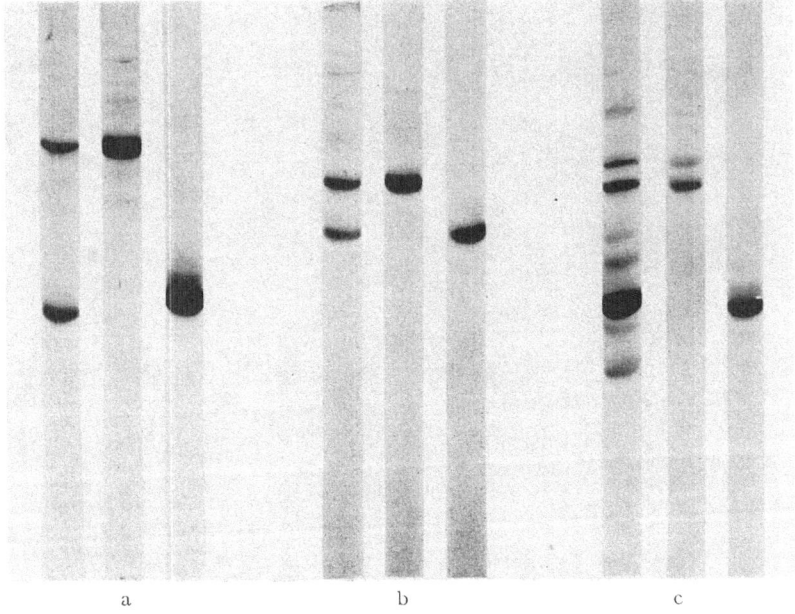

a                          b                          c

Fig. 6. Polyacrylamide Gel Analysis of CMC Fractions Further Purified on Sephadex G 100. In each group of three gels the starting material is on the left, the less retarded Sephadex peak in the middle and the more retarded peak on the right. (a) Proteins $30_3$ and $30_{10b}$. The Sephadex elution profile is given in Fig. 7. (b) Proteins $30_{11}$ and $30_{10a}$. (c) Proteins $30_g$ and $30_{4b}$. The solvent in fractionations (a) and (c) was 1 M propionic acid. The solvent in fractionation (b) was 6 M urea, 0.25 M sodium acetate, pH 5.6

fractions are still rather complicated mixtures of several proteins. Many proteins which have about the same electrophoretic mobility on the gels are separated into different peaks on the CMC column, as for example protein $50_7$ occuring in fractions 132 and 173 or protein $50_{16}$ present in fractions 28 and 85.

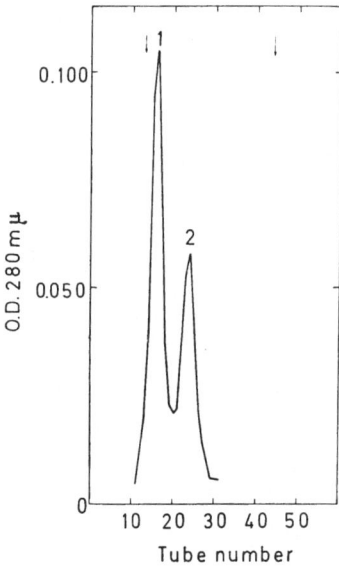

Fig. 7. Elution Profile of a CMC Fraction Containing Proteins $30_3$ and $30_{10b}$ Chromatographed on Sephadex G 100 in 1 M Propionic Acid. The void volume and the hold-up volume are indicated by arrows

The further purification was done mainly by chromatography on Sephadex G 100 in 1 M propionic acid according to FLEISCHMAN, PORTER, and PRESS (1963). Fig. 6 demonstrates the efficiency of this method: The first gel of each of the three sets was loaded with protein fractions from the CMC column. The other two gels show protein samples recovered from these mixtures after the Sephadex chromatography. Fig. 7 shows a typical elution profile of one of the 30 S CMC peaks. Fig. 8 depicts heavily loaded polyacrylamide gels of the purified 50 S proteins. The purity of the fractions was measured by densitometry of the longitudinally sliced and amido schwarz stained gels. Previous work had shown the agreement between color intensity and the radioactivity of uniformly labelled proteins (TRAUT, 1966). Most of the proteins are about 90 % pure by this criterion as is indicated in the lower line. Fig. 9 shows split gels of the purified 30 S proteins against 30 S total protein. The 30 S proteins appear to be somewhat more pure than the isolated 50 S proteins by visual inspection.

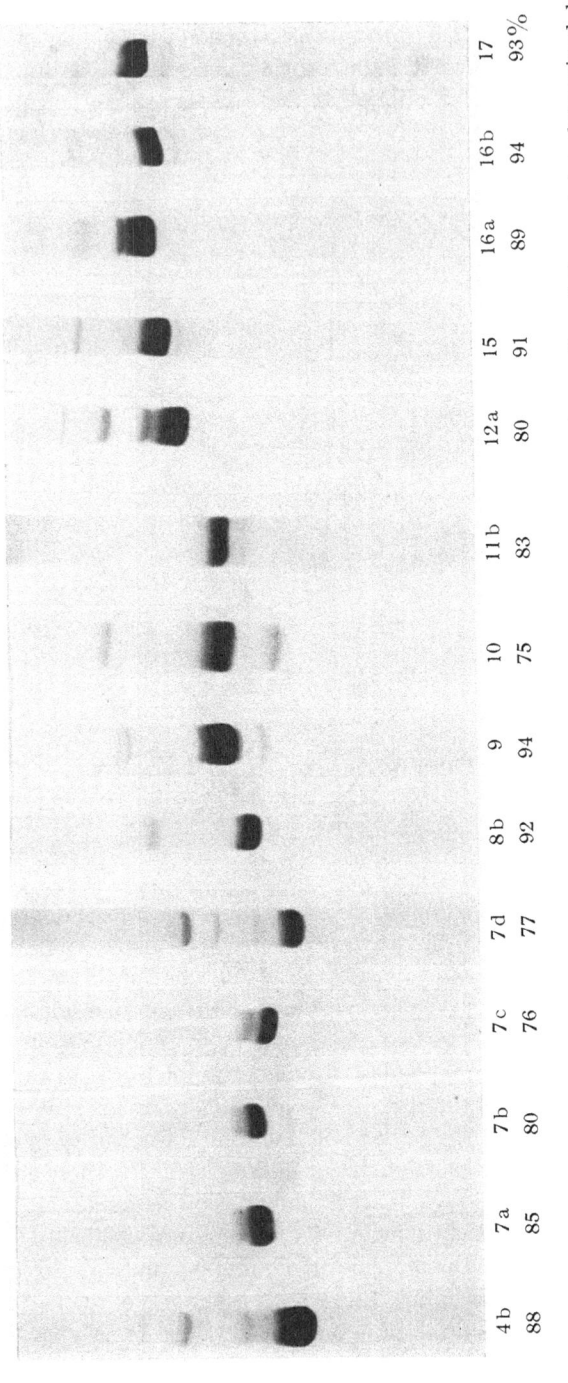

Fig. 8. Polyacrylamide Gel Analysis of Purified 50S Proteins. The percentages refer to the purity of the proteins determined by densitometry of the sliced gels

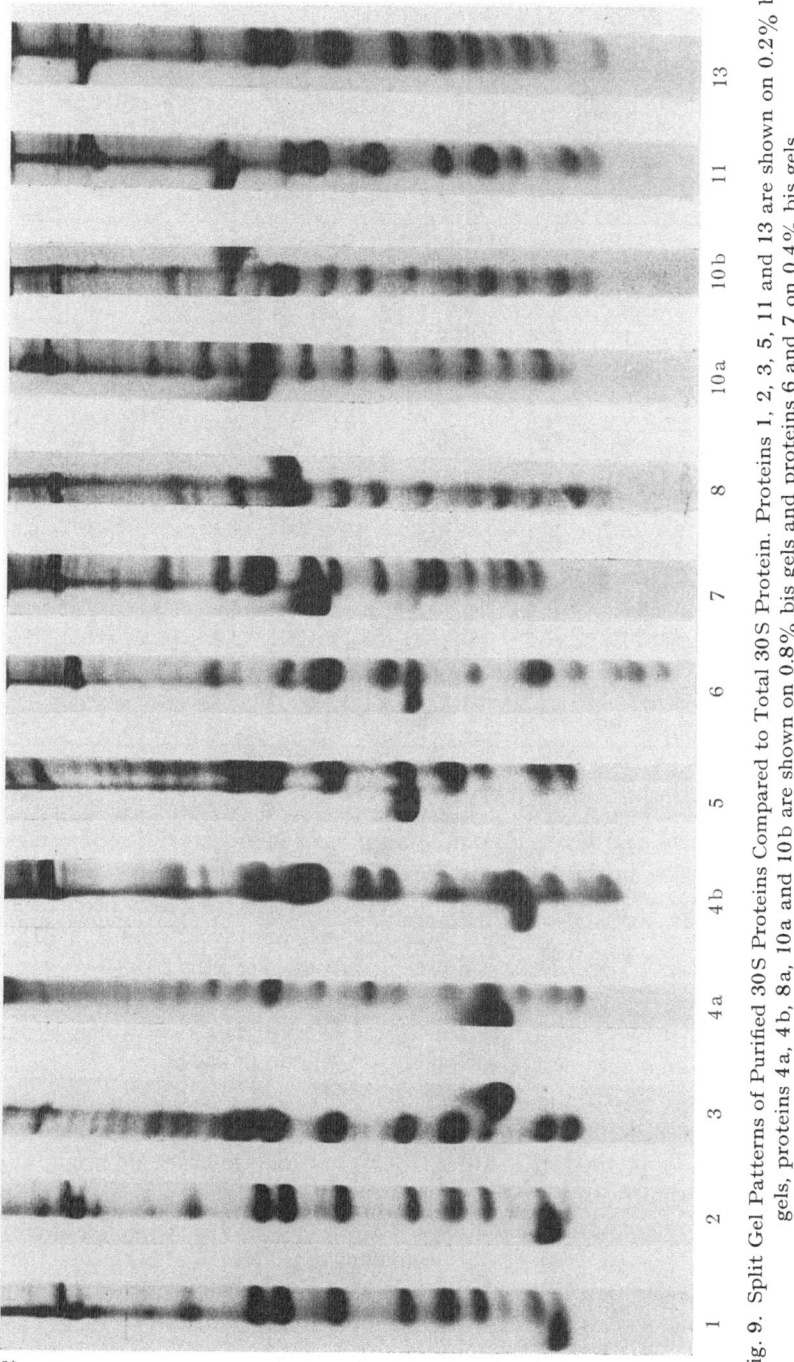

Fig. 9. Split Gel Patterns of Purified 30S Proteins Compared to Total 30S Protein. Proteins 1, 2, 3, 5, 11 and 13 are shown on 0.2% bis gels, proteins 4a, 4b, 8a, 10a and 10b are shown on 0.8% bis gels and proteins 6 and 7 on 0.4% bis gels

3*

## Chemical and Physical Analysis of the Purified Proteins

There are major differences in the amino acid compositions of nearly all the proteins isolated. As an example all the purified proteins of the 30 S und 50 S protein which have the same electrophoretic mobility in gel electrophoresis are listed in the Table 1. Most of them can be distinguished

Table 1. *Amino acid composition of selected proteins*

|      | $30_{4a}$ | $30_{4b}$ | $50_{4b}$ | $30_5$ | $50_{7c}$ | $50_{7d}$ | $30_{8a}$ | $30_{10a}$ | $50_{11b}$ | $50_{12d}$ | $30_{13}$ | $50_{22}$ |
|------|------|------|------|------|------|------|------|------|------|------|------|------|
| Lys  | 5.8  | 7.9  | 12.6 | 8.0  | 8.1  | 11.8 | 6.9  | 5.0  | 10.4 | 9.2  | 7.5  | 8.4  |
| His  | 2.4  | 4.1  | 1.0  | 1.2  | 1.0  | 3.2  | 2.2  | 4.2  | 3.0  | 2.4  | 1.7  | 0.6  |
| Arg  | 12.1 | 11.7 | 14.1 | 13.5 | 14.2 | 11.8 | 5.9  | 9.2  | 11.8 | 6.0  | 5.5  | 6.1  |
| Asp  | 10.2 | 8.4  | 6.3  | 6.1  | 5.9  | 8.0  | 8.5  | 11.0 | 11.4 | 18.3 | 11.8 | 9.5  |
| Thr  | 3.8  | 5.9  | 2.2  | 4.2  | 4.5  | 5.2  | 5.0  | 5.0  | 3.0  | 4.5  | 4.6  | 4.1  |
| Ser  | 4.2  | 5.8  | 3.9  | 5.4  | 5.8  | 5.5  | 4.1  | 3.8  | 2.4  | 3.4  | 4.7  | 3.4  |
| Glu  | 9.5  | 11.7 | 8.1  | 12.7 | 12.6 | 7.7  | 10.5 | 16.0 | 6.7  | 8.1  | 14.2 | 17.6 |
| Pro  | 2.6  | 1.6  | 0    | 2.5  | 3.0  | 2.6  | 3.0  | 3.8  | 7.0  | 5.4  | 2.0  | 3.1  |
| Gly  | 7.4  | 7.1  | 5.7  | 10.1 | 10.6 | 10.4 | 12.4 | 4.5  | 11.6 | 10.7 | 9.7  | 8.0  |
| Ala  | 13.4 | 9.1  | 17.6 | 8.5  | 9.7  | 9.4  | 11.9 | 10.9 | 8.4  | 11.5 | 8.7  | 9.6  |
| Val  | 9.8  | 6.4  | 7.4  | 7.6  | 8.0  | 8.3  | 11.1 | 8.0  | 8.9  | 10.7 | 12.0 | 10.1 |
| Met  | 1.0  | 0.7  | 0    | 2.0  | 1.5  | 0.6  | 4.0  | 4.0  | 1.3  | 0    | 0.6  | 1.7  |
| Ile  | 7.3  | 4.1  | 7.8  | 5.3  | 5.5  | 4.3  | 5.2  | 5.1  | 4.2  | 5.3  | 5.1  | 5.6  |
| Leu  | 4.9  | 11.0 | 5.0  | 7.1  | 7.5  | 6.2  | 5.2  | 4.3  | 6.1  | 6.5  | 8.3  | 6.8  |
| Tyr  | 1.2  | 2.2  | 4.4  | 3.0  | 1.7  | 0.9  | 1.1  | 2.4  | 1.8  | 2.0  | 0.6  | 1.5  |
| Phe  | 4.6  | 2.4  | 4.1  | 3.0  | 3.4  | 4.6  | 2.4  | 3.3  | 2.0  | 2.9  | 3.1  | 3.9  |

Amino acid compositions are given in moles/100 moles amino acids.

by differences in amino acid composition. Some obviously different values are underlined in the table. An exception are the proteins $30_5$ and $50_{7c}$ which show only minor differences. Since they are also eluted from the CMC column at about the same ionic strength, are rather soluble in 10 % TCA and have no detectable amino terminal end group they may represent a protein which is part of both the 30 S and the 50 S subunit. However the differences between other proteins show clearly that two proteins can have the same electrophoretic mobility without having identical structures.

Amino end group analysis was done according to Waller and Harris (1961) by the method of Sanger (1945), but on a micro scale using ³H-labelled F-dinitro-benzene in the reaction. Table 2 shows the results of the end group determination. So far 10 met end groups, 4 ala, 1 thr, and 1 ser have been detected. From the amount of protein used in the reaction and the yield of labelled end groups a molecular weight of the different proteins was calculated. The last column shows the ratio of these end group molecular weights to the physical molecular weights described below. The values lie between 1,5 and 2,5. The proteins never reacted to 100 % under the conditions employed; therefore the calculated molecular weights are always higher than the physical molecular weights.

Table 2. *End groups and molecular weights by end group yields of ribosomal proteins*

|  | End group | End group Molecular weight | EG/SE |
|---|---|---|---|
| 30S Proteins |  |  |  |
| 3 | Thr | 16,000 ± 15% | 1.5 |
| 4a | Met | 14,000 | 1,6 |
| 7 | B | 150,000 | 10.0 |
| 10a | Met | 8,500 | 1,5 |
| 11 | Ser | 22,000 | 1.6 |
| 50S Proteins |  |  |  |
| 4b | Ala | 33,100 | — |
| 5 | Ala | — | — |
| 7a | Met | 16,500 | 1,3 |
| 7b | Met | 20,000 | — |
| 7c | B | 111,000 | — |
| 7d | B | 82,000 | — |
| 8b | B | 128,000 | 9.7 |
| 9 | Met | 34,000 | 1.9 |
| 10 | Met | 32,000 | — |
| 11b | Ala | 56,400 | 1.9 |
| 12a | B | 71,000 | — |
| 15 | Met | 48,500 | 2.3 |
| 16a | Met | 32,000 | 1.6 |
| 16b | Met | 42,500 | 2.5 |
| 17 | Ala | 49,500 | 2.2 |
| 22 | Met | 59,000 | 1.3 |

10 × Met; 4 × Ala; 1 × Thr; 1 × Ser; 5 × No End group.
The last column gives the ratio of molecular weights determined by end group analysis to the molecular weights determined by sedimentation equilibrium.

There are 5 proteins which yielded only very small amounts of labelled end group, usually with no single amino acid being predominant. End group molecular weights calculated for two of them were found to be 10 times higher than the physical molecular weight. These proteins we assume to have blocked end groups. The nature of the blocking agent has not been determined.

Fingerprints were made of the tryptic digests of 13 of the purified 30 S proteins and of nine of the purified 50 S proteins. The 30 S fingerprints are shown in Fig. 10 together with the fingerprint of the total 30 S protein. All these fingerprints show clearly different patterns, and the same observation was made for the 50 S proteins. From 10 different 30 S

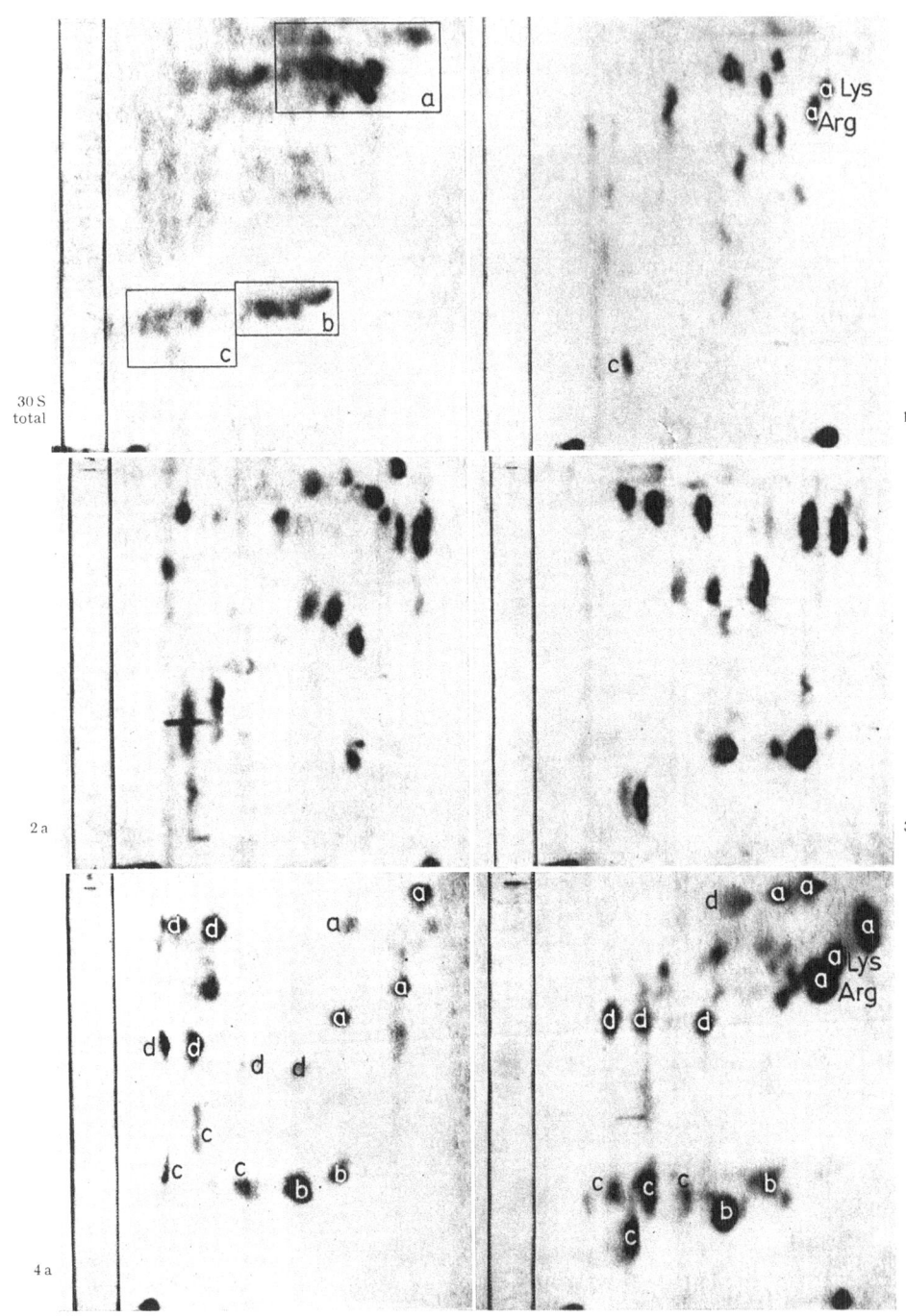

Fig. 10a

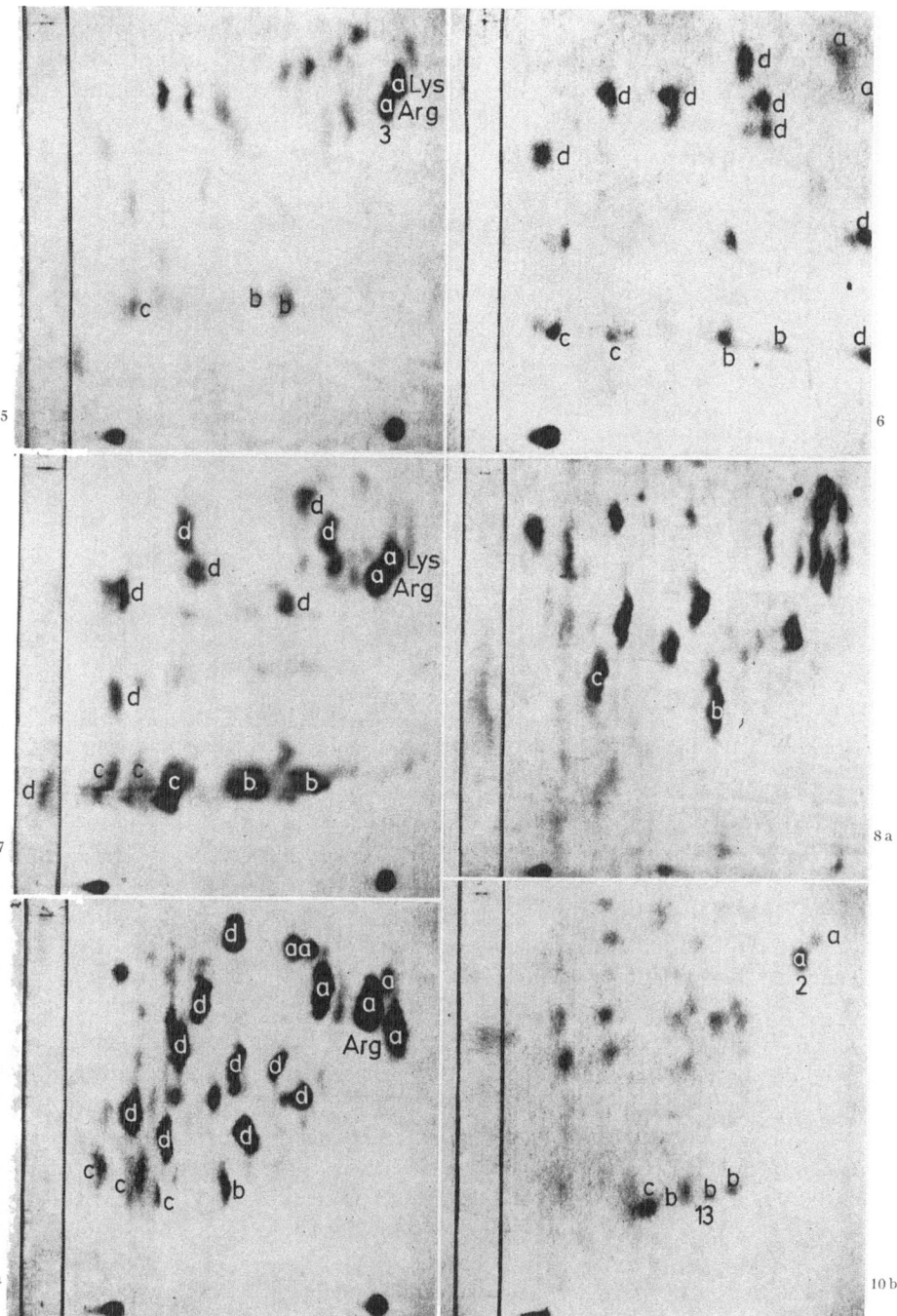

Fig. 10 b

fingerprints 75 spots located in regions which are heavily stained in the total 30 S fingerprint were eluted and submitted to amino acid analysis. No common tryptic peptides were found. Therefore it seems improbable to us that there exist long regions of identical sequence in the different proteins analysed so far. But a definite answer will be given only by a complete sequence determination of the pure proteins.

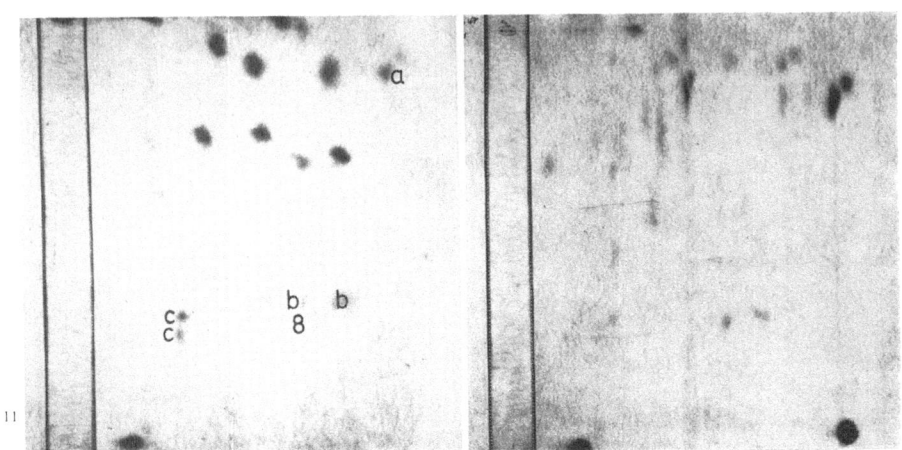

Fig. 10 c

Fig. 10a—c. Tryptic Fingerprints of Total and Purified 30 S Ribosomal Protein. The fingerprints are oriented so that the vertical direction is the chromatographic dimension and the horizontal, that of electrophoresis. The origin is in the upper left of the fingerprints and the cathode is at the right. Three regions containing intense spots in the total 30 S fingerprint are indicated by a, b, and c. Spots in fingerprints of individual proteins which fall into these regions are designated by the letter corresponding to their region.

The physical molecular weights of the proteins were determined by sedimentation equilibrium runs in the ultracentrifuge (Yphantis, 1960). The samples were dialysed against a buffer containing 6 M urea, 0.1 M NaCl, 0.01 M sodium acetate, pH 5.6, and 0.01 M beta-mercaptoethanol. Bovine serum albumin and lysozyme were run as standards in this buffer and the results were in agreement with published values. In Table 3 the molecular weights of some of the 30 S and 50 S proteins are listed. They range from 4,500 to about 40,000, most of them being between 10- and 20,000. The weight average of the 13 isolated proteins of the 30 S subunit is 16,600, a little higher than the measured molecular weight of the total 30 S protein, found to be 14,100.

The total 70 S protein gives a molecular weight of 21,000, similar to the value of 23,000 determined by Möller and Chrambach (1967). This is definitely higher than the average molecular weight determined for the 30 S and 50 S protein separately. Also the mixture of one part of 30 S protein plus two parts of 50 S protein gives a molecular weight of 21,000, and the value is not changed by the use of 6 M guanidine-HCl instead of the 6 M urea normally used for the centrifuge runs. We believe that this is caused by a strong and possibly specific interaction between 30 S and 50 S proteins.

Table 3. *Sedimentation equilibrium molecular weights of ribosomal proteins*

| 30 S Proteins | | 50 S Proteins | |
|---|---|---|---|
| 1 | 4,500 ± 7% | 7a | 12,700 ± 15% |
| 2a | 13,000 | 8b | 13,200 |
| 3 | 10,700 | 9 | 18,000 |
| 4a | 9,000 | 11b | 30,200 |
| 4b | 5,500 | 15 | 20,800 |
| 5 | 12,800 | 16a | 19,800 |
| 6 | 13,700 | 16b | 17,100 |
| 7 | 14,500 | 17 | 22,900 |
| 8a | 20,000 | 22 | 44,300 |
| 10a | 5,600 | | |
| 10b | 27,600 | | |
| 11 | 14,000 | | |
| 13 | 23,000 | | |
| Total | 14,100 | Total | 15,200 |

70 S Total 21,000

## Quantitation

The ratio between the amount of a given protein present on the ribosome and its molecular weight gives the number of copies of that protein which are present. This calculation has been made for the 13 30 S proteins which have been isolated.

The quantities of the different species in the total 30 S protein mixture were determined by means of polyacrylamide gel electrophoresis of radioactive uniformly labelled 30 S protein. The gels were stained and cut into slices according to the visible band pattern. The sections of the gels were solubilized and the amount of radioactivity was determined by scintillation counting. Fig. 11 shows the amount of radioactivity in the different sections of the gels expressed as percent of the total amount of radioactivity found in the gels. From the percentage of the radioactivity found in each protein band the amount of this protein species can be calculated taken the total radioactivity to be equal to the 30 S protein weight of 330,000 dalton units (Tissières, Watson, Schlessinger, and Hollingworth, 1961).

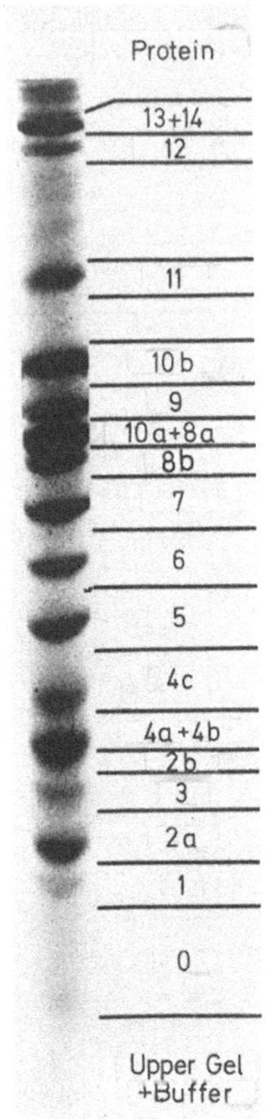

| Protein | % | Total CPM[1] | Daltons Protein/ 30 S Subunit | Sed Eqlbm M W | Copies |
|---|---|---|---|---|---|
| | 10.1 | | 24,400 | | |
| 13+14 | | 7.4 ± 1.6 | 10,200 | | |
| 12 | | 3.1 ± .38 | | | |
| 11 | 5.3[2] | 4.6 ± .48[2] | 15,200 | 14,000 | 1.1 |
| 10b | 1.5 | 6.3 ± .80 | 20,800 | 27,600 | 0.76 |
| 9 | | 6.0 ± 1.0 | 19,800 | | |
| 10a+8a | | 8.6 ± 1.4 | 28,400 | 25,600 | 1,1 |
| 8b | | 4.7 ± .68 | 15,500 | | |
| 7 | | 5.2 ± .80 | 17,200 | 14,500 | 1.2 |
| 6 | | 4.8 ± .12 | 15,800 | 13,700 | 1.2 |
| 5 | | 4.8 ± .65 | 15,800 | 12,750 | 1.2 |
| 4c | | 4.2 ± .81 | 14,800 | | |
| 4a+4b | | 5.0 ± .75 | 16,500 | 14,500 | 1.1 |
| 2b | | 2.0 ± .20[3] | 6,600 | | |
| 3 | | 2.7 ± .70[4] | 8,900 | 10,700 | 0.83 |
| 2a | | 4.3 ± .80[2] | 14,200 | 13,700 | 1.0 |
| 1 | | 1.6 ± .14 | 5,300 | 4,550 | 1.2 |
| 0 | | 0.8 ± .20 | 2,600 | | |
| Upper Gel +Buffer | | 7.0 | 2,600 | | |

Fig. 11. Distribution of [14]C 30 S Ribosomal Protein Among the Bands Resolved on Polyacrylamide Electropherograms. The gels were stained with amido schwarz and then cut into sections as indicated by the black horizontal lines. The gel sections were dissolved and counted. The radioactivity contained in each section is expressed as a percent of the total recovered, including the upper gel and the buffer compartment between the spacer gel and a piece of dialysis tubing held over the top of the gel column during the run. The gel is made from 7.5% acrylamide and 0.8% bisacrylamide. 1 average of 8 gels. 2 average of 7 gels, 3 average of 2 gels. 4 average of 6 gels

About 25 % of the label is not found in the major bands but is found in the upper gel, the top part of the gel and in the space between the major protein bands. This "minor band" material is also found as a background in the CMC column elution pattern. It seems to be related to the presence of many different proteins present only in small amounts in the ribosomes as for example the initiation factors (F. GROS, personal communication), or the ribosomal RNase which was shown to be present as only one molecule per 10 ribosomes (SPAHR and HOLLINGWORTH, 1961). Since the minor band material is presumed to be present in the ribosome isolated by the method of TISSIÈRES (TISSIÈRES, WATSON, SCHLESSINGER, and HOLLINGWORTH, 1959) used both here and in his determination of the ribosomal molecular weight, it is included in our calculation.

Seventy-five percent of the label is found in the major bands. The amount of the individual bands is calculated and listed in dalton units. The ratio of the amount of protein found in the gel to the molecular weight determined for the purified proteins should indicate how many copies of each of these protein species are present in the 30 S protein mixture. These ratios are listed in the last column of Fig. 11. Within the experimental error in all cases a ratio of one is found. Thus this group of proteins are present in equimolar amounts, and there is one copy of each per 330,000 daltons of protein. The most probable conclusion is that there is one copy of each per 30 S ribosomal subunit, and that all 30 S ribosomes have exactly the same composition with one copy of all of the major proteins.

Assuming the presence of one copy per 30 S subunit for all major proteins one can equate the daltons of protein indicated by the radio-activity data with the molecular weight and thereby predict the molecular weight of those proteins not yet isolated but only identified on gels. These new molecular weights combined with the molecular weights of the known proteins lead to a calculated weight average molecular weight of 15,900 which is in rather good agreement with the measured value of 14,100.

## Summary

1. Proteins of 30 S and 50 S E. coli ribosomal subunits were separated by chromatography on carboxymethyl cellulose columns and they were further purified to about 90% purity by chromatography on Sephadex G 100.

2. They show differences in amino acid composition, amino end groups, fingerprints of tryptic peptides, composition of isolated tryptic peptides and molecular weights.

3. The quantitative distribution of the 30 S proteins was determined and compared to the molecular weights of 13 proteins which had been isolated. One copy of each of these 13 proteins was found per 330,000 daltons of 30 S protein.

4. This strongly suggests that ribosomes are homogeneous and that the protein of each subunit contains one copy of each of the major protein species.

## References

ATSMON, A., P. SPITNIK-ELSON, and D. ELSON: J. molec. Biol. 25, 161 (1966).
BOLTON, E. T.: In: CANTONI, G. L., and D. R. DAVIES, (Eds.): Procedures in Nucleic Acid Research, New York: Harper & Row 1966.
BRAKKE, M. K.: Arch. Biochem. 107, 388 (1964).

Cammack, V. A., and H. E. Wade: Biochem. J. **96**, 621 (1965).
Fleischmann, J. B., R. R. Porter, and E. M. Press: Biochem. J. **88**, 220 (1963).
Fraenkel-Conrat, H.: Virology **4**, 1 (1957).
Leboy, P. S., E. C. Cox, and J. G. Flaks: Proc. nat. Acad. Sci. (Wash.) **52**, 1367 1964).
Möller, W., and A. Chrambach: J. molec. Biol. **23**, 377 (1967).
Sanger, F.: Biochem. J. **39**, 507 (1945).
Spahr, P. F., and B. R. Hollingworth: J. biol. Chem. **236**, 823 (1961).
Tissières, A., J. D. Watson, D. Schlessinger, and B. R. Hollingworth: J. molec. Biol. **1**, 221 (1959).
Traut, R. R.: J. molec. Biol. **21**, 571 (1966).
—, P. B. Moore, H. Delius, H. Noller, and A. Tissières: Proc. nat. Acad. Sci. (Wash.) **57**, 1294 (1967).
Waller, J. P.: J. molec. Biol. **10**, 319 (1964).
—, and J. I. Harris: Proc. nat. Acad. Sci. (Wash.) **47**, 18 (1961).
Yphantis, D. A.: Ann. N. Y. Acad. Sci. **88**, 586 (1960).

## Discussion

*Stöffler:* We want to demonstrate a new method for electrophoretic separation of ribosomal proteins on gelatinized cellulose acetate. Only $5-40$ $\mu$g ribosomal protein are applied. The whole electrophoretic procedure (including staining and destaining) takes only about 2 h. The buffers used were 0.03 M Na-phosphate, pH 6.7, and a tris-EDTA-boric acid buffer system, pH 8.9 both containing, 8 M urea and 0.05 M $\beta$-mercapto-ethanol. The separation pattern of E. coli B and E. coli K ribosomal proteins is shown in following figure:

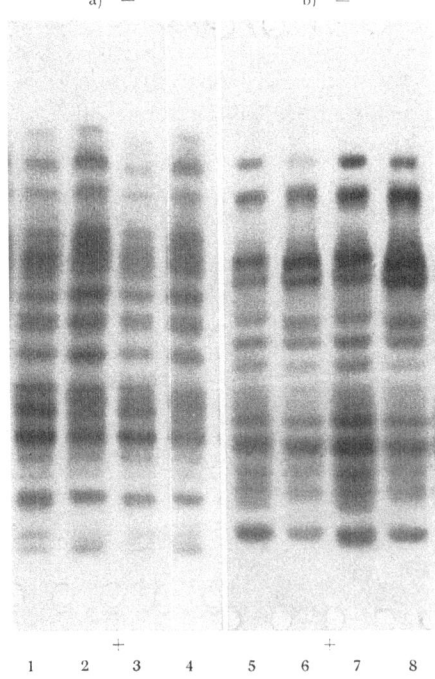

a) Separation of ribosomal protein E. coli B and E. coli K on gelatinized cellulose acetate (Cellogel-Chemetron Milan); 0.03 M Na-phosphate buffer, pH 6.7, 8 M urea, 0.05 M mercapto-ethanol; 200 V, $3-5$ mA running time 70 min; staining with amido black. 2, 4 Coli B ribosomal protein; 1, 3 Coli K 12 ribosomal protein

b) Separation pattern in a tris-EDTA-boric acid buffer at pH 8.9, 8 M urea and 0.05 M $\beta$-mercapto-ethanol. 200 V, $1.5-3.0$ mA; running time **85** min; staining with amido black. 5, 7 Coli B ribosomal protein; 6, 8 Coli K 12 ribosomal protein. Origins are indicated by arrows. For more details see: G. Stöffler, molec. gen. Genetcis **100**, 374 (1967)

Electrophoretic patterns are very similar to those obtained by starch gel electrophoresis. In contrast to polyacrylamide electrophoresis only little or no sieving effects is responsible for the separation.

*Delius:* May I ask if you analysed the protein samples which you isolated from the preparative polyacrylamide-electrophoresis by this method ?

*Stöffler:* So far we have checked only five isolated proteins with this method. Four of these proteins gave single bands in this electrophoretic system, too. One protein, homogeneous in disc electrophoresis showed two bands in cellulose acetate electrophoresis. This particular one was ommitted in our paper.

# Structural Studies on Ribosomal Proteins:
# Peptides from Cyanogen Bromide Cleavage
# and Tryptic Digestion

By

W. Rombauts*

Virus Lab., Univ. of California, Berkeley, Calif., U.S.A.

Ribosomal proteins were prepared from *E. coli* B and A 19 ribosomes and ribosomal subunits following Spitnik-Elson's method [Biochem. biophys. Res. Commun. *19*, 557 (1965)]. Spectral properties indicated that less than 1 % residual RNA was present. Amino acid analysis of proteins derived from whole ribosomes showed 2.5 % of the residues to be methionine. The ribosomal proteins were subjected to cleavage at the C-terminal side of the methionine residues by treatment with cyanogen bromide in 0.1 N HCl for 30 h at 30°C [Gross and Witkop: J. biol. Chem. *237*, 1856 (1962)]. Under these conditions only 30 % of the material became dializable, as checked by the recovery of amino acids and by the distribution of counts when $^{14}$C labeled proteins were used. This fraction contained about 60 % of the original methionine recovered as homoserine and its lactone. The remainder was a high molecular weight substance which showed a similar sedimentation pattern as the untreated material. It contained approximately 35 % methionine, recovered as homoserine and its lactone. The total extend of the cleavage reaction approached, therefore, 100 %. On polyacrylamide gel run at pH 4.5 in 8 M urea [Leboy et al.: Proc. nat. Acad. Sci. (Wash.) *52*, 367 (1964)] the high molecular weight material showed only about 15 bands as opposed to the 35 bands observed with untreated protein (Fig. 1 and 2). A similar reduction of the number of bands was observed when proteins from the 30S or 50S subunits only were used.

Ribosomal proteins were also reduced, carboxymethylated and digested with trypsin. A small fraction of the digest was insoluble in pyridinium acetate buffer of pH 3.5. Aliquots of the supernatant corresponding to 0.1 μM of total ribosomal protein were chromatographed on

* Present address: Department of Biochemistry, Faculty of Medicine, University of Leuven, Leuven, Belgium

Spinco PA-35 resin following the method of JONES (Technicon Symposium, New York, 1967) with a pyridine acetate gradient.

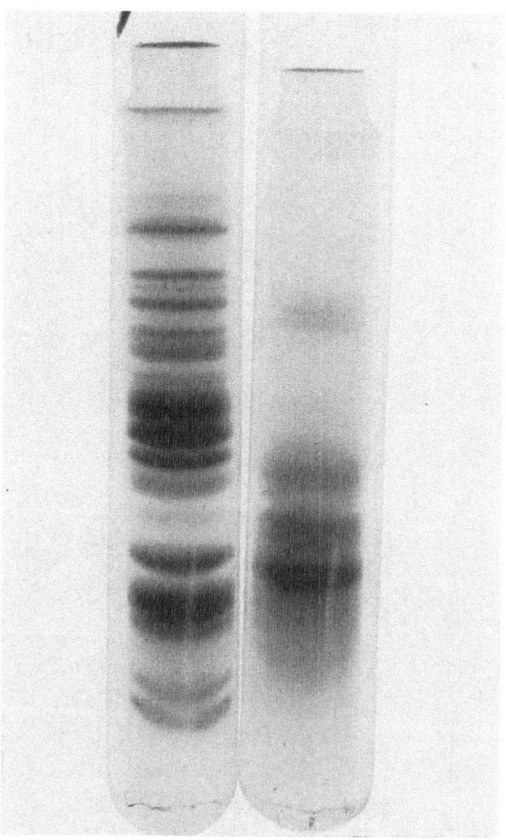

Fig. 1. Acrylamide gel electrophoresis patterns of normal ribosomal proteins (left gel) and CNBr-treated ribosomal proteins (right gel), stained with naphtol blueblack. The electrophoresis was done at pH 4.5 in 8 M urea, according to LEBOY et al.: Proc. nat. Acad. Sci. (Wash.) 52, 367 (1964). The site of application is on top of the spacer gel, at the top of the picture. Migration is towards the anode, at the bottom of the picture. About 100 μg was applied in the case of the untreated proteins, and about 70 μg, the amount remaining after dialysis, in the case of the CNBr-treated proteins

Peptide patterns were obtained showing about 40 peaks in amounts ranging from 0.05 to 0.2 μM. Peptide patterns of proteins from whole ribosomes and from the 50S subunit were very similar. Those of the proteins from the 30S subunit showed some differences although a general

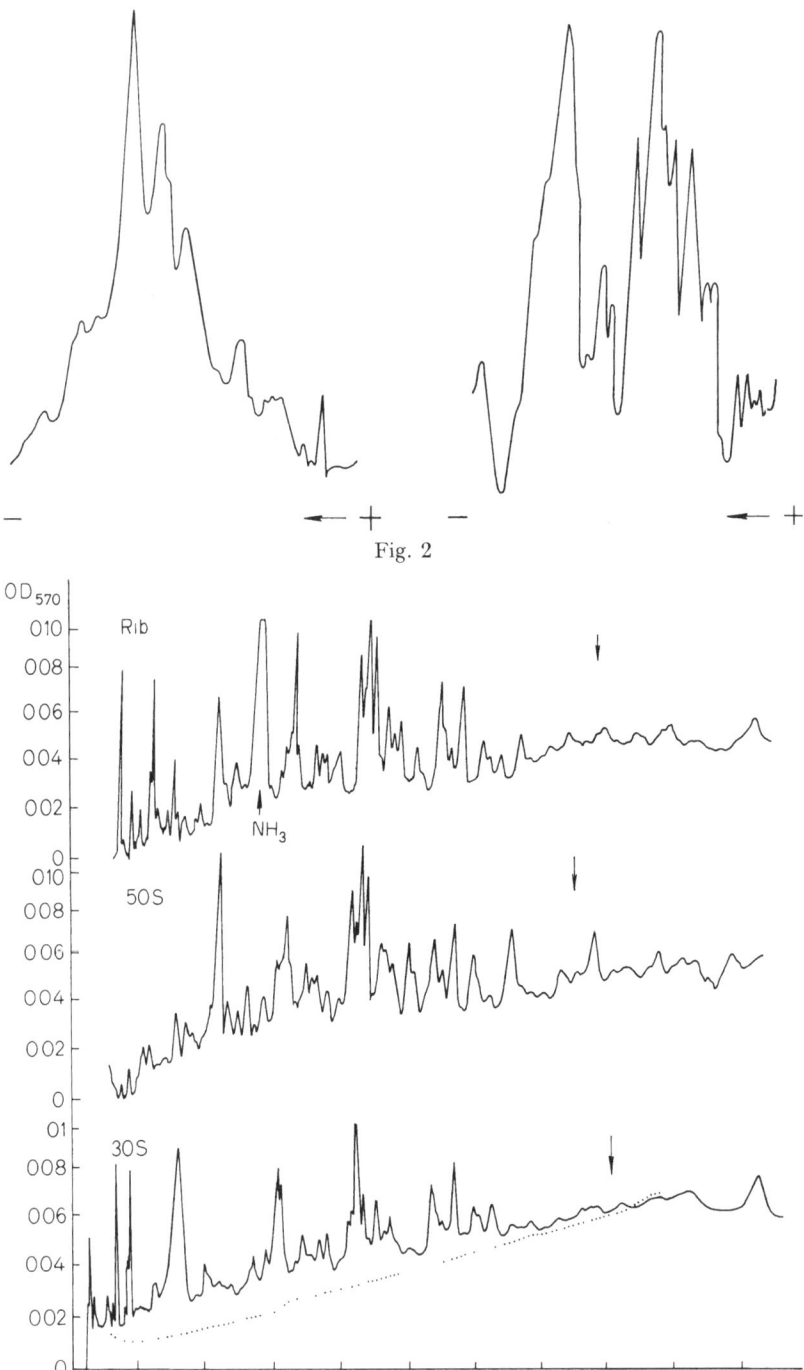

Fig. 2

Fig. 3

correspondence with the 50 S and whole ribosome patterns is obvious. In addition, a few peaks are specific for the 30 S or the 50 S pattern (Fig. 3). If all ribosomal proteins had grossly different sequences, a much larger number of tryptic peptides should be found in far smaller yield. Similarly, cyanogen bromide cleavage should produce a number of different peptides larger than the number of chains originally present.

On the basis of opposite evidence from the cyanogen bromide and tryptic digestion experiments, it is therefore suggested that ribosomal proteins may possess a large degree of homology in their sequences.

Supported by NSF grant GB-3107 and in part by the Nat. Fund for Scientific Research, Belgium

## Discussion

*Zillig:* Is the resolution power of the method used for separation of peptides high enough to ensure, that bands are not corresponding to still complex mixtures of similar peptides ?

*Rombauts:* The method I have used for separating the tryptic peptides from ribosomal protein is at least equal to and often more powerful than the combination chromatography/electrophoresis ("fingerprinting"). It has frequently been used to separate the abnormal tryptic peptides from hemoglobin variants which often differ from the normal protein by only a single amino acid exchange. However, the major peptide peaks from the chromatogram are now being analysed with additional separation methods to check the possible occurence of mixtures of similar but different peptides.

---

Fig. 2. Densitometer tracings of acrylamide gels such as in Fig. 1. The right tracing is from a run with untreated ribosomal proteins and the left one from a run with CNBr-treated and dialized ribosomal proteins. Planimetry of the tracings shows that the surface under the left curve is about 70% of the surface under the right tracing

Fig. 3. Tryptic peptide patterns obtained with ribosomal proteins from whole ribosomes and the 50 S and 30 S particles. Peptide separation was done on PA-35 Beckman Custom Research Resin in a pyridine-acetate gradient pH 3.5 → 5.0 following BENSON et al.: Anal. Biochem. 16, 91 (1966). In each case about 1.5 mg peptides were applied. The dashed curve represents a blank run to show the increase of the base line due to ninhydrine positive impurities present in the pyridine. The arrows indicate the end of the gradient and the start of elution with limiting buffer pH 5.0

# The Role of 30 S Ribosomal Subunits in Initiation of Protein Synthesis

M. Nomura

University of Wisconsin, Laboratory of Genetics, Madison, Wisconsin, U.S.A.

## Background

N-formylmethionyl-tRNA (fMet-tRNA$_F$)[1] is involved in the initiation of protein synthesis in *E. coli*, and the triplet for fMet-tRNA$_F$, AUG (and possibly GUG and others), is most likely the signal for initiation of synthesis of polypeptide chains [1–8]. The most widely entertained hypothesis of the mechanism of chain initiation is that 70S ribosomes bind with a high affinity to the initiator codon in mRNA and/or fMet-tRNA$_F$, leading to formation of a 70S-mRNA-fMet-tRNA$_F$ complex at the starting site on mRNA. An essential feature of this hypothesis is that the postulated affinity of 70S ribosomes for the initiator codon (and/or fMet-tRNA$_F$) is so high that it *excludes* all possible interactions of 70S ribosomes with other codons leading to binding of aminoacyl tRNA's other than fMet-tRNA$_F$ (see, for example, ref. [6]). However, the vast majority of experiments have shown that 70S ribosomes can make stable mRNA-70S-aminoacyl-tRNA complexes using a variety of synthetic mRNA's and corresponding aminoacyl-tRNA's. Thus, it is difficult to believe this hypothesis in its simplest form unless we make a further assumption that only the initiator codon is "exposed" and *all* other codons are masked by a secondary structure of RNA in the natural mRNA.

## Alternate Model

These considerations led us to turn our attention to the ribosomal subunits, the 30S and 50S particles, and to formulate a specific alternative hypothesis [9, 10].

Our model proposes the following: (a) *In vivo* there exist only free 30S and 50S particles, and polysomes, no free 70S ribosomes. (The paper by MANGIAROTTI and SCHLESSINGER [11] suggesting that all the ribosomes

---

[1] tRNA$_F$ refers to the methionine-tRNA species that can be formylated and tRNA$_M$ refers to the one that cannot be formylated (see ref. [2]).

in *E. coli* exist either as a part of polysomes or as free 30S and 50S particles was available to us when the present model was formulated.) (b) 30S particles have a high affinity for the initiator codon (AUG or related nucleotide sequences) and/or fMet-tRNA$_F$, and form a 30S-mRNA-fMet-tRNA$_F$ complex ("initiation complex") but do not respond to any other codons to bind other aminoacyl tRNA's. (c) 30S−50S complexes, that is 70S particles, are able to respond to all the codons and bind corresponding aminoacyl tRNAs, but are unable to bind fMet-tRNA$_F$ without prior dissociation into free 30S particles. (d) *In vivo*, 30S−50S complexes can be formed only via the formation of an initiation complex (30S-mRNA-fMet-tRNA$_F$ complex) as an obligatory intermediate. (e) Polypeptide chain termination causes dissociation of the 30S−50S complex into the two subunit particles simultaneously with the release of polypeptidyl-tRNA.

Thus, according to the proposed model, ribosomal particles are present in the pool *in vivo* as free 30S and 50S particles, not as 70S particles, and the only site where they are able to bind aminoacyl-tRNA is at the initiator codon and the first complex is the initiation complex consisting of 30S, fMet-tRNA$_F$ and mRNA. Following the formation of the initiation complex, a 50S particle joins and the resultant 30S−50S-mRNA aggregate now acquires the capacity to bind ordinary aminoacyl-tRNA with high efficiency and will bind the aminoacyl-tRNA specified by the second codon. Peptide bond formation will then follow.

At the end of the cistron, the chain termination codon will cause dissociation of the 30S−50S complex as stated above, and the liberated 30S particles, either still on mRNA or close to the mRNA, would then be ready to recognize the initiator AUG codon of the next cistron.

## Experimental Evidence for New Model

The experimental results obtained in our laboratory to support the present model are as follows [9, 10]: (a) Using purified ribosomal particles (not treated with high concentration of salts) RNA from bacteriophage f2 was shown to stimulate, in the absence of "initiation factor" [12−15], binding of fMet-tRNA$_F$ to 30S particles, but not to 50S particles or 70S particles. The presence of 50S particles inhibited the binding of fMet-tRNA$_F$ to 30S particles. The stimulation of fMet-tRNA$_F$ binding to the 30S particles by f2 RNA is specific to this tRNA alone. Binding of no other tRNA tested was stimulated by f2 RNA. (b) The 30S−50S complex or the 70S particle, on the other hand, was more efficient than the 30S particle in the binding of tRNA (including Met-tRNA$_M$; see Footnote) other than fMet-tRNA$_F$, when synthetic polynucleotides containing appropriate codons were used as messengers. (c) A synthetic polynucleotide

4*

52     M. NOMURA:

containing A, U and G (1:1:1) in random sequence ("random" poly AUG)
stimulated binding of fMet-tRNA$_F$ to 30S particles in the absence of
initiation factor. The presence of 50S particles inhibited the binding of
fMet-tRNA$_F$ to ribosomal particles. (d) The initiation complex consisting
of the 30S particle, mRNA, and fMet-tRNA$_F$ was isolated. The addition
of 50S particles to this initiation complex converted it to a complex
consisting of the 30S particle, the 50S particle, mRNA and fMet-tRNA$_F$.
This joining of 50S particles to the initiation complex was shown not to
involve dissociation of fMet-tRNA$_F$ from the initiation complex. (e) It
was observed that, in the presence of initiation factor, but not in its
absence, phage f2 RNA stimulates the binding of fMet-tRNA$_F$ to 70S
particles, but that this binding is also specifically inhibited by added 50S
particles. The poly U directed binding of Phe-tRNA to 70S particles was
not inhibited by 50S particles.

## Discussion of Experimental Results

These results indicate that the 30S ribosomal structure permits interaction of the AUG codon in mRNA (either f2 RNA or "random" poly AUG)
with the anticodon on fMet-tRNA$_F$, but not interaction of other codons
with the anticodons of the corresponding tRNA's. On the other hand, the
opposite is true with the 30S—50S complex or 70S ribosomal structure.

According to our view, the weak binding of ordinary tRNA such as
Phe-tRNA to 30S particles directed by synthetic polynucleotides [16—18]
is an *in vitro* non-physiological binding reaction which may utilize the
binding site restricted only to the initiation tRNA under *in vivo* physiological conditions. In fact, binding of several other tRNA's takes place
only on the 30S—50S complex and not on the 30S particle [9].

Convincing evidence for the above conclusions is provided by
the experiments done using "random" poly AUG. This poly-nucle-

Table 1. *Binding of fMet-tRNA$_F$ and Val-tRNA to ribosomes stimulated by random*
*poly AUG*

| Aminoacyl-tRNA | Ribosomal particles | tRNA bound ($\mu\mu$ moles) | | |
|---|---|---|---|---|
| | | — polymer | + poly AUG | Stimulation |
| f-Met-tRNA$_F$ | 30S | 0.10 | 1.10 | 1.00 |
| | 50S | 0.04 | 0.03 | 0 (−0.01) |
| | 30S + 50S | 0.23 | 0.30 | 0.07 |
| Val-tRNA | 30S | 0.12 | 0.12 | 0 |
| | 50S | 0.09 | 0.13 | 0.04 |
| | 30S + 50S | 0.23 | 0.68 | 0.45 |

Data are taken from the paper by NOMURA and LOWRY [9].

otide contains the initiator codon, AUG, and in addition, codons for several ordinary amino acids. It is particularly rich in codons for valine (GUA, GUU, GUG). Therefore, the binding of val-tRNA and of fMet-tRNA$_F$ was compared. As shown in Table 1, 30S particles in the absence of 50S particles showed good binding of fMet-tRNA$_F$, but none of Val-tRNA. Addition of 50S particles to 30S particles caused good binding of Val-tRNA, but very little binding of fMet-tRNA$_F$. Thus, the 30S ribosomal structure permits interaction of the AUG codon with the anticodon on fMet-tRNA$_F$ selectively, whereas the 70S (30S−50S complex) structure interferes with this interaction, and permits other kinds of codon-anticodon interaction.

In the presence of initiation factor, the inhibition of fMet-tRNA binding by 50S particles is diminished, and phage f2-RNA-directed binding of fMet-tRNA can be observed with 70S particles. The effect of the initiation factor can be interpreted in the following way: the 70S particle preparation always contains a small amount of "free" 30S particles in equilibrium with the 70S particles. The initiation factor accelerates the binding of fMet-tRNA$_F$ to these free 30S particles, and this, in turn, results in further dissociation of 70S particles. Thus, the sequence of reactions would be as follows:

(1) $70S \rightleftharpoons 30S + 50S$

(2) $30S + f2 \ RNA + fMet\text{-}tRNA_F$

$\xrightarrow{\text{Initiation factor}} 30S-f2 \ RNA\text{-}fMet\text{-}tRNA$ complex

(3) $30S-f2 \ RNA\text{-}fMet\text{-}tRNA$ complex $+ 50S$
$\rightarrow 30S-50S\text{-}f2 \ RNA\text{-}fMet\text{-}tRNA$ complex.

The presence of excess amounts of 50S particles overcomes, at least partially, the effect of the initiation factor by decreasing the fraction of free 30S particles in equilibrium with the 70S particles.

## Other Work Related to the Model

GHOSH and KHORANA have recently confirmed our model using a polymer with the repeating sequence AUG as a mRNA [19]. They first isolated the initiation complex consisting of 30S particles, poly AUG and H³-fMet-tRNA$_F$ in the absence of 50S particles, and then added stepwise 50S particles and C¹⁴-Met-tRNA$_M$ to the initiation complex. Isolation of the resultant complex and mixing of this complex with the supernatant protein fraction resulted in formation of a dipeptide, H³-formylmethionyl C¹⁴-methionine. Thus, under these experimental conditions, the initiation of formation of peptide bonds can certainly take place on *free* 30S particles.

The cyclic dissociation and reassociation of ribosomal particles concurrent with peptide chain termination and with initiation has also been proposed by other workers on the basis of other experimental results. SCHLESSINGER, MANGIAROTTI, and APIRION examined conditions necessary for *in vitro* formation of 70S particles from "native" 30S and 50S particles [20]. They found that these native 30S and 50S particles do not associate spontaneously, but can form 70S particles in the presence of mRNA, tRNA, and $K^+$ and $Mg^{++}$ ions, and that mRNA containing the AUG sequence is more effective than poly U, in promoting association. Furthermore, they showed dissociation of 70S particles into 30S and 50S particles as a result of chain termination with puromycin following polyphenylalanine synthesis. These results do not prove that the initial step in the protein synthesis is the selective binding of fMet-rRNA$_F$ to the *30S particles* rather than to the 70S particles, but the results are certainly consistent with the model described here.

The other work pertinent to the present model is that done by KAEMPFER, MESELSON, and RASKAS [21]. These workers examined the distribution of isotopic labels among 70S particles as well as 50S and 30S ribosomal subunits following transfer of a growing bacterial culture from a heavy to a light isotopes medium. They have demonstrated that (a) 30S and 50S particles remain intact, and (b) 70S particles undergo subunit exchange. This exchange occurs presumably by dissociation into their 30S and 50S subunits and reformation from a pool of these subunit particles. Their experimental results are convincing and subunit exchange *in vivo* has been firmly established. However, these experiments do not prove the hypothesis that ribosomes must dissociate into their subunits after *every* round of protein synthesis. It would be obviously important to perform similar experiments using an *in vitro* protein synthesizing system to establish the present model more firmly.

BISHOP has presented data suggesting that the *de novo* chain initiation of hemoglobin synthesis in the rabbit reticulocyte system is accomplished mainly by ribosomal subunits [22, 23]. Although in animal cells no "initiation tRNA" corresponding to fMet-tRNA$_F$ in bacterial systems has yet been found, these results hint at the presence of a general initiation mechanism common for both animal and bacterial systems.

It may also be interesting to note that phage MS2 RNA injected into bacterial cells appears to associate first with 30S ribosomal subunits *in vivo* [24]. It has also been reported that newly synthesized vaccinia virus mRNA is bound to small ribosomal subunits before it enters polysomes [25]. These observations are all consistent with the present model.

### Further Discussion of the Model

Several lines of evidence strongly indicate the existence of polycistronic mRNA and its sequential reading starting from the first cistron near the "operator" site [26−29]. The sequential reading of polycistronic mRNA appears to be also operating in *in vitro* translation of phage RNA [*30, 31*]. Thus, the first site to which ribosomes attach is presumably the initiation codon of the first cistron of a polycistronic mRNA. Yet, the products of all the cistrons synthesized *in vitro* on phage RNA messenger seem to have formylmethionine at one end [*32*]. Therefore, all the cistrons appear to have the codon (AUG) for fMet-tRNA at their starting point. The question now arises why ribosomes start only at the initiator codon of the first cistron and not at the initiator codon of the other cistrons. Another related problem is how an AUG codon for an internal methionine residue, which is inside a cistron can be distinguished from the initiator AUG codons, especially those initiators in the middle of polycistronic mRNA.

The present model does not explain why the 30 S particles start at a particular AUG codon (the initiator codon of the first cistron) and not at some other codons corresponding either to internal methionines or to the initiator codons of the cistrons subsequent to the first one. Clearly, the initiator codon of the first cistron must have some special feature distinct from all other AUG codons. This special feature could be some specific nucleotide sequences adjacent to AUG, or merely the proximity of the AUG codon to the end of the mRNA. Alternatively, it is also possible that all other AUG codons are masked by some secondary structure of mRNA and that only the AUG of the first cistron is "exposed" to 30 S particles. The final answer to this question must perhaps await the determination of the base sequence in natural mRNA responsible for the initiation of the first cistron.

However, the model does explain why the ribosomes pick up Met-$tRNA_M$ and reject fMet-$tRNA_F$ at the internal AUG codons, and vice versa at the initiator codon of the second cistron. The basic feature of the proposal is that fMet-$tRNA_F$ binds only to 30 S particles, whereas met-$tRNA_M$ binds only to the 30 S–50 S complex. The 30 S particle makes the initiation complex at the initiator codon of the first cistron, and the 50 S particle joins the initiation complex as described before. Any AUG codon for internal methionine is then encountered by the protein synthesizing 30 S–50 S complex, which is now capable of binding only met-$tRNA_M$ and not fMet-$tRNA_F$. The initiator codon AUG of the second cistron will be encountered by ribosomes which have just finished translation of the first cistron. According to the present model, the chain termination at the end of the first cistron causes dissociation of the 30 S–50 S complex and the ribosomal particle which interacts with the nearby initiator codon of the next cistron is the 30 S particle, which is now capable of binding only fMet-$tRNA_F$ and not Met-$tRNA_M$. It is quite possible that, upon chain termination, both the 50 S particle and the finished polypeptide are released from the mRNA, but the 30 S particle remains attached to mRNA if it is in the middle of a polycistronic mRNA. This picture may make it easier to understand how the 30S particles which have just finished translation of the first cistron are able to interact easily with the initiator codon of the second cistron, whereas the free 30S particles in the pool are not.

The above discussion is also pertinent to the mechanism of polarity. Nonsense codons (amber, ochre or UGA codons) created by mutation inside a cistron cause peptide chain termination at the site of mutation and the release of unfinished polypeptide fragments. The premature chain termination causes reduction of translation of the cistrons of an operon on the operator-distal side of the mutated cistron. The degree of the reduction (polarity) depends on the distance from the nonsense codon to the start of the next cistron [29]. In order to explain such a relationship between the degree of polarity and the distance, it has been proposed that ribosomes travel along mRNA without synthesizing polypeptides, and the ribosomes will fall off mRNA with a probability which is dependent on the distance between the nonsense codon and the start of the next cistron [33]. Alternatively, the ribosome may fall off at the site of nonsense mutation, but have a chance of reattachment to the initiator site of the next cistron, the probability of this reattachment depending on the distance between the mutational site and the reattachment site. Previously, it was difficult to imagine why ribosomes do not make peptides or do not make stable aminoacyl tRNA-mRNA-ribosome complexes in response to codons after the nonsense codon. According to the present model, 70S ribosomes dissociate into 30S and 50S particles upon chain termination and, therefore, cannot respond to ordinary codons except to a nearby initiator codon (AUG). Any ribosomes that continued to travel along mRNA past the site of nonsense codon would be 30S particles rather than 70S particles.

Recently, Sarabhai and Brenner studied the problem of chain reinitiation within cistrons after premature chain termination caused by nonsense mutation [34]. They have succeeded in making reinitiation signals ("starters") within the rII B cistron by mutation, and examined conditions necessary for the reinitiation. They have found that the presence of a chain termination signal within a certain distance from the starter is essential for the starter to function, but the termination signal can be either to the left or to the right of the starter. According to the present model, the starter can respond only to 30S particles and not to 70S particles, and thus, dissociation of 70S particles by a chain termination signal near the site of the starter would be necessary.

Examination of ribosomes from several different sources in nature has always revealed the presence of two different dissociable subunits. Previously, it was difficult to explain why ribosomes must contain two dissociable subunits. With the large variety of ribosomal proteins, it was thought that all the ribosomal functions could be performed on a single particle. According to the present model, the dissociation of the 70S ribosomes into the 30S and 50S subunits is a property obligatory for the initiation of protein synthesis, and serves the important regulatory function already discussed.

## Acknowledgements

The author wishes to thank Mr. C. Lowry and Mrs. C. Guthrie for their participation in the present work, Drs. H. Temin and M. Susman for reading the manuscript. The work described in this paper was supported, in part, by Public Health Service Research Grant No. GM-15422 from the National Center for Urban and Industrial Health, and by the National Science Foundation Grant No. GB-6594.

## References

1. Marcker, K. A., and F. Sanger: J. molec. Biol. 8, 835 (1964).
2. Clark, B. F. C., and K. A. Marcker: J. molec. Biol. 17, 394 (1966).
3. Adams, J. M., and M. R. Capecchi: Proc. nat. Acad. Sci. (Wash.) 55, 147 (1966).
4. Webster, R. E., D. L. Engelhardt, and N. D. Zinder: Proc. nat. Acad. Sci. (Wash.) 55, 155 (1966).
5. Capecchi, M. R.: Proc. nat. Acad. Sci. (Wash.) 55, 1517 (1966).
6. Sundararajan, T. A., and R. E. Thach: J. molec. Biol. 19, 74 (1966).
7. Bretscher, M. S., and K. A. Marcker: Nature (Lond.) 211, 380 (1966).
8. Ghosh, H. P., D. Söll, and H. G. Khorana: J. molec. Biol. 25, 275 (1967).
9. Nomura, M., and C. Lowry: Proc. nat. Acad. Sci. (Wash.) 58, 946 (1967).
10. — —, and C. Guthrie: Proc. nat. Acad. Sci. (Wash.) 58, 1487 (1967).
11. Mangiarotti, G., and D. Schlessinger: J. molec. Biol. 20, 123 (1966).
12. Stanley, W. M. jr., Salas, A. J. Wahba, and S. Ochoa: Proc. nat. Acad. Sci. (Wash.) 56, 290 1966)(.
13. Salas, M., M. B. Hille, J. A. Last, A. J. Wahba, and S. Ochoa: Proc. nat. Acad. Sci. (Wash.) 57, 387 (1967).
14. Revel, M., and F. Gros: Biophys. Biochem. Res. Commun. 25, 124 (1966).
15. Eisenstadt, J., and G. Brawerman: Biochemistry 5, 2777 (1966).
16. Suzuka, I., H. Kaji, and A. Kaji: Proc. nar. Acad. Sci. (Wash. 55, 1483 (1966).
17. Pestka, S., and M. Nirenberg: J. molec. Biol. 28, 145 (1966).
18. Matthaei, J. H., H. P. Voigt, G. Heller, R. Neth, G. Schoch, H. Kubler, F. Amelunxen, G. Sander, and A. Parmeggiani: Cold Spr. Harb. Symp. quant. Biol. 31, 25 (1966).
19. Ghosh, H. P., and H. G. Khorana: Proc. nat. Acad. Sci. (Wash.) 58, 2455 (1967).
20. Schlessinger, D., G. Mangiarotti, and D. Apirion: Proc. nat. Acad. Sci. (Wash.) 58, 1782 (1967).
21. Kaempfer, R., M. Meselson, and H. J. Raskas: J. Mol. Biol., 31, 277 (1968).
22. Bishop, J. O.: Biochim. biophys. Acta (Amst.) 119, 130 (1966).
23. — J. molec. Biol. 17, 285 (1966).
24. Godson, G. N., and R. L. Sinsheimer: J. molec. Biol. 23, 495 (1967).
25. Joklik, W. K., and Y. Becker: J. molec. Biol. 13, 511 (1965).
26. Jacob, F., and J. Monod: J. molec. Biol. 3, 318 (1961).
27. Ames, B. N., and P. E. Hartman: Cold Spr. Harb. quant. Biol. 28, 349 (1966).

28. Martin, R. G., D. F. Silbert, D. W. E. Smith, and H. J. Whitfield: J. molec. Biol. 21, 357 (1966).
29. Newton, W. A., J. R. Beckwith, D. Zipser, and S. Brenner: J. molec. Biol. 14, 290 (1965).
30. Ohtaka, Y., and S. Spiegelman: Science 142, 493 (1963).
31. Engelhardt, D. L., R. E. Webster, and N. D. Zinder: J. molec. Biol. 29, 45 (1967).
32. Vinuela, E., M. Salas, and S. Ochoa: Proc. nat. Acad. Sci. (Wash.) 57, 729 (1967).
33. Yanofsky, C., and J. Ito: J. molec. Biol. 21, 313 (1966).
34. Sarabhai, A., and S. Brenner: J. molec. Biol. 27, 145 (1967).

## Discussion

*Starlinger:* Doesn't one have to distinguish between an entry step at the beginning of each individual gene? If this were true, the selection of the proper reading frame might occur at the entry step rather than at the initiation step.

*Nomura:* First initiation signal at the beginning of an operon must be different from the initiation signal of the 2nd and 3rd cistrons, and this first entry must determine the correct reading frame. However, upon any sort of chain termination, ribosomes dissociate and the liberated free 30 S particles recognize any nearby signal with proper base sequence (e.g. AUG) irrespective of the frame the 70 S ribosome complex has been following.

*Stöffler:* Ochoa's group has found two initiation factors in the 70 S ribosomes, which can be dissociated by high salt concentrations (0.5 M $NH_4Cl$). J. E. Allende and H. Weissbach (Biochem. biophys. Res. Commun. 28, 82 (1967) showed more recently a GTP requirement for factor II.

Do you have this GTP requirement for N-formyl-Met-tRNA binding in your system, too?

*Nomura:* We have shown a stimulatory effect of the "initiation factor" in our system, but we have never studied a GTP requirement carefully. However, I have learned that Dr. Wahba's group has recently shown a GTP requirement in a system using the 30 S ribosomal subunits and synthetic oligonucleotides as a mRNA.

*Matthaei:* 30 S-subunits specifically selecting an F-Met-codon in $f_2$-RNA is a lovely finding. Under the ionic ratio of 9 mM $Mg^{++}$ to 25 mM $Tris^{++}$ + 40 mM $NH_4^+$, there is apparently no binding of various other amino acid adaptors to 30 S-particles under direction of this mRNA. This specificity of recognition *in vitro* is rather extraordinary and may concern the initiator-codons or longer sequences in mRNA. Do you agree, that after a translocation step moving F-Met-tRNA and its codon to the next (= ? donor-) site, the same checking-site on the 30 S-particle may well be the site for recognition of the adapters for chain-internal amino acids?

We found this checking site first in 1964, but clearly more evidence is required to prove this explanation of its function and to clarify the parts which the two subunits may play in the process of recognition of the amino acid adapters.

*Nomura:* We believe that the weak binding of ordinary tRNA such as phe-tRNA to 30 S particles directed by synthetic polynucleotides is an *in vitro* nonphysiological binding reaction which may utilize the binding site restricted to fMet-tRNA$_F$ under *in vivo* physiological conditions in the presence of natural messenger. We still do not know whether this "initiation" site on 30 S is converted to the site for other aminoacyl-tRNA after joining of 50 S, or alternatively, this initiation site is converted to the so-called "peptidyl site" and the site for other aminoacyl-tRNA on 30 S−50 S complex has no relation to the initiation site.

# Structure and Function of Ribosomes: Studies on Dissociation and Reconstitution of Ribosomal Particles

By

P. Traub and M. Nomura

Laboratory of Genetics, University of Wisconsin, Madison, Wisconsin U.S.A.

## 1. Introduction

The ribosome is a cell organelle that is essential for protein synthesis. That the ribosomes provide sites for the binding of mRNA and the amino-acyl tRNA has been established (see reviews by WATSON [1, 2]). In addition, the structural complexity of ribosomes suggests greater complexity in function and, in fact, there are several indications that the structure of ribosomes is profoundly involved in the translation of the genetic code [3, 4, 5]. Recent experiments also suggest that the peptidyl transferase is a part of 50S ribosomal proteins [6, 7]. However, the detailed function of ribosomes in protein synthesis is still unclear. It appears that a comprehensive understanding of the mechanism of protein synthesis must await an elucidation of the structure and function of ribosomes.

In this article we shall review our recent work on structure and function of *E. coli* ribosomes. The principal objective of our work is to identify all of the molecular components of ribosomes and to relate their properties to the overall function of the ribosomes in protein synthesis. The approach we have taken to achieve this goal is to prepare artificial ribosome derivatives deficient in one specific component, and examine the alteration of the functions in these ribosome derivatives. Such a specific deletion of one component from the ribosome particle can be made by dissociating the ribosome into its structural components and reconstituting the particle with the omission of the desired component. This type of approach has been made feasible by the success of reconstitution of functionally active ribosomal particles from inactive protein-deficient sub-ribosomal particles and ribosomal proteins [8, 9].

## 2. Partial Fractionation of the Ribosomal Proteins and Reconstitution of Ribosomes and Subribosomal Particles

When purified 30S and 50S ribosomal particles are centrifuged to equilibrium in 5 M CsCl in the presence of 0.04 M Mg$^{++}$, they dissociate

into smaller ribonucleoprotein particles and free proteins, called "split"
proteins [10]. The smaller particles are recovered from the band at the
density of about 1.65 in the CsCl gradient, while the free proteins are
recovered from the top of the gradient. The 30S particles yield 23S "core"
particles and the 50S particles yield 40S "core" particles. Both of these
core particles are inactive in protein synthesis. Both MESELSON's group
[8] and our group [9] have independently demonstrated that mixing the
split proteins with these core particles results in the formation of particles
which are similar to the original ribosomal particles in sedimentation
properties and are active in *in vitro* polypeptide synthesis. It has been
shown that functionally active 30S particles can be reconstituted by the
combination of 23S core particles and the split proteins from 30S (SP30),
but not by the combination of 23S core particles and the split proteins
from 50S (SP50). Conversely, active 50S ribosomes can be reconstituted
by the combination of 40S core particles and the SP50 split proteins, but
not by the combination of 40S core particles and the SP30 split proteins
[8, 9]. It has also been shown that the proteins obtained from 23S cores
or 40S cores cannot replace the respective split proteins (SP30 or SP50)
in the reconstitution of functional ribosomal particles [11]. These results
suggest that at least some of the split proteins from 30S particles are
distinct from split proteins from 50S particles and from proteins in the
23S or 40S core particles, and that the restoration of synthetic activity
to core particles requires the addition of some specific proteins. Analysis
of proteins by polyacrylamide gel electrophoresis has also revealed that
proteins in the four fractions (SP30, SP50, 23S cores, and 40S cores)
are distinct from each other [11, 12].

We then started fractionation of split proteins [13, 14, 15]. The split
proteins recovered from the top of the CsCl density gradient are soluble
in the presence of a high concentration of salts or urea, but are insoluble
upon removal of the salts. Thus, we had to use urea during the fractionation in order to keep proteins in solution. The fractionation we have
achieved so far is summarized in Fig. 1.

Split proteins from 50S particles (SP50) have been fractionated by
adsorption to and elution from DEAE in the presence of 6M urea. Two
fractions were obtained: the basic fraction (SP50B) which was not adsorbed by the DEAE column at pH 8, and the acidic fraction (SP50A) which
was adsorbed by the DEAE column at pH 8 and then eluted with 2 M
LiCl at pH 4.5. Similarly, the split proteins from 30S particles yielded
2 fractions, SP30B and SP30A. The basic fraction, SP30B, represents
proteins which are not adsorbed on DEAE, and the acidic fraction,
SP30A, represents proteins which are adsorbed on DEAE. Once separated,
the four protein fractions (SP50A, SP50B, SP30A and SP30B) are
soluble in buffers at neutral pH and moderate salt concentrations and this

made it possible to study the function of these proteins in a variety of systems for the assay of ribosomal functions as will be described below.

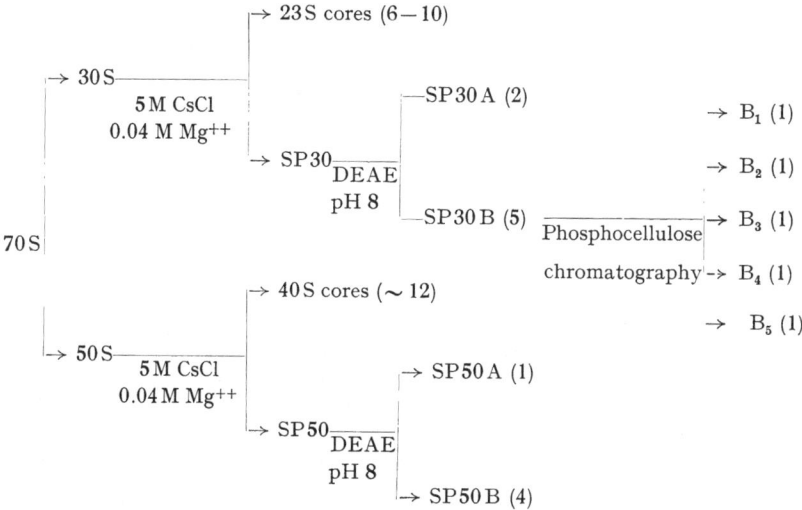

Fig. 1. Dissociation of ribosomal particles and fractionation of subribosomal components. (Numbers within parentheses indicate the numbers of major protein bands observed on polyacrylamide gel electrophoresis.)

In the initial stage of our work [13, 14], we prepared four different subribosomal particles in addition to 40S and 23S core particles; (a) [40, A] particles from 40S core particles and SP50A proteins, (b) [40, B] particles from 40S core particles and SP50B proteins, (c) [23, A] particles from 23S core particles and SP30A proteins, and (d) [23, B] particles from 23S core particles and SP30B proteins. In addition, control reconstituted particles, [40, A, B] or [23, A, B] particles, were prepared from 40S cores, SP50A and SP50B proteins or from 23S cores, SP30A and SP30B proteins. These reconstituted particles were isolated by centrifugation and analyzed for their protein composition. It was found that the control particles contain all the original ribosomal proteins, whereas the four subribosomal particles are deficient in the specific protein components which were omitted from the reconstitution mixture. These particles were then analyzed for their several known ribosomal functions, and each activity of the artificial subribosomal particles was compared with that of the control particles. The activity of the control reconstituted "30S particles", that is [23, A, B] particles, was generally 50 to 100 % of that of the "native" 30S particles; and the control reconstituted "50S particles" showed about 30 to 60 % of the activity of the "native" 50S particles, when they were assayed for their poly U-dependent phenylalanine incorporation activity in the presence of native 50S and 30S particles, respectively. Activity of artificial subribosomal particles was compared with that of the control reconstituted particles.

Table 1. *Summary of the functional capacity of various subribosomal particles derived from 50S particles*

| Particles | Poly U-directed phenylalanine synthesis |
|---|---|
| [40, A, B] | + + |
| [40S] | − |
| [40, A] | + |
| [40, B] | − |

Details are described in the original paper [*14*].

It has been found that [40, B] particles, which lack SP50A proteins, are completely inactive as are 40S core particles in poly U directed polyphenylalanine synthesis, whereas [40, A] particles, which lack SP50B proteins, are partially active in polyphenylalanine synthesis. [23, A] par-

Table 2. *Summary of the functional capacity of various subribosomal particles derived from 30S particles*

| | Polypeptide synthesis | | tRNA binding | | | poly U binding |
|---|---|---|---|---|---|---|
| | poly U-phenyl-alanine | f2-valine | poly U-Phe-tRNA | | f2 RNA-Met-tRNA | |
| | (+50S) | (+50S) | (+50S) | (−50S) | (−50S) | (−50S) |
| [23, A, B] | + + | + + | + + | + + | + + | + + |
| 23S | − | − | − | − | − | − |
| [23, A] | − | − | − | − | − | − |
| [23, B] | + | + | + | + | + | + + |
| [23, A, B (1, 2, 3, 4, 5)] | + + | + + | + + | + + | + + | + + |
| [23, A, B (−, 2, 3, 4, 5)] | + | + | + | + | + | + + |
| [23, A, B (1, −, 3, 4, 5)] | + | + | + | + | + | + + |
| [23, A, B (1, 2, −, 4, 5)] | − | − | − | − | − | + + |
| [23, A, B (1, 2, 3, −, 5)] | + + | + + | + + | + | + | + + |
| [23, A, B (1, 2, 3, 4, −)] | − | − | − | − | − | + + |

Summary of results of several experiments are given. Details are given in original papers [*14, 15*]. Particles reconstituted without B₁, but with all other components present, are designated as [23, A, B (−, 2, 3, 4, 5)], and other particles in corresponding ways.

ticles, which lack SP30B proteins, are completely inactive in both poly U-directed polyphenylalanine synthesis and poly U-directed Phe-tRNA binding, whereas [23, B] particles, which lack SP30A proteins, are partially active in these two functions. The ability of the various 30S ribosomal derivatives to bind a synthetic mRNA, poly U, was also studied. Both 23S and [23, A] particles were inactive, whereas both [23, B] particles and [23, A, B] particles were able to bind poly U. These results are summarized in Table 1 and 2.

The difference in activity between [40, A] and [40, A,B] particles is due to the lack of the protein component of SP50B in the [40, A] particle, since the addition of SP50B proteins stimulated the activity of [40, A] particles. The inactivity of [40, B] particles and 40S particles is not due to an inactivation during the preparation or storage of the particles, because the addition of SP50A proteins and SP50A plus SP50B proteins, respectively, restored the activity to the same level as that of [40, A, B]. Similarly, the partially active [23, B] particles can be stimulated by the missing SP30A protein components, and the inactive [23, A] particles and 23S core particles can be stimulated by SP30B proteins and SP30B plus SP30A proteins, respectively [13, 14].

It is concluded from these experiments that some protein(s) in the SP50A fraction are indispensable for the synthesis of polypeptide, and

Table 3. *Stimulation of poly U-directed phenylalanine incorporation activity of various defective subribosomal particles by various split protein fractions*

| Subribosomal particles | Split protein fraction tested | Stimulation of phenylalanine incorporation (%) |
|---|---|---|
| [23, A] | SP30A | 0 |
|  | SP30B | > 1,000 |
|  | SP50A | 0 |
|  | SP50B | 0 |
| [23, B] | SP30A | 130 |
|  | SP30B | 0 |
|  | SP50A | 3 |
|  | SP50B | 0 |
| [40, A] | SP30A | 0 |
|  | SP30B | 0 |
|  | SP50A | 0 |
|  | SP50B | 86 |
| [40, B] | SP30A | 0 |
|  | SP30B | 0 |
|  | SP50A | 462 |
|  | SP50A | 0 |

Detailed data are described in the original paper [14].

that the proteins in the SP 50 B fraction may be dispensable, but have a stimulatory effect. Similarly, it is concluded that some proteins in the SP 30 B are indispensable for both the specific tRNA binding function and the incorporation activity, while the SP 30 A proteins may be dispensable, but have a stimulatory action in some way.

Systematic studies [14] showed that each defective subribosomal particle can be stimulated only by the protein components which are missing in that subribosomal particle, demonstrating the functional specificity of these four protein fractions (see Table 3). These four protein fractions have also been shown to have protein band patterns different from each other in polyacrylamide gel electrophoresis [13, 14]. Thus, it is concluded that the ribosomal proteins are physically and functionally heterogeneous.

### 3. Possible Functions of "Dispensable" Ribosomal Proteins in SP 30A Fraction

The presence of ribosomal proteins (SP 30 A and SP 50 B) which appear to be functionally dispensable poses questions as to their possible functions. In an effort to uncover functions for these proteins, we have examined the differences between the functional ability of [23, B] particles and of [23, A, B] particles in detail [16]. Specifically, three questions were asked: (a) whether [23, B] particles can respond to natural messenger RNA, (b) whether the absence of SP 30 A proteins in [23, B] particles results in increased error frequency in reading of the genetic code, and (c) whether the stimulation of the activity of [23, B] particles by the SP 30 A proteins is due to the stimulation of the activity of each of the partially active [23, B] particles, or alternatively, is due to an increase in the number of active particles in the preparation.

Our studies have shown the following results: (a) the [23, B] particles are partially active in polypeptide synthesis directed by a natural mRNA, RNA formyl phage f2. (b) The [23, B] particles are partially active in the binding of formyl Met-tRNA directed by the triplet AUG, random poly AUG, or phage f2 RNA. (c) The possibility that incomplete ribosomes would make translation errors was examined in the system with the [23, B] particles and the native 50 S particles using poly U or copolymer r-UUC with a repeating triplet sequence as mRNA. There was no significant increase in translation errors relative to the system with the native 30 S particles. Thus, the deletion of the SP 30 A proteins from the 30 S particle does not change the translation fidelity of the particle. (d) The specific tRNA binding experiments, done with either poly U-Phe-tRNA or f2 RNA-formyl Met-tRNA systems, indicate that the stimulation by SP 30 A proteins is due to an increase in the number of specific binding sites and therefore, probably in the number of active particles in the preparation.

It is suggested that [23, B] particles can assume several different structures, some active and others inactive, and that, in the presence of the SP 30 A proteins, only the active form of the [23, B] particle is structurally permitted in [23, A, B] particles.

The same argument could be applied to the stimulatory action of the SP 50 B proteins in the 50 S ribosomes. The SP 50 B is dispensable for the *in vitro* polypeptide formation *per se* and yet has a specific stimulatory action. Thus, the ribosome contains many different proteins, some of which at least may not have any functional role directly related to polypeptide bond formation, but have a structural role to make the functional configuration a stable one.

However, the above argument does not exclude the possibility that these proteins which have a structural role may also have some specific function which cannot be detected in the *in vitro* assay system used in our studies, such as a role in the chain termination or a role in some unknown mechanism regulating ribosome function *in vivo*.

## 4. Isolation of Five Individual Ribosomal Proteins and their Functional and Chemical Characterization

Knowing that the proteins in the SP 30 B fraction are essential for the 30 S function, we next undertook to fractionate proteins in this fraction into individual protein components [15].

Separation of ribosomal proteins in SP 30 B fraction was done using a phosphocellulose column. Five major protein peaks were observed. Analysis of the protein composition in each of the separated peaks by polyacrylamide gel electrophoresis revealed that the five proteins ($B_1$ to $B_5$) obtained from the column are essentially pure and correspond to the five major protein components revealed by electrophoresis of the original SP 30 B fraction.

"30 S" particles ([23, A, B (1, 2, 3, 4, 5)]) were prepared from 23 S core particles, SP 30 A proteins and five purified ribosomal proteins ($B_1$ to $B_5$) and their functional capacities in protein synthesis were examined [15]. The reconstituted particles, [23, A, B (1, 2, 3, 4, 5)], showed activity comparable to the original undissociated 30 S or the particles ([23, A, B]) reconstituted from 23 S core particles, SP 30 A fraction and SP 30 B fraction. Thus, the SP 30 B fraction can be replaced by a mixture of five purified proteins.

Next, particles were prepared using 23 S core particles, the SP 30 A fraction, and various combinations of the five SP 30 B proteins, so that five kinds of subparticles were produced, each particle being deficient in a different SP 30 B protein.

These various particles were then analyzed for their activity in performing the various known 30 S ribosomal functions, and the following conclusions were obtained [15]: (a) Particles deficient in $B_3$ or $B_5$ are almost completely inactive both in specific tRNA binding and in amino acid incorporation, although they are able to bind $C^{14}$-poly U to a great extent. (b) Particles deficient in $B_1$ have a greatly reduced activity (10 to 40 % of the activity of control [23, A, B (1, 2, 3, 4, 5)] particles remained) and (c) particles deficient in $B_2$ retain partial activity (50 to 80 % of the activity of the control particles), in both tRNA binding and amino acid incorporation. (d) Particles which are deficient in $B_4$ retain full activity in both amino acid incorporation and Phe-tRNA binding in the presence of 50 S particles. However, the particles deficient in $B_4$ show a definite reduction in the f 2 RNA dependent binding of formyl-Met-tRNA (activity varies from 20 to 40 % of the control). This result suggests that $B_4$ is not needed for chain elongation, but has some role related to chain initiation. (e) All the particles studied here are able to bind $C^{14}$-poly U to a great extent (more than 50 % of the control particles). Thus, particles deficient in $B_3$ and $B_5$ fail to bind Phe-tRNA, although they are able to bind poly U.

The inactivity or reduction of activity of particles deficient in $B_1$, $B_2$, $B_3$ or $B_5$ is not due to an inactivation during the preparation or storage of the particles, but is due to the lack of the respective specific proteins, because addition of the missing proteins to these inactive particles almost completely restored the activity.

These results clearly demonstrate that two of the split proteins from 30 S particles, $B_3$ and $B_5$, and possibly a third ($B_1$), are essential for amino acid incorporation as well as for specific amino acyl-tRNA binding. $B_4$ is entirely dispensable for amino acid incorporation and for aminoacyl tRNA-binding directed by the synthetic messenger poly U. However, $B_4$ may have a role in chain initiation. On the other hand, a definite conclusion was not obtained with $B_2$, although it was clearly shown to have at least a stimulatory activity in polypeptide synthesis. These results are also summarized in Table 2.

The five isolated proteins were also studied chemically [15]. Both amino acid compositions and peptide patterns after trypsin digestion of the five purified proteins were determined. The data clearly show that these five proteins are chemically different from each other.

Chemical heterogeneity of ribosomal proteins has been shown by many other workers [17—21]. Tissières and his coworkers [22, 23], Wittmann and his coworkers [24], and Kurland and his coworkers [25] have now independently isolated a number of purified ribosomal proteins and demonstrated the chemical heterogeneity of these proteins. Our studies not only confirm the chemical heterogeneity of ribosomal proteins, but also demonstrate that chemically distinct proteins are also functionally distinct. Thus, the physical heterogeneity of the ribosomal proteins reflects the very complex but highly organized functional structure of the ribosomes.

## 5. Concluding Remarks

A few years ago, the study of the complex structure of ribosomes seemed to be a formidable one, and the ribosomes were treated as a black box in discussions of mechanism of protein synthesis. Partly because of our ignorance and partly because of our over-emphasis of functions of mRNA and tRNA, we regarded the role of ribosomes in protein synthesis as a passive one. As already described in the introduction, several recent developments have changed our concept of the ribosome from an inert passive structure to an active structure with fundamental roles in protein synthesis. Although our studies are still in their infancy, we are now beginning to see the inside of this black box, and are in a position to be able to design experiments to study in detail the functions of ribosomes in protein synthesis. Thus, for example, we now have several purified ribosomal proteins (basic split proteins from 30S, $B_1 - B_5$), and the activity of these purified proteins can be assayed by using appropriate defective artificial particles which are deficient in one pertinent protein. It is now possible to apply a variety of known techniques in protein chemistry to these proteins and to study their structure and function individually.

On the other hand, an obviously important next task would be an extension of our present type of approach to the 23S and 40S core particles, that is, an attempt to dissociate core particles to free RNA and mixture of proteins, and then to reconstitute functional ribosomal particles.

Finally, it should be mentioned that the present reconstitution system provides an experimental system in which the mode of assembly of the ribosome from subribosomal components can be studied. Our experiments done on this subject [26] have already shown that under proper conditions the association of split proteins with core particles is spontaneous and reasonably rapid, supporting the hypothesis of self-assembly of ribosome structure from subribosomal components.

## Acknowledgements

The authors wish to thank Drs. K. Hosokawa, G. R. Craven, and D. Söll for their participation in various stages of the present work, and Dr. M. Susman for reading the manuscript. The work described in this paper was supported, in part, by Public Health Service Research Grant No. GM-15422 from the National Center for Urban and Industrial Health, and by the National Science Foundation Grant No. GB-6594.

5*

## References

1. Watson, J. D.: Science **140**, 17 (1963).
2. — Bull. Soc. Chim. Biol. **46**, 1399 (1964).
3. Gorini, L., and E. Kataja: Proc. nat. Acad. Sci. (Wash.) **51**, 487 (1964).
4. — — Proc. nat. Acad. Sci. (Wash.) **51**, 995 (1964).
5. Davies, J., W. Gilbert, and L. Gorini: Proc. nat. Acad. Sci. (Wash.) **51**, 883 (1964).
6. Traut, R. R., and R. E. Monro: J. molec. Biol. **10**, 63 (1964).
7. Monro, R. E.: J. molec. Biol. **26**, 147 (1967).
8. Staehelin, T., and M. Meselson: J. molec. Biol. **16**, 245 (1966).
9. Hosokawa, K., R. Fujimura, and M. Nomura: Proc. nat. Acad. Sci. (Wash.) **55**, 198 (1966).
10. Meselson, M., M. Nomura, S. Brenner, C. Davern, and D. Schlesinger: J. molec. Biol. **9**, 696 (1964).
11. Traub, P., M. Nomura, and L. Tu: J. molec. Biol. **19**, 215 (1966).
12. Gesteland, R. F., and T. Staehelin: J. molec. Biol. **24**, 149 (1967).
13. Nomura, M., and P. Traub: Organizational Biosynthesis. Ed. by Vogel, H. T., Lampen, T. O., and Bryson, V. New York-London: Academic Press, P. 459 (1967).
14. Traub, P., and M. Nomura: J. molec. Biol. in press (1968).
15. —, K. Hosokawa, G. R. Craven, and M. Nomura: Proc. nat. Acad. Sci. (Wash.) **58**, 2430 (1967).
16. —, D. Söll, and M. Nomura: J. molec. Biol. in press (1968).
17. Waller, J. P., and J. J. Harris: Proc. nat. Acad. Sci. (Wash.) **49**, 538 (1961).
18. — J. molec. Biol. **10**, 319 (1964).
19. Leboy, P. S., E. C. Cox, and J. G. Flaks: These Proceedings **52**, 1367 (1964).
20. Traut, R. E.: J. molec. Biol. **21**, 571 (1966).
21. Spitnik-Elson, P.: Biochim. biophys. Acta (Amst.) **74**, 105 (1963); **80**, 594 (1964).
22. Traut, R. R., P. B. Moore, H. Delius, H. Noller, and A. Tissieres: Proc. nat. Acad. Sci. (Wash.) **57**, 1294 (1967).
23. Delius, H., R. R. Traut, P. B. Moore, H. Noller, and P. Pearson: in this volume (1967).
24. Kaltschmidt, E., V. Rudloff, G. Stöffler, A. Chersi, M. Dzionara, D. Donner, and H. G. Wittmann: in this volume (1967).
25. Hardy, S. J. S., G. Craven, and C. G. Kurland: (1967) in preparation.
26. Nomura, M., and P. Traub: J. molec. Biol. in press (1968).

## Discussion

*Matthaei:* Would you like to tell us a bit more on how you isolated functioning ribosomal proteins and under which conditions you reconstitute these to active ribosomal particles?

*Nomura:* Ribosomes are centrifuged in 5 M CsCl for about 36 h. The "cores" are obtained from the A band. "Split proteins" are recovered from the top of the gradient. The split proteins are mixed with cores in about 5 M CsCl and then dialyzed against Tris (0.01 M, pH 7.4) $-MgCl_2$ (0.01 M) $-NH_4Cl$ (0.03M) $-\beta$-mercaptoethanol $(6 \times 10^{-3}M)$. Reconstituted particles are recovered by centrifugation. When fractionated proteins, which are soluble in buffer of moderate salt concentration, are used, proteins and cores were mixed in suitable buffers without high concentration of salts.

# The Formation of the Peptide Bond on Ribosomes

By

I. Rychlík

Institute of Organic Chemistry and Biochemistry
Prague, Czechoslovakia

Polypeptide chains of proteins are synthesized on ribosomes by a cyclic reaction during which the nascent polypeptide is transferred from polypeptidyl-tRNA to the neighbouring aminoacyl-tRNA. In each completed cycle of the reaction the polypeptidyl-tRNA is extended by one aminoacid residue, tRNA which has lost the polypeptide is excluded from the ribosome and a new aminoacyl-tRNA elected by the codon-anticodon interaction enters the reaction [1−3]. In addition to the

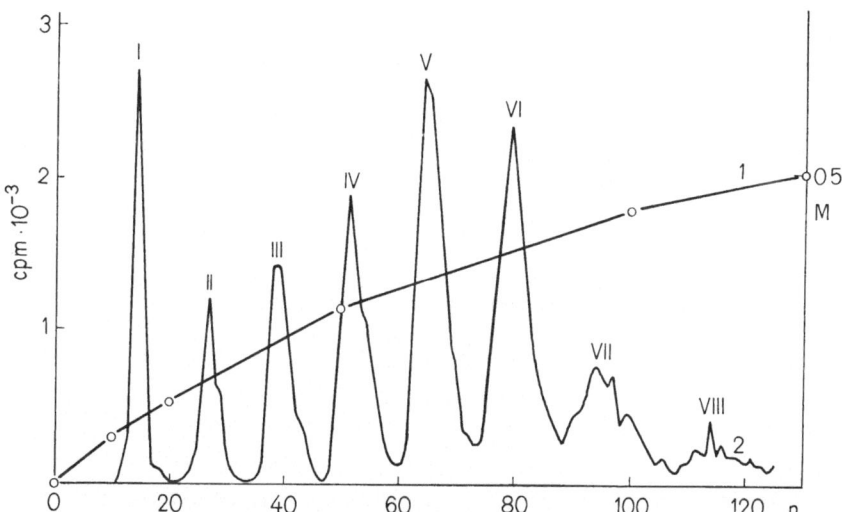

Fig. 1. Synthesis of polylysyl-tRNA in the complete system. The system contained in a final volume of 0,3 ml: 1 mg sRNA, 30 μmol Tris acetate pH 7.8, 30 μmol magnesium acetate, 30 μmol ammonium acetate, 0.1 μmol ATP, 1 μmol phospho-enolpyrruvate, 2 μg pyruvate kinase, 30 μg polyadenylic acid, 5 μg lysyl-tRNA ligase, ribosomes 0.4 mg of protein, SE 100 supernatant enzymes 0.5 mg of protein and lysine-$^{14}$C 0.2 μC. After incubation for 15 min at 35° C polylysyl-tRNA was isolated, the lysine peptides released from tRNA by 0.2 M KOH and chromato-graphed on CM-cellulose as described [7]. I dilysine, II trilysine, III tetralysine, IV pentalysine, V hexalysine, VI heptalysine

ribosome and the two tRNA derivatives soluble enzymes from the postribosomal supernatant are involved in the cycle. The mutual relations between the ribosome and supernatant enzymes are clearly shown in the poly A directed cell free Escherichia coli system synthesizing lysine peptides [4, 5]. Figure 1 shows the course of the synthesis of lysine peptides attached to tRNA in the complete system. A series of intermediates of polylysine synthesis is formed ranging from dilysyl-tRNA up to octalysyl-tRNA [6, 7]. When supernatant enzymes are omitted from the medium only dilysyl-tRNA is formed with a minor trace of trilysyl-tRNA the amount of which depends on the procedure used for washing the ribosomes [7] (Fig. 2).

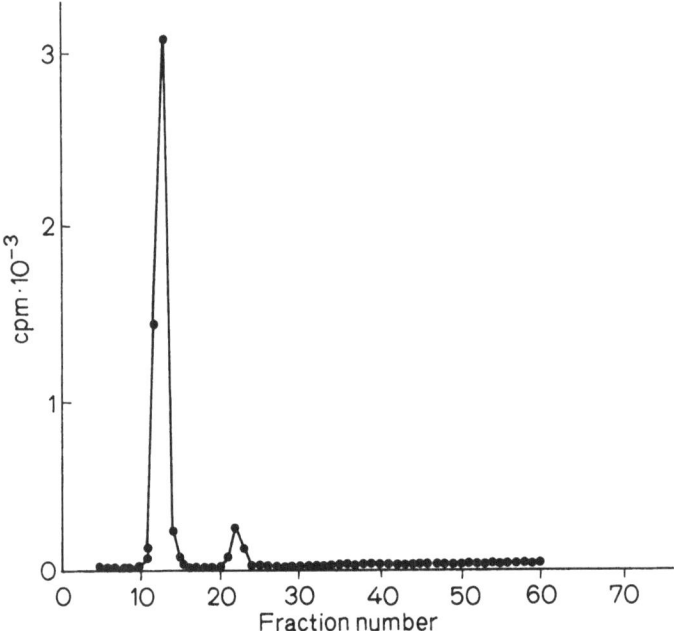

Fig. 2. Synthesis of dilysyl-tRNA in absence of SE 100 enzymes. Conditions as in Fig. 1. Proteins of the SE 100 supernatant were omitted

These experiments indicate that the ribosome catalyzes the formation of the peptide bond between two lysine residues leading to the synthesis of dilysyl-tRNA in the absence of supernatant enzymes and GTP. In order that the whole cycle could be repeated, and peptides higher than dilysine formed, the action of enzymes of the postribosomal supernatant and GTP are required in addition to the ribosomes.

More direct evidence for the ribosomal location of the peptide bond forming step, comes from studies of the puromycin reaction. Puromycin inhibits the synthesis of proteins by replacing aminoacyl-tRNA [8–10]. The nascent peptide is transferred to puromycin instead of the next incoming aminoacyl-tRNA. TRAUT and MONRO [11] were the first to use the puromycin reaction for analysing the growth of the peptide chain. We have succeeded in simplifying their system further in that we used individually obtained components each of which was purified independently [12, 13]. As peptidyl-tRNA we utilized polylysyl-tRNA labeled with [14]C in the peptide moiety the isolation of which we described earlier [6, 13].

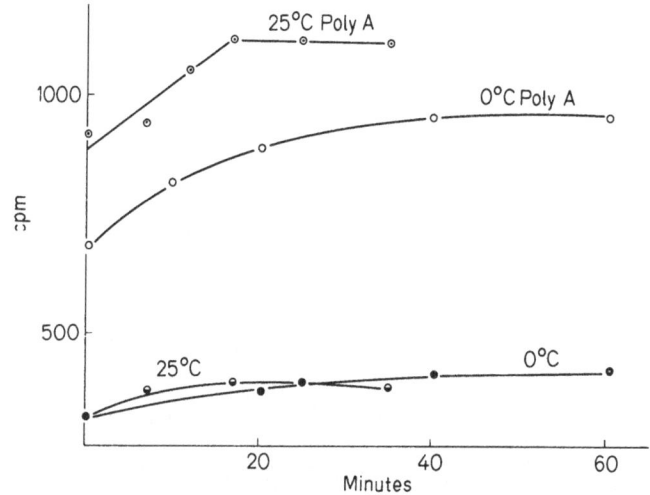

Fig. 3. Time course of the binding reaction of polylysyl-tRNA to the ribosomes. Ribosomes (160 $\mu$g protein), poly A (20 $\mu$g), polylysyl-tRNA (40 $\mu$g, 2,050 counts/ per min) were incubated in a total volume of 0.2 ml buffer (0.1 M ammonium acetate, 0.01 M magnesium acetate, 0.1 M Tris-acetate, pH 7.2) at 0° C or 25° C. At the given time intervals the amount of radioactivity attached to the ribosomes was determined

Polylysyl-tRNA is bound to the ribosome when added to a ribosome suspension in a buffer at neutrality and containing ammonium and magnesium ions (Fig. 3). The binding is messenger RNA specific and apparently a ternary complex of polylysyl-tRNA, ribosome and messenger RNA is formed [13] (Table 1). When puromycin is added to such a complex, a rapid transfer of lysine peptides from polylysyl-tRNA to puromycin occurs [12, 14]. Figure 4 shows the course of the puromycin reaction at 20°C and 34°C. The reaction kinetics is rather complex. The total amount of peptides split off depends on the amount of polylysyl-tRNA which was bound to the ribosomes under the experimental conditions used. No peptides are split off if puromycin is incubated with polylysyl-tRNA only or with polylysyl-tRNA in the presence of postribosomal enzymes (Table 2). The puromycin reaction with the polylysyl-tRNA-ribosome complex does not require supernatant enzymes. If these are added to the system, the overall amount of peptide split off is increased, but the difference between the amount of peptides split off with puromycin and without puromycin is not changed.

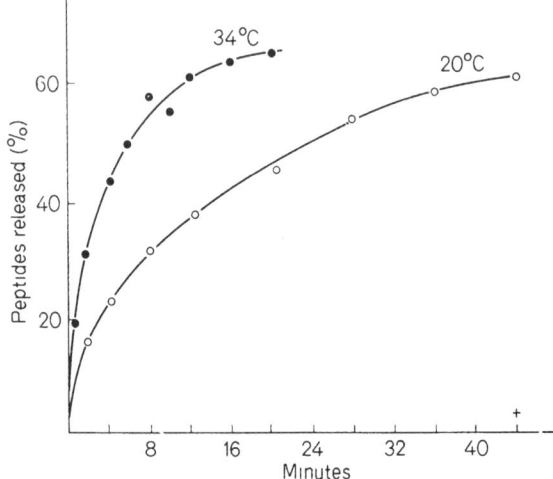

Fig. 4. Release of peptides by puromycin. Ribosomes (300 μg protein), poly-A (20 μg), polylysyl-tRNA (15 μg, 3,600 counts/min) and puromycin $10^{-4}$ M were incubated at 20° C or 34° C in 0.1 M ammonium acetate, 0.01 M magnesium acetate, 0.1 M Tris-acetate buffer (pH 7.2) in a final volume of 0.2 ml. At intervals shown, samples were withdrawn, precipitated and washed with 5% trichloroacetic acid and counted

Table 1. *Messenger specificity of the interaction of polylysyl-tRNA with ribosomes*

| Messenger polynucleotide | Polylysyl-tRNA bound to ribosome counts/min |
|---|---|
| — | 320 |
| Poly A | 1,200 |
| Poly U | 312 |
| Poly C | 446 |

Ribosomes (160 μg), poly A (20 μg), polylysyl-tRNA (40 μg, 2,050 counts/min) were incubated in a total volume of 0.2 ml buffer (0.1 M ammonium acetate, 0.01 M magnesium acetate, 0.1 M Tris-acetate, pH 7.2) at 25° C for 15 min and the radioactivity attached to ribosomes measured [13].

The puromycin reaction proceeding in this simple system displays many characteristics of the more complex process of the formation of proteins de novo. The transfer of lysine peptides from polylysyl-tRNA to puromycin depends on $Mg^{2+}$ ions which can be replaced by $Ca^{2+}$ ions. The reaction proceeds only in the presence of $NH_4^+$ or $K^+$ ions, the ammonium kation being the more effective [7]. The antibiotics erythromycin and other macrolides as well as chloramphenicol and tetracyclines which inhibit the synthesis of proteins de novo inhibit equally well the puromycin reaction [11, 14, 15].

Table 2. *Release of lysine peptides by puromycin*

Polylysyl-tRNA (43 μg, 1,257 counts/min) and the additions indicated were incubated at 35° C for 30 min under conditions described in Fig. 4

| Polylysyl-tRNA incubated with | Radioactivity bound to tRNA after incubation (counts/min) | | Radioactivity released by puromycin counts/min |
|---|---|---|---|
| | Without puromycin | With $10^{-4}$ M puromycin | |
| No additions | 1,160 | 1,120 | 40 |
| S-100 | 960 | 950 | 10 |
| Ribosomes | 1,080 | 942 | 138 |
| Ribosomes + poly A | 1,110 | 710 | 400 |
| Ribosomes + poly A + S-100 | 900 | 490 | 410 |
| Zero time control | 1,210 | — | — |

The additions were: poly A, 20 μg; ribosomes, 300 μg protein; S-100, enzymes from the 100,000 × g supernatant 200 μg protein.

These results were confirmed in other laboratories. GOLDBERG and MITSUGI [16] and COUTSOGEORGOPOULOS [17] made use of the polylysyl-tRNA-ribosome-puromycin system while studying the effects of antibiotics on the ribosomal peptide forming process. In 1966 BRETSCHER and MARCKER [18] and also LEDER and BURSZTYN [19] described the transfer of N-formylmethionine to puromycin from N-formylmethionyl-tRNA in the simple ribosomal system. LEDER [20] reported that N-formylamino-acids are transferred from tRNA to puromycin. We observed the transfer of the peptidyl residue to puromycin from benzyloxycarbonylvalyl-phenylalanyl-tRNA and from N-formylglycylphenylalanyl-tRNA. The last two derivatives were prepared by chemical synthesis from phenyl-alanyl-tRNA and nitrophenyl esters of protected aminoacids.

The activity catalysing the puromycin reaction is firmly bound to the ribosomes. It cannot be removed by washing the ribosomes with buffers containing 0.5 M $NH_4$Cl or 3 M KCl which are used for removing adsorbed enzymes from ribosomes [21, 22]. This indicates that the active enzyme site catalysing the transfer of the growing polypeptide to aminoacyl-tRNA represents an integral component of the ribosome. Monro, Maden and Traut [3] arrived at the same conclusion and later also Goldberg and Mitsugi [16], Coutsogeorgogoulos [17] and others. Monro, Maden and Traut [3] suggested to call the enzyme polypeptidyl-transferase.

There is a close analogy between the molecular mechanism of the formation of the peptide bond on the ribosome as first formulated by Nathans and Lipmann [23] and transpeptidation reactions catalysed by proteolytic enzymes in which polypeptides are formed from aminoacid or peptide esters [24, 25]. Both are displacement reactions of the general type

$$RCO - X + B \qquad RCO - B + X$$

where X is $-OR'$ and B-nucleophiles as $NH_2-R''$. The proteolytic enzymes operate generally via a double displacement mechanism whereby an acyl enzyme intermediate is formed. Brattsten, Synge, and Watt [26] proposed a similar mechanism also for proteosynthetic reactions. They postulated that, in the course of adding each amino acid residue, the growing polypeptide chain is transferred from ester linkage with tRNA to linkage with ribosomal protein through its carboxyl group, perhaps by ester linkage to an alcoholic group.

This analogy led us to explore the effect of typical inhibitors of proteolytic enzymes on the puromycin reaction. We tested diisopropylfluorophosphate which inhibits serine proteinases by blocking the serine hydroxyl in the active center, further SH-group blocking compounds active with SH-proteinases and chloromethylketone derivatives of aminoacids which block the histidine of the active site [27, 28]. From all compounds tested here only the chloromethylketones acted as inhibitors of the puromycin reaction [29]. We tested N-tosyl-L-phenylalanyl chloromethylketone (1-chloro-4-phenyl-3-tosylamino-2-butanone, Tos-L-Phe-CH$_2$Cl), N-acetyl-L-phenylalanyl chloromethylketone (1-chloro-4-phenyl-3-acetylamino-2-butanone, Ac-L-Phe-CH$_2$Cl) and α-N-tosyl-L-lysyl-chloromethylketone (7-amino-1-chloro-3-tosylamino-2-heptanone, α-Tos-L-Lys-CH$_2$Cl). The first two compounds are specific chymotrypsin inhibitors, the last one a trypsin inhibitor. All three compounds inhibit the puromycin reaction to approximately the same degree (Table 3). In respect to the puromycin reaction these inhibitors do not show the aminoacid residue specificity typical of the inhibition of proteolytic enzymes [27]. The chloromethylketones were found by Pulkrábek and Rychlík [29] to inhibit also the synthesis of lysine peptides de novo in the cell free poly A directed system of Escherichia coli. If compared with chloramphenicol, a typical inhibitor of proteosynthesis, the chloromethyl ketones are a little more active on molar basis (Table 3). When ribosomes only were treated with the chloromethylketones and thereafter the inhibitor removed by dialysis, the ribosomes retained full activity in catalysing the puromycin reaction whereas proteolytic enzymes are

Table 3. *Inhibitors of the puromycin reaction*

As polypeptidyl-tRNA N-acetyl-phenylalanyl-tRNA was used. The reaction was performed under conditions described by MONRO and MARCKER [35]. With 40% of ethanol in the incubation medium

| Inhibitor | Concentration (M) | Inhibition (%) |
|---|---|---|
| Tos-L-Phe-CH$_2$Cl | $10^{-4}$ | 58 |
| Ac-L-Phe-Ch$_2$Cl | $10^{-4}$ | 64 |
| α-Tos-L-Lys-CH$_2$Cl | $10^{-4}$ | 70 |
| Z-L-Phe-ONP | $10^{-3}$ | 76 |
| Z-Gly-Phe-NA | $10^{-3}$ | 40 |
| Chloramphenicol | $10^{-4}$ | 40 |
| no inhibitor | | 0 |

inhibited irreversibelly. This weakens the interpretation that chloromethylketones inhibit the puromycin reaction by substituting a histidine residue in peptidyl-transferase. An alternative explanation of the mode of action could make use of the remote analogy between the structure of the chloromethylketones and the peptidyl terminus of peptidyl-tRNA with which they could compete for peptidyl-transferase. In Table 3 the inhibitory effect of benzyloxycarbonyl-L-phenylalanyl-p-nitrophenylester (Z-L-Phe-ONP) and of benzyloxycarbonyl-glycyl-L-phenylalanyl nitroanilide (Z-Gly-Phe-NA) is shown. However, these compounds resembling to some degree the peptidyl terminus of tRNA are much weeker inhibitors of the puromycin reaction than chloromethylketones.

By analogy with other enzymes we assumed that the ribosomal peptidyl transferase contains an active site catalysing the transfer of the nascent peptide chain proper and two binding sites which determine the substrate specificity of the reaction, one for peptidyl-tRNA, the other for aminoacyl-tRNA. We studied [30] the binding site for aminoacyl-tRNA using instead of whole aminoacyl-tRNA molecules simpler compounds (Fig. 5), namely synthetic glycyldinucleoside phosphates of defined structure and related compounds modelled on the glycine terminal sequence of glycyl-tRNA (i.e. RNA-CpA-Gly). These compounds were prepared by CHLÁDEK and ŽEMLIČKA [31]. The effect of these compounds was examined in the polylysyl-tRNA-ribosome system used above for testing the puromycin reaction [12, 14]. The release of lysine peptides from polylysyl-tRNA was a measure of the biological activity of the compounds. The simplest derivatives represent the terminal aminoacid of tRNA

I. RYCHLÍK:

either free or bound in ester or amide linkage. This group comprised free
glycine and phenylalanine, their amide, ethylester, nitrophenylester and
nitroanilide. Besides these we tested the amide and ethyl ester of tyrosine,
leucine, valine and alanine. All the compounds mentioned were fully
inactive in releasing lysine peptides from polylysyl-tRNA bound to the
ribosome. From compounds more related to the aminoacid terminus of

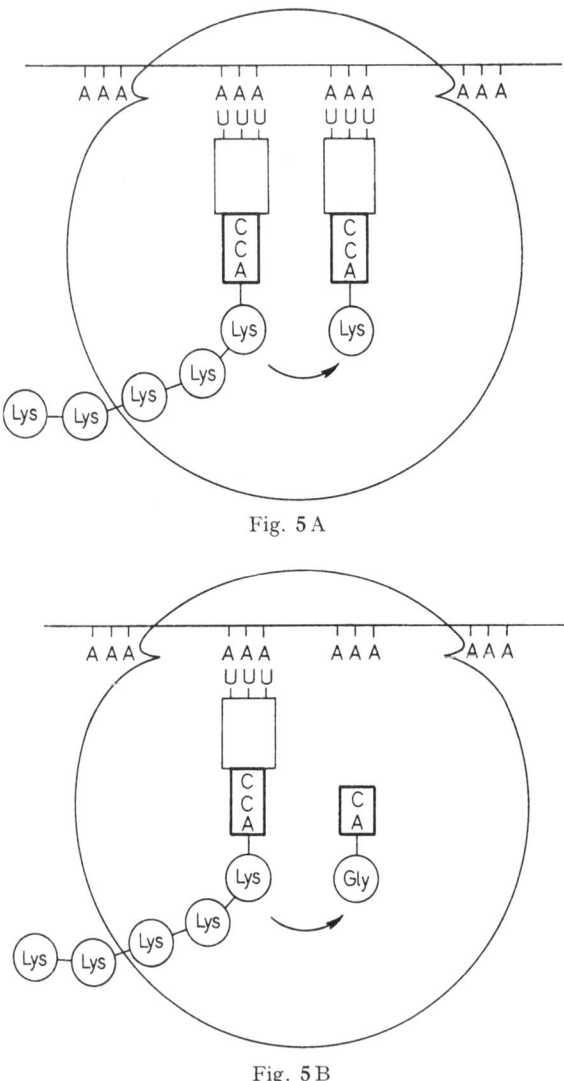

Fig. 5. A) Schematic drawing of the complex between polylysyl-tRNA, lysyl-tRNA,
poly A and ribosome. B) Schematic drawing of the complex between CpA-Gly,
polylysyl-tRNA, poly A and ribosome

tRNA (Table 4) tested here A-Gly as well as pA-Gly equivalent to terminal
aminoacyladenosine and aminoacyladenosine-5′-phosphate of tRNA,
respectively, did not release peptides from the polylysyl-tRNA ribosome

Table 4. *Release of lysine peptides from polylysyl-tRNA by CpA-Gly and related
compounds*

[14]C Polylysyl-tRNA (10 $\mu$g, 1,900 counts/min (Expt. A); 2,400 counts/min (Expt. B),
E. coli ribosomes[13] (120 $\mu$g of protein) and polyadenylic acid (10 $\mu$g) were incubated at 35° C for 20 min in 0.1 M ammonium acetate, 0.01 M magnesium acetate,
0.1 M Tris-maleate buffer (pH 6.5) in a final volume of 0.1 ml. The samples were
precipitated with trichloroacetic acid, washed and counted

| Additions | Concentration (M) | Radioactivity bound to tRNA after incubation (counts/min) | Radioactivity released by the compound tested (counts/min) |
|---|---|---|---|
| Expt. A | | | |
| None | — | 1,870 | — |
| A-Gly | $10^{-4}$ | 1,850 | 20 |
| A-Gly | $10^{-3}$ | 1,840 | 30 |
| pA-Gly | $10^{-4}$ | 1,845 | 25 |
| CpA-Gly | $10^{-4}$ | 1,380 | 490 |
| CpA | $10^{-4}$ | 1,880 | none |
| CpA-(ZGly) | $10^{-4}$ | 1,820 | 50 |
| Puromycin | $10^{-4}$ | 880 | 990 |
| Expt. B | | | |
| None | — | 2,390 | — |
| UpA-Gly | $10^{-5}$ | 2,360 | 30 |
| UpA-Gly | $6.10^{-4}$ | 2,270 | 120 |
| UpA-Gly | $10^{-3}$ | 2,310 | 80 |
| UpU-Gly | $10^{-5}$ to $10^{-3}$ | 2,390 | none |
| dCpA-Gly | $5.10^{-4}$ | 2,210 | 180 |
| CpA-Gly | $10^{-4}$ | 1,880 | 510 |

complex. These results are in accordance with experiments using glycyl-
puromycin which shows no effect on the synthesis of proteins in cell free
systems [32]. However, A-Leu and A-Tyr as the leucyl or tyrosyl derivatives of puromycin depress the synthesis of proteins on ribosomes [32, 33].
Only when A-Gly is attached to the cytidine 3′-phosphate residue which
represents the natural neighbour of adenosin in the terminal sequence
RNA-CpA-Gly, is an active compound formed (CpA-Gly) which splits off
peptides from polylysyl-tRNA attached to ribosomes.

The following evidence is advanced that CpA-Gly causes release of
lysine peptides in that these peptides are transferred to it in the same way
as to puromycin or to aminoacyl-tRNA. The only effective compound is
CpA-Gly, which contains a free amino group in the glycine residue. No

peptides are split off if the polylysyl-tRNA-ribosome complex contains, instead of the above compound, either the parent dinucleoside phosphate CpA or CpA-(ZGly) in which the glycine amino group is substituted by a benzyloxycarbonyl group.

The release of peptides is very specific with respect to the dinucleoside phosphate to which glycine is attached. Peptides are split off by CpA-Gly, which represents a part of the natural terminal sequence of glycyl-tRNA. Its close analogue, UpA-Gly, has a very small effect; the more remote UpU-Gly is fully inactive. dCpA-Gly shows also a low effect equivalent to UpA-Gly.

Monro [34] and Monro and Marcker [35] explored the peptidyl-tRNA binding site of the ribosomal peptidyl-transferase using a similar approach [34, 35]. He prepared by digesting with $T_1$ ribonuclease the terminal hexonucleotide fragment from N-formylmethionyl-tRNA and observed the transfer of formylmethionine from the fragment to puromycin. The reaction proceeds only with 30 % to 60 % ethanol in the medium. Černá and Monro [35] found that the shortest fragment from which N-formylmethionine is transferred to puromycin is CpCpA-(FMet).

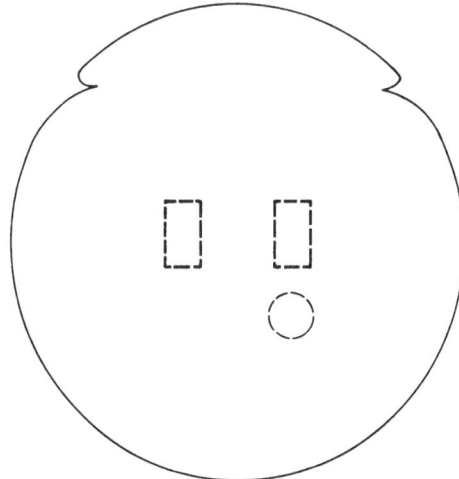

Fig. 6. Schematic drawing of the ribosome binding sites for -CpCpA termini of aminoacyl- and polypeptidyl-tRNA's, respectively

The results of experiments described here can be summarised as follows. The peptidyl-transferase which catalyses the transfer of the growing polypeptide chain from peptidyl-tRNA to aminoacyl-tRNA is a ribosomal enzyme. The active site of the enzyme is composed of a site catalysing the transfer of the polypeptide chain proper and of two binding sites determining the specificity of the reaction (Fig. 6). One of them binds the terminal sequence − CpA-aa by a specific interaction with nucleotides of that sequence; the other binding site interacts in a similar manner with the − CpCpA-terminus of peptidyl-tRNA. This interaction directs the growing polypeptide and the aminoacid to be attached into proper positions so that the transfer of the peptide residue can occur. It is not the whole molecule of tRNA that is involved in the reaction but only the terminal sequences −CpA-aa and −CpCpA-peptide, respectively. The general occurence of the CpCpA sequence in all transfer ribonucleic acids is probably the expression of the general function of this sequence in the transfer reaction.

## References

1. WATSON, J. D.: Bull. Soc. Chim. biol. (Paris) 46, 1399 (1964).
2. WETTSTEIN, F. O., and H. NOLL: J. molec. Biol. 11, 35 (1965). − NOLL, H.: Science 151, 1241 (1966).
3. MONRO, R. E., B. F. MADEN, and R. R. TRAUT: In: Genetics elements properties and function, p. 179. Ed. by D. SHUGAR. London: Academic Press 1967.
4. NIRENBERG, M. W., and J. H. MATTHAEI: Proc. nat. Acad. Sci. (Wash.) 47, 1588 (1961).
5. GARDNER, R. S., A. J. WAHBA, C. BASILIO, R. S. MILLER, R. LENGYEL, and J. F. SPEYER: Proc. nat. Acad. Sci. (Wash.) 48, 2087 (1962).
6. RYCHLÍK, I.: 6th Internat. Congr. of Biochemistry, Abstracts I p. 83, New York 1964. Coll. Czech. Chem. Commun. 30, 2259 (1965).
7. PULKRÁBEK, P., and I. RYCHLÍK: Biochim. biophys. Acta (Amst.) 155, 219 (1968).
8. YARMOLINSKY, M. B., and G. L. DE LA HABA: Proc. nat. Acad. Sci. (Wash.) 45, 1721 (1959).
9. HULTIN, T.: Biochim. biophys. Acta (Amst.) 51, 219 (1961).
10. ALLEN, D. W., and P. C. ZAMECNIK: Biochim. biophys. Acta (Amst.) 55, 865 (1962).
11. TRAUT, R. R., and R. E. MONRO: J. molec. Biol. 10, 63 (1964).
12. RYCHLÍK, I.: Abstracts of the 2nd Meeting of the Federation of European Biochemical Societies, p. 299, Vienna 1965.
13. − Coll. Czech. Chem. Commun. 31, 2583 (1966).
14. − Biochim. biophys. Acta (Amst.) 114, 425 (1966).
15. ČERNÁ, J., and I. RYCHLÍK: Coll. Czech. Chem. Commun. (in press).
16. GOLDBERG, H. J., and K. MITSUGI: Biochemistry 6, 383 (1967).
17. COUTSOGEORGOPOULOS, C.: Biochem. biophys. Res. Commun. 27, 46 (1967).
18. BRETSCHER, M. S., and K. A. MARCKER: Nature (Lond.) 211, 380 (1966).
19. LEDER, P., and H. BURSZTYN: Biochem. biophys. Res. Commun. 25, 233 (1966).
20. − Fed. Proc. 26, 457 (1967).
21. SMITH, T.: J. molec. Biol. 8, 772 (1964).
22. LEDERBERG, S., B. ROTMAN, and V. LEDERBERG: Biochem. biophys. Res. Commun. 12, 324 (1963).
23. NATHANS, D., and F. LIPMANN: Proc. nat. Acad. Sci. (Wash.) 47, 497 (1961).
24. WALLEY, S. G., and J. WATSON: Biochem. J. 57, 529 (1954).
25. RYCHLÍK, I., K. I. DANCHEVA, and M. CERHOVÁ: Coll. Czech. Chem. Commun. 30, 138 (1965).
26. BRATTSTEN, I., R. L. M. SYNGE, and W. B. WATT: Biochem. J. 97, 678 (1965).

27. SCHOELLMANN, G., and E. SHAW: Biochemistry 2, 252 (1963).
28. JAROŠKOVÁ, L., and J. RUDINGER: in preparation.
29. PULKRÁBEK, P., and I. RYCHLÍK: Coll. Czech. Chem. Commun. (in press).
30. RYCHLÍK, I., S. CHLÁDEK, and J. ŽEMLIČKA: Biochim. biophys. Acta (Amst.)
      138, 642 (1967).
31. CHLÁDEK, S., and J. ŽEMLIČKA: Coll. Czech. Chem. Commun. 32, 1776 (1967).
32. NATHANS, D., and A. NEIDLE: Nature (Lond.) 197, 1067 (1963).
33. WALLER, J. P., T. ERDÖS, F. LEMOINE, S. GUTTMANN, and E. SANDRIN: Biochim.
      biophys. Acta (Amst.) 119, 566 (1966).
34. MONRO, R. E.: J. molec. Biol. 26, 147 (1967).
35. —, and K. A. MARCKER: J. molec. Biol. 25, 437 (1967).
36. ČERNÁ, J., and R. E. MONRO: personal communication.

# Synthesis of Polyphenylalanine in a Cell-free System from E. coli A19: Ribosomal Binding Sites, Initiation with Formyl-methionyl-tRNA, and Codon-Interaction of Aminoacyl-tRNA

By

H. MATTHAEI, K. LØVE, M. MILBERG, G. SANDER,
D. SWAN, and H.-P. VOIGT

*Abbreviations used:* AA = amino acid; 19a-tRNA = tRNA charged with 19 amino acids; F = formate; Met = methionine; mRNA = messenger-RNA; Phe = phenylalanine; poly-Phe = polyphenylalanine; Poly-U = poly-uridylic acid; tRNA = transfer-RNA.

## Introduction: Individual Steps in Protein Synthesis

Molecular biology deals with specificity relationships between the molecular members of the various functional systems of the cell. The *decoding system* is the center of gene action and is outlined for the bacterial

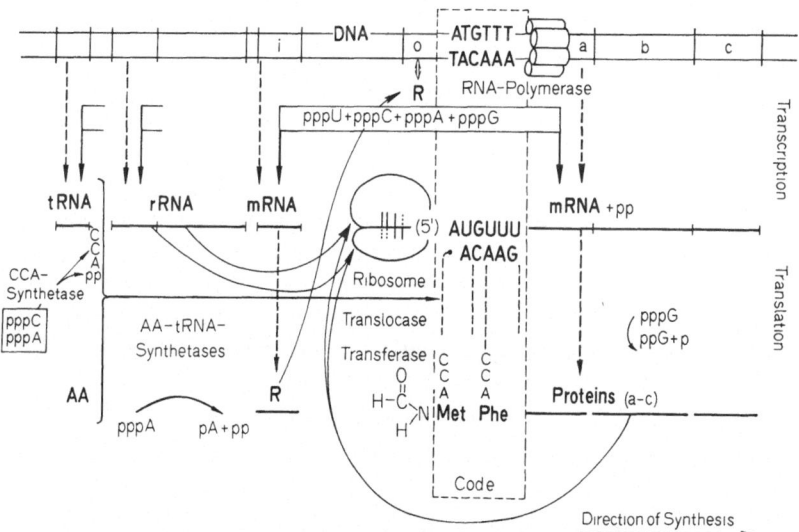

Fig. 1. Diagram illustrating the decoding system of the bacterial cell: Molecular constituents and major functions (from [5], improved)

cell in Fig. 1. This system involves, (1) *Template-controlled enzymatic processes* of information transfer: The synthesis of the different species of RNA and of protein. (2) *Nontemplate-controlled enzymic reactions:* The synthesis of aminoacyl-tRNA. (3) *Formation of structural complexes by macromolecules:* The spontaneous combination of ribosomal RNA and proteins to form ribosomes, and the interaction of the repressor (R) and the operator (o) region of DNA to form the repressed state of an operon. The crucial and typical functions in the decoding system are those bringing together biochemical specificities which reside in molecules belonging to different classes of substance: RNA-polymerase transcribes DNA into RNA and aminoacyl-tRNA-synthetases recognize the specific relationships between aminoacids and their corresponding transfer-RNAs. The final translation-problem is thus reduced, from an overall recognition of correspondence between DNA-triplets and aminoacids by the whole decoding system, to a presumed checking of the specific interaction between RNA-triplets, i. e. of codons in mRNA, and of anticodons in aminoacyl-tRNA. This final translation most certainly takes place on the ribosome.

After the deciphering of the genetic code, current interest in protein synthesis is concentrated in an effort towards an understanding of the ribosomal mechanism. The individual steps in the mechanism can be postulated and partially described as indicated in Fig. 2. The degree of detail in this kind of working hypothesis is limited to consideration of several sites on the ribosome, three types of functionally different codons (initiation, – internal- and termination-signals), and three types of transfer-RNAs (initiator-, internal- and terminator-tRNA)[1]. Apart from special problems in initiation and termination, there are three major steps to be considered in the incorporation of each amino acid into a polypeptide (= repeating unit): (1) Binding of aminoacyl-tRNA, (2) transfer of the esterified carboxyl end of the growing peptide to the amino group of the next incoming aminoacyl-tRNA, and (3) translocation of the resulting peptidyl-tRNA. In the initial steps, the question of the sequence of events in macromolecular interaction with ribosomal sites (and enzymes) is due to be analyzed. Once this is known, more detailed working hypotheses will be needed, which take into consideration the specific substructures and conformational changes within the macromolecular participants, in order to understand their interaction chemically.

Our present report deals with the following questions:

1. How many binding sites for aminoacyl-tRNA are there on the ribosome?

2. Which site does initiator-tRNA enter first?

---

[1] see addendum

3. How is the synthesis of poly-phenylalanine initiated?

4. Can one demonstrate interference by amino acid adapters, which do not fit the codon to be read?

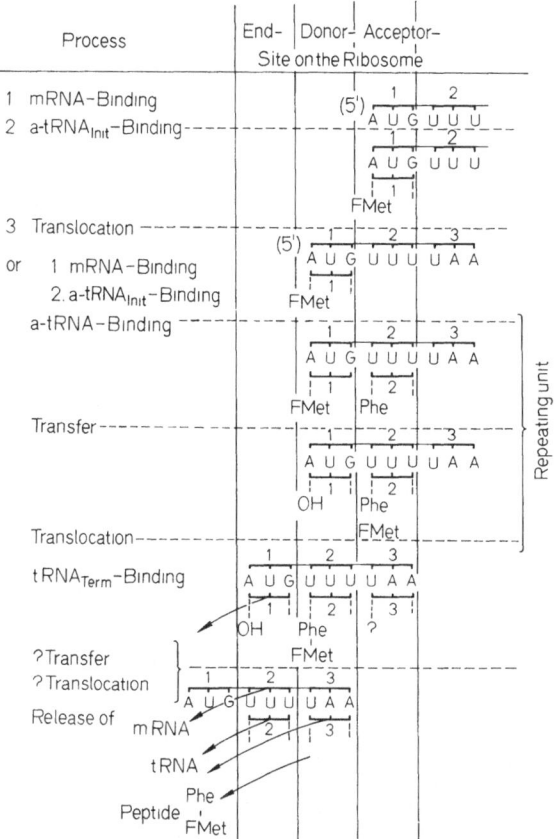

Fig. 2. Our current working hypothesis on subsequent steps in protein synthesis: The initiation triplet AUG could be bound first into the acceptor- or donor-site (from [5], improved)

The new observations should be helpful in designing new assay systems for partial functions like those of individual factors of, or on, the ribosome.

For experimental details and pertinent literature the reader is referred to the publications quoted.

## Experimental Results

*1. The Number of Binding Sites for Amino Acyl-tRNA on E. coli-Ribosomes*

The minimum number of ribosomal binding sites for tRNA appears to be *two* for a logical reason: The coded synthesis of peptide bonds occurring on ribosomes [9] should require sites for binding *of both* substrates, the peptidyl-donating tRNA and the amino acyl-tRNA accepting this peptidyl-group in a transfer reaction. In Fig. 2, our current working hypothesis is shown, which illustrates possible functions of these sites. Amino acyl-tRNA is bound under the coding function of mRNA to both 70S ribosomes and their isolated 30S subunits [4]. Poly-Phe-tRNA, however, is bound to 50S subunits [11]. A *third* site apparently binds non-aminoacylated tRNA and is characterized by a Mg-requirement which is higher than that for the other sites [15].

All amino acid adapters for chain-internal amino acids may be bound primarily to the "acceptor"-site, which is probably located on the 30S-ribosome (ref. in [13]). It appears to be a separate question, however, whether the initiator-adapter F-Met-tRNA is bound primarily to the same site on the 30S-subunit as the adapters for chain-internal amino acids (see section No. 2); and furthermore, whether this first-entered site is indeed the acceptor site. The functional definition of this site, i.e. binding of an amino acyl-tRNA which accepts the carboxyl-group of a peptidyl-residue, could not of course apply in the special case of N-formyl-methionine, which has the amino group blocked.

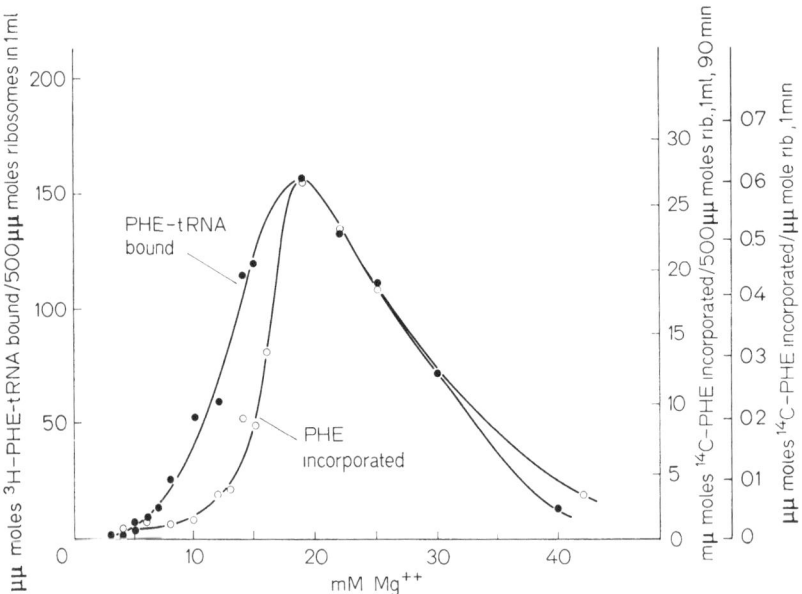

Fig. 3. $Mg^{++}$-dependence of the poly-U-dependent binding of phenylalanyl-tRNA to E. coli A 19-ribosomes (left ordinate) and of the synthesis of polyphenylalanine (right ordinate), ribosomes are prepared from E. coli A 19 [3, 5], washed three times with ammonium chloride [13], and tested as described (1); conditions in table 1

Synthesis of the model peptide poly-Phe requires relatively high $Mg^{++}$-concentrations [12]. In the lower $Mg^{++}$-concentration range the $Mg^{++}$-requirement for the peptide synthesis is considerably higher than for the binding of Phe-tRNA (Fig. 3). The optima for the binding- and peptidization-systems, however, are the same. In the experiment shown in Fig. 3 it was 19 mM. This optimum varies between 14 and 19 mM with different ribosome preparations. REVEL and HIATT [12] have suggested that the high $Mg^{++}$ reguirement might be due to some initial step in the synthesis of polyphenylalanine. In fact they have shown that after preincubation with a high $Mg^{++}$-concentration (15 mM), the polymerization took place

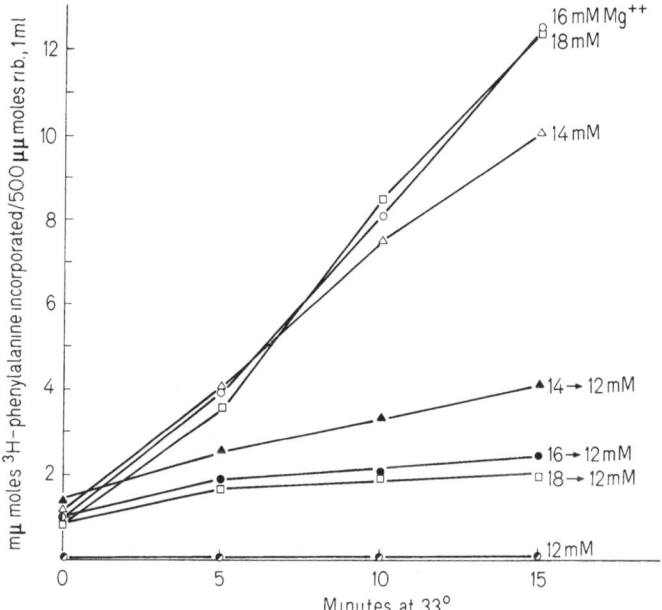

Fig. 4. $Mg^{++}$-requirement of the synthesis of polyphenylalanine and the influence of higher $Mg^{++}$-concentrations during a preincubation period of 7 min at 25° C upon incorporation during further incubation for the time period given on the abscissa, at 33° C and the $Mg^{++}$-concentrations indicated on the curves; e.g. 14 → 12 mM means $Mg^{++}$-concentration reduced from 14 mM during preincubation to 12 mM for incubation at 22° C [1]

just as well after lowering this concentration to one half (7.5 mM). Fig. 4 represents some similar experiments with our cell-free system from E. coli A 19, which after reduction of $Mg^{++}$-concentrations from 14, 16 or 18 mM to 12 mM shows at least some incorporation continuing at 12 mM, where no peptidization of phenylalanine would have occurred without preincubation at a higher $Mg^{++}$-concentration. During the last few years, this phenomenon was generally assumed to result from a *mis-initiation* of this model peptide synthesis. Due to the absence of any initiator-codons in poly-U, a molecule of phenylalanyl-tRNA would presumably have to be bound to the donor-site and this erroneous binding would require extra $Mg^{++}$. We have tried to check whether Phe-tRNA is bound to all the different ribosomal sites, and to what extent this binding occurs even at $Mg^{++}$-concentrations far below the minimum required for synthesis of poly-Phe.

Saturation of poly-U-programmed ribosomes with the substrate Phe-tRNA in the binding-equilibrium follows a curve with three poly-U-dependent slopes [7]. Extrapolation of the first two slopes to a Phe-tRNA-concentration of zero indicates an equal filling of the two, presumably corresponding, sites at various $Mg^{++}$-concentrations (in low-salt, see Table 1) and, moreover, at very different states of purification. We recently found that beyond the end of the third slope, saturation-curves are horizontal straight lines.

Table 1. *Composition of the cell-free systems employed*

| Component | Binding system | | Peptidizing system |
|---|---|---|---|
| | "high" — "low salt" | | |
| Tris-HCl pH 7.8 | 100 mM | 10 mM | 100 mM |
| KCl | 80 | 6 | 80 |
| $MgCl_2$ | 17 | 17 | 17 |
| $\beta$-Mercaptoethanol | 6 | — | 6 |
| (MET) | 6 | | |
| ATP-K | 1 | — | 1 |
| GTP-K | — | — | 0.2 |
| CTP-K | 0.5 | — | 0.5 |
| PEP-K | 8 | — | 8 |
| Phospho-enol- | 20 $\gamma$/ml | — | 20 $\gamma$/ml |
| pyruvate-kinase | | | |
| Phenylalanine | — | — | 0.05 or one |
| 20 Amino acids | 0.05 | — | 0.05 |
| minus Phe | | | |
| Poly-U (Up) | 0.15 | 0.15 | 0.15 |
| tRNA | ³H-phe-tRNA: | | 0.75 mg/ml |
| | 0.75 mg/ml | | tRNA or ³H- |
| | | | phe-tRNA |
| Ribosomes | 500 $\mu\mu$moles/ml | | 500 $\mu\mu$moles/ml |
| S 100 | — | — | 100 $\mu$l/ml |

Under appropriate conditions, the total amount of substrate bound to all three binding sites for Phe-tRNA reaches an average of 1.5 moles Phe-tRNA per ribosome. This is considered as further evidence for the binding of more than one molecule of Phe-tRNA to some, or all, of the ribosomes active in the binding of this substrate. We now postulate *three* sites on the ribosome which bind Phe-tRNA.

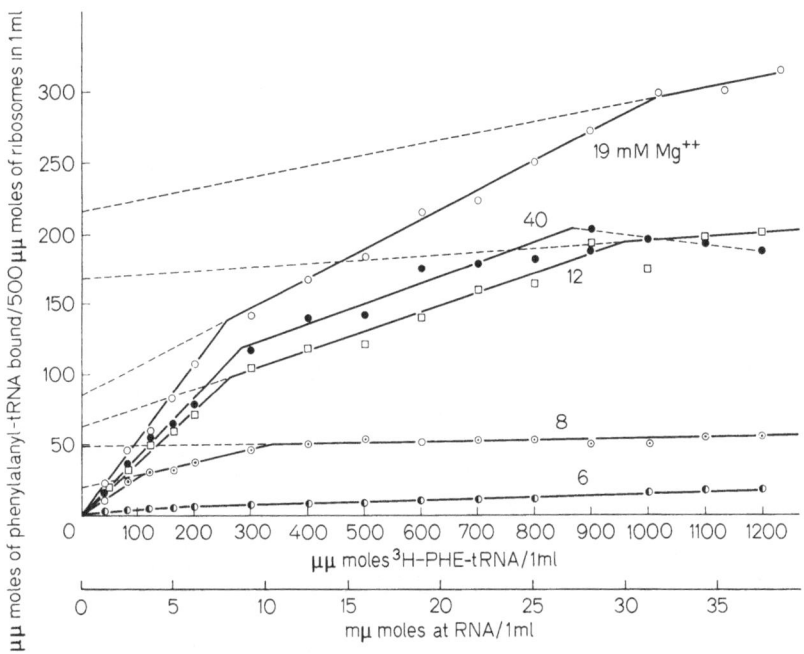

Fig. 5. $^3$H-phenylalanyl-tRNA-saturation curves: Each curve shows three subsequent poly-U-dependent slopes, which may indicate three different ribosomal binding sites, demonstrable in both low [7] and high concentrations of monovalent cations. The high-salt-binding system used here contained 100 mM Tris-HCl pH 7.2, 80 mM KCl, and the $Mg^{++}$-concentrations indicated on the curves. They were incubated for 30 min at 33° C and filtered through Millipore-filters HAWP 0.25 μ

To answer the question of how many sites bind Phe-tRNA in the peptidizing system, saturation-curves were also obtained at the higher salt-concentration (see Table 1): Saturation curves are similar, displaying again three slopes before full saturation. Fig. 5 shows a much smaller extent of binding below 12 mM $Mg^{++}$; Fig. 6, however, proves that even down to 4 mM $Mg^{++}$, there are three slopes to be seen. If the reason for the failure of A 19-ribosomes to synthesize poly-Phe, normally below 10 mM $Mg^{++}$, is related somehow to binding of Phe-tRNA to the donor site, it could be due to the low frequency or short life-span of complexes, but not complete absence of Phe-tRNA from any one of the three sites indicated by these slopes. We wish also to stress the point that we do not yet know if spontaneous binding of Phe-tRNA from the dissolved state

to the donor site can be correlated to any one of the three slopes, or
whether Phe-tRNA, if bound to the donor site, just does not lead to the
initiation of a polypeptide synthesis.

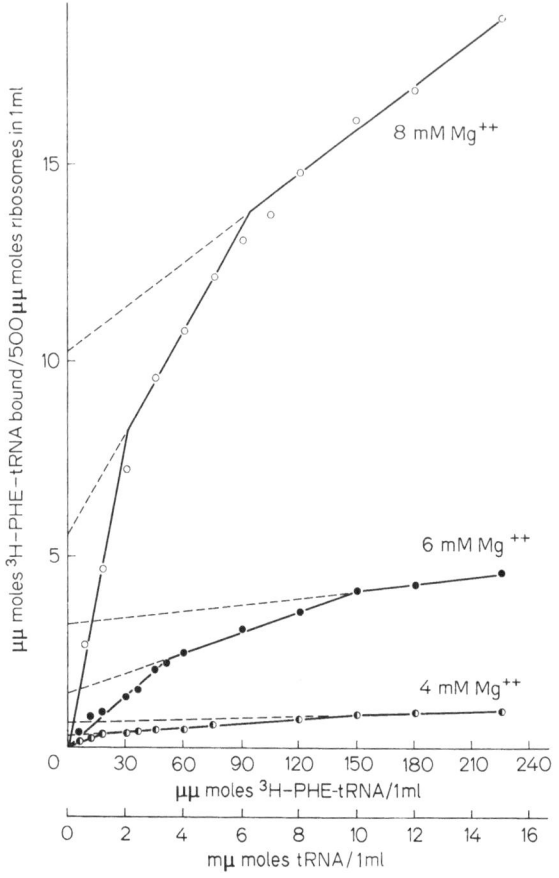

Fig. 6. Phenylalanyl-tRNA-saturation curves: Three slopes are seen, even at
Mg++-concentrations far below the level required for polyphenylalanine-synthesis
(see fig. 3). Only a few of the presumed sites corresponding to these slopes seem to
be filled, however. Conditions as in fig. 5

So far we can tentatively correlate at least one of the three slopes to a
separate ribosomal subunit. The first and steepest slope, and apparently
also the third slope are seen with 30S particles, both in the presence and
absence of a 50S particle. 50S-subunits do not bind Phe-tRNA to any
greater extent, with or without poly-U. When 50S subunits are added in
stoichiometric amounts to 30S-subunits, however, a second slope appears
and the site seemingly represented by this slope is filled to the same extent
as the site corresponding to the first slope (Fig. 7).

The requirement of both subunits for the display of the second slope may suggest that this site is related to peptide synthesis. Since binding of aminoacyl-tRNA is also accomplished by the 30S-subunit, apparently even to two sites, the second slope may correspond rather to the donor site, being located on the 50S-subunit [11]. It remains to be explained by further experiments why poly-Phe-synthesis does not occur below 10 mM $Mg^{++}$ in spite of the fact that the donor-site may have bound phenylalanyl-tRNA to some extent. It would be rather unexpected if the site corresponding to the third slope were the end-position for tRNA, which would normally receive a non-charged molecule of tRNA after transfer and translocation (see Fig. 2). Thus in subsequent investigation we shall have to look into the possibility of whether there is more than one site in front of the donor site, for checking of specific correspondence between codons and tRNA. The proposed possible existence of several sites for simultaneous checking may promote the speed and high fidelity reached in translation.

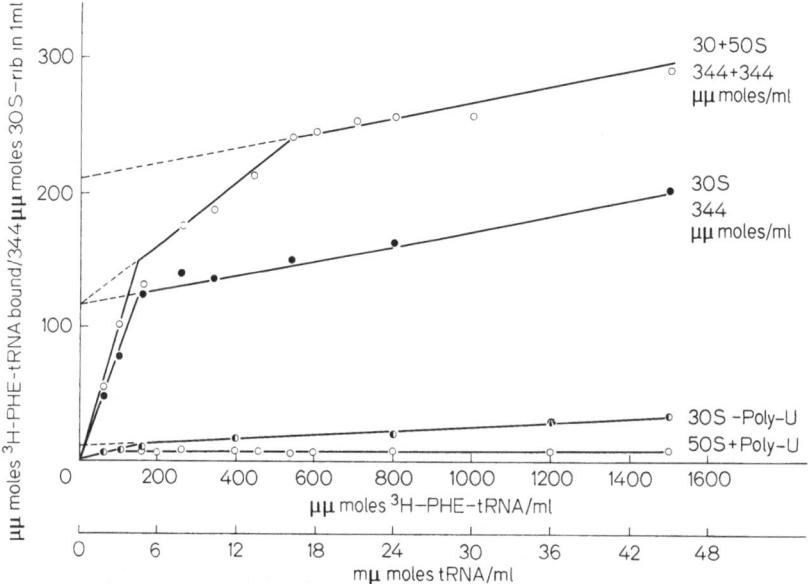

Fig. 7. Phenylalanyl-tRNA-saturation curves: Binding to 30S-subunits displays only the first and third slopes of the curve obtained for either 30S- plus 50S-, or 70S-ribosomes. The low-salt-binding system (Table 1) containing 10 mM Tris-HCl pH 7.2 − 13.5 mM KCl − 19 mM $MgCl_2$ was incubated for 30 min at 35° with the concentrations of ³H-phenylalanyl-tRNA given on the abscissa [7]

Another open question remains to be answered by further work: Whether the large fraction of purified ribosomes, active in binding of Phe-tRNA (up to 50 %), may be considered representative of the probably much smaller fraction of particles which is active in poly-Phe-synthesis. At the moment we can only contribute a kinetic argument, which would rather disfavour such an assumption. The incorporation of Phe into the polypeptide takes place at a rate of about one molecule Phe per average ribosome per minute [1]. In the high-salt ionic environment of the peptidizing system, however, the maximum initial velocity of binding of Phe-tRNA to the ribosome is far below this value.

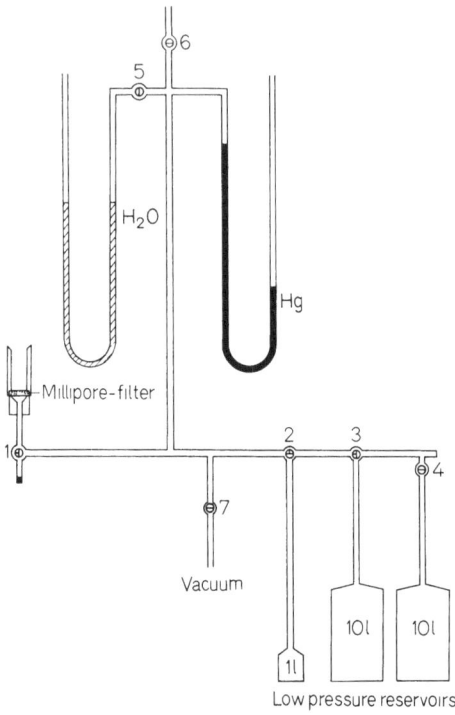

Fig. 8. Apparatus for fast kinetic measurements of the filling of ribosomal binding sites using Millipore-filters

By a special modification of the cellulose nitrate filter technique, reported first by NIRENBERG and LEDER [10], we can measure binding to ribosomes in the second- and even millisecond-range (see Fig. 8) [6]. It is only in the special low-salt environment indicated in Fig. 9, that we can see an initial velocity high enough to account for the binding in a good peptidizing system. This speed decreases after the first second of exposure of filter-bound poly-U-saturated ribosomes to a solution of Phe-tRNA. The extrapolated first section of this time curve shows a much smaller binding by these ribosomes than would be expected from the saturation curve slopes. Therefore ribosomes should bind Phe-tRNA much faster in the presence of supernatant factors than in their absence. Much further work will be needed in order to understand the function of those ribosomal sites which seem to be indicated by Phe-tRNA-saturation curves.

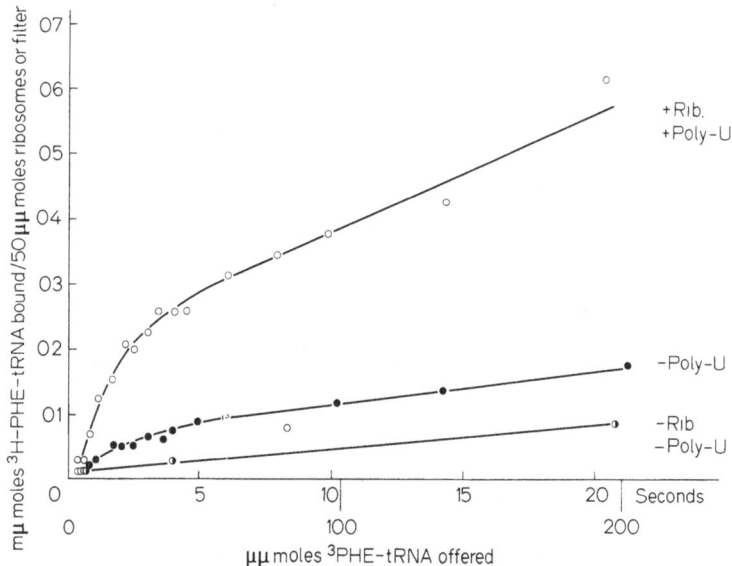

Fig. 9. Binding of phenylalanyl-tRNA: Fast kinetics in a low-salt-system with 10 mM K-phosphate buffer pH 5.2 — 60 mM $MgCl_2$ and 310 $\mu\mu$ moles tRNA (including 10 $\mu\mu$ moles [3]H-Phe-tRNA) per ml. This solution flows through the Millipore-filter containing 50 $\mu\mu$ moles of poly-U-charged ribosomes

## 2. The Binding of Initiator-tRNA to Ribosomes

The initiator-adapter in E. coli (F-Met-tRNA) was shown to bind to ribosomes at an optimum of $3-5$ mM $Mg^{++}$, and hardly at all above 10 mM. Washing of ribosomes with 0.5 M $NH_4Cl$ largely reduces this binding capability [14]. There is another $Mg^{++}$-optimum for Met-tRNA$_{Met}$ at around 16 mM (Fig. 10). This observation could be explained as being due to either two binding sites, or one binding site which assumes different specificity with varied Mg-concentration. From the viewpoint of structure one could also have expected the initiator-adapter, carrying the peptide N-formyl-methionine, to enter firstly the donor rather than the acceptor site (see Fig. 2). One could even have asked which of these sites the initiator triplet AUG would enter first. This problem seems to be solved rather easily. Table 2 shows results from two different experiments in which the triplet ApUpG*-[3]H was bound to both 30S and 50S subunits;

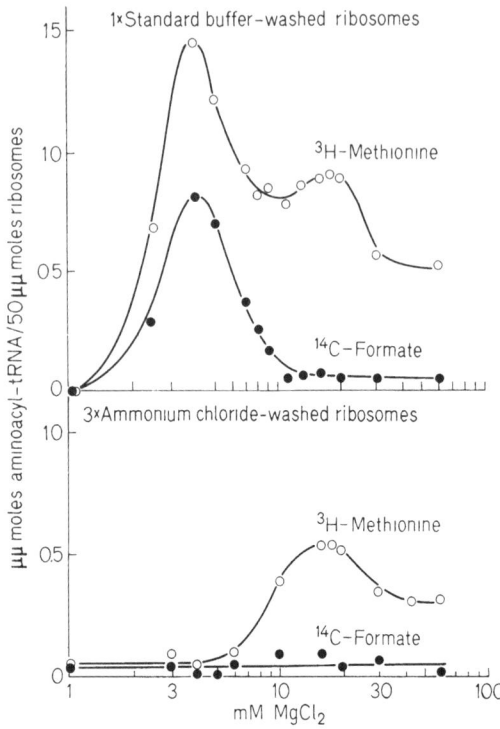

Fig. 10. Binding of $^{14}$C-formyl-$^{3}$H-methionyl-tRNA to ribosomes coded with poly-A, U, G: Mg$^{++}$-optima for the formylatable and non-formylatable methionine-adaptors are different. Ammoniumchloride-washed ribosomes have essentially lost their ability to bind formyl-methionyl-tRNA [14]

Table 2. *Binding of ApUpG\*-$^{3}$H and $^{14}$C-formyl-methionyl-tRNA by ribosomes and their subunits from E. coli A 19*

| Expt. I | Rib. concn. | ApUpG\*-$^{3}$H | $^{14}$C-F\*-Met-tRNA |
|---|---|---|---|
| | μμMoles/ml | μμMoles/ml | μμMoles/ml |
| 30 S | 3,000 | 21.0 | 5.2 |
| 50 S | 3,000 | 19,4 | 0.4 |
| 30 + 50 S | 1,500 + 1,500 | 67.3 | 35.1 |
| 70 S (original) | 3,000 | 107.0 | 84.6 |
| Expt. II | | | |
| 30 S | 3,000 | 7.3 | 4.0 |
| 50 S | 3,000 | 30.0 | 1.7 |
| 30 + 50 S | 1,500 + 1,500 | 52.5 | 51.3 |
| 70 S (original) | 3,000 | 127.5 | 90.0 |

Expts. I and II were done with 3 and 5 times standard buffer-washed ribosomes, respectively.

$^{14}C-F^*$-Met-tRNA, however, was bound only to 30 S and not to 50 S particles.

These experiments, favouring the idea that 30 S subunits are the entrance for F-Met-tRNA as well for all other amino acyl-tRNAs, are coincident with experiments by Dr. NOMURA (this volume). He has shown that only 30 S-particles bind f2-RNA and thereupon also F-Met-tRNA.

## 3. The Initiation of the Poly-U-Directed Synthesis of Poly-Phenylalanine

According to the first section of this paper we could explain the failure of ribosomes to synthesize poly-Phe below 10 mM $Mg^{++}$, by the following hypotheses:

(1) Phe-tRNA never enters the donor site directly.

(2) Phe-tRNA binds to the donor site but is never transferred to initiate a polypeptide-chain.

(3) Phe-tRNA binds to the donor site, but never stays there long enough to have a chance of transfer to a second molecule of Phe-tRNA bound in the acceptor site.

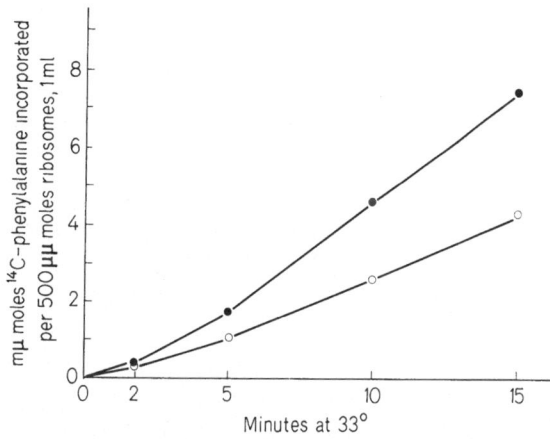

Fig. 11. Synthesis of polyphenylalanine from phenylalanine: Stimulation by methionine. Systems (Table 1) with (●——●), and without (○————○) 0.54 μmoles/ml L-methionine were preincubated for 10 min at 0° [1]

In none of these cases would there be a chance for *mis-initiation*. Should initiation occur, perhaps, by a *mis-coding* of an initiator-tRNA with the Phe-triplets in poly-U? Free methionine can apparently help in overcoming the lag-period in poly-Phe synthesis and lead to a peptidizing system which incorporates Phe faster after the lag (Fig. 11). F-Met is incorporated into acid precipitable material in systems starting from either ¹⁴C-formyl-tetrahydrofolate and methionine, or from F-Met-tRNA. In the latter case incorporation of ¹⁴C is stimulated more significantly by poly-U. The poly-U-stimulated incorporation of ¹⁴C-formate is 1/73 of the

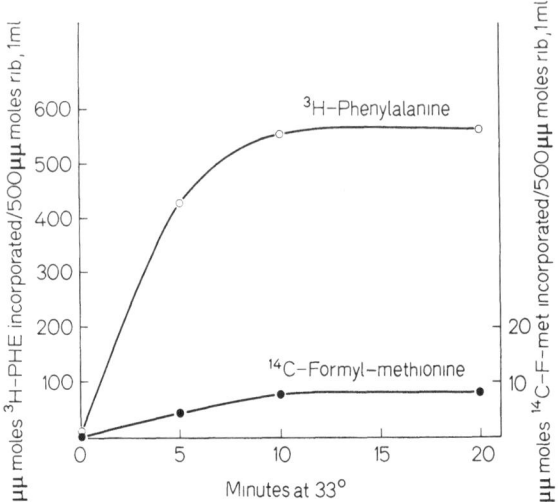

Fig. 12. Synthesis of polyphenylalanine: Incorporation of ¹⁴C-formyl-methionine coded with poly-U. Peptidizing systems (see Table 1) contained (¹⁴C-formyl-methionyl + ³H-phenylalanyl minus 18 aminoacyl)-tRNA, and 120 mμmoles/ml of the 20 protein amino acids [1]

³H-Phe incorporation from (¹⁴C-F-Met + ³H-Phe)-tRNA (Fig. 12). ¹⁴C-incorporation was stimulated by 100% above the minus-poly-U blank. These experiments were fully reproducible. The product should be ¹⁴C-F-Met-(³H-Phe)$_n$. It is under further investigation.

Another artificial initiation of the synthesis of poly-Phe does occur by "mis-initiation". LUCAS-LENARD and LIPMANN have demonstrated that N-acetyl-Phe-tRNA acts as an initiator-tRNA at only 4 mM Mg$^{++}$ in the presence of GTP and two more factors in addition to the T- and G-factors [2].

### 4. A Demonstration of Interference by Charged Amino Acid Adapters which do not fit the Codon to be Read

In an investigation of the Mg-dependent initial rates of different steps in the synthesis of poly-Phe [1], it seemed curious to us that the complete system starting from free Phe was usually about 10-times as fast as the partial system starting from (Phe + 19a)-tRNA. Although the mixture of the 19 amino acids other than Phe may also *stimulate* the final rate of Phe-incorporation if it is added during a preincubation period, *later* addition of the other amino acids may inhibit the synthesis of poly-Phe. This inhibition by the unrequired amino acids can be demonstrated in

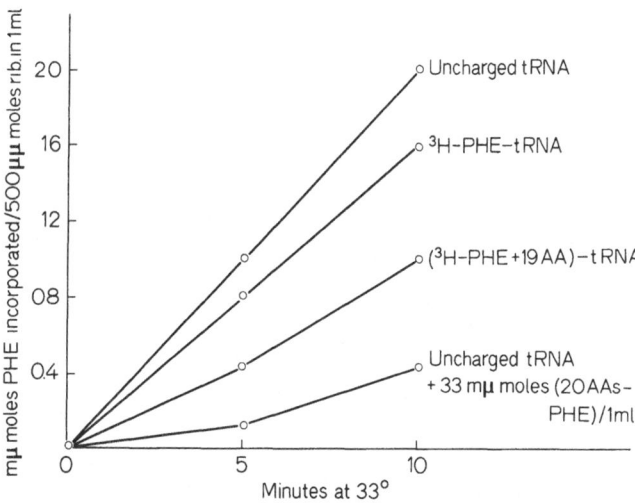

Fig. 13. Synthesis of polyphenylalanine, starting from uncharged tRNA and free phenylalanine or phenylalanyl-tRNA: In both cases, the initial rate of incorporation is inhibited by the 19 other amino acids. Reaction mixtures were preincubated for 5 min at 33° C. At zero-time, ³H-Phe (± 19 aminoacyl)-tRNA or the 19 amino acids other than phenylalanine were added. ¹⁴C-phenylalanine was present in the cases of uncharged tRNA during preincubation. Zero-time-blanks are subtracted [1]

systems starting from either Phe or Phe-tRNA (see Fig. 13). Therefore it appears likely that the *charged* unrequired amino acid adaptors interfere somewhere in the synthetic process. We propose that this interference occurs at the checking stage. Surprisingly enough, this phenomenon could only be demonstrated with the poorer of our S 100-supernatant preparations from E. coli. Dr. A. Parmeggiani found recently that our supernatants are extraordinarily rich in both transfer-factors, due to the culture conditions [3, 5]. We hope to develop this system into an assay for a still hypothetical "checking-factor", and an investigation of the checking process.

**Addendum:** We now differentiate "aminoacyl-" and "anticodon-checking" [8]. The inhibition by charged unrequired adaptors can be overcome, partly at least, by T-factor and therefore seems to indicate interference in amino acyl-checking. — After the observation of Capecchi the existence of a termination-tRNA appears unlikely [see 8].

## Summary

This progress-report refers to some recent work on the polyphenylalanine synthesizing system and on ideas derived from it:

1. Three slopes in the Phe-tRNA-saturation curve may indicate three ribosomal binding sites with different sorption-equilibria.

2. The sites corresponding to the first and third slopes require the 30S-subunit, whereas the second slope is only displayed by the combined 30S- and 50S-subunits.

3. F-Met-tRNA is coded by the triplet ApUpG and bound to the 30S- but not the 50S-subunit. Therefore one general entrance site for initiator- and chain-internal-tRNAs is assumed to exist on the 30S-particle. The hypothesis is put forward that several checking sites may be located ahead of the donor site. In the simplest case the three sites found in saturation curves for binding equilibria may correspond to the three sites discussed so far in the literature.

4. The high $Mg^{++}$-requirement for the synthesis of polyphenylalanine may be explained by the need for a miscoding of an initiator-tRNA by poly-U. Methionine stimulates the synthesis of poly-Phe, and the incorporation of $^{14}C$-F-Met is stimulated with poly-U by 100% above the minus-polymerblank. The product obtained should be on the average $^{14}C$-F-Met-(Phe)$_{73}$.

## Acknowledgement

The excellent assistance of Wiltrud Ludewig and Dorothea Schwarz is gratefully acknowledged; stud. math. Ute and Gerlind Rahmel helped us a great deal in both experimentation and evaluation of data. The Deutsche Forschungsgemeinschaft supported this work.

## References

1. Løve, K.: Die $Mg^{++}$-Abhängigkeit der Geschwindigkeiten einzelner Schritte der Synthese von Polyphenylalanin in einem zellfreien System aus Escherichia coli. Dissertation (Tübingen 1967).

2. Lucas-Lenard, J., and F. Lipmann: Initiation of Polyphenylalanine Synthesis by N-Acetylphenylalanyl-tRNA. Proc. nat. Acad. Sci. (Wash.) 57, 1050—1057 (1967).

3. Matthaei, J. H., and M. W. Nirenberg: Characteristics and stabilization of DNAase-sensitive protein synthesis in E. coli extracts. Proc. nat. Acad. Sci. (Wash.) 47, 1580—1588 (1961).

4. —, F. Amelunxen, K. Eckert u. G. Heller: Zum Mechanismus der Proteinbiosynthese I. Die Bindung von Matrizen-RNS und Aminoacyl-RNS an Ribosomen. Ber .Bunsenges. Phys. Chem. 68, 735—742 (1964).

5. —, G. Heller, H. P. Voigt, R. Neth, G. Schöch, and H. Kübler: Analysis of the genetic code by amino acid adapting. Symposium "Structure and function of genetic elements" (D. Shugar, Ed.). Proceedings of the 3rd Meeting of European Biochemical Societies; p. 233—249 (Warsaw 1966).

6. —, H. P. Voigt, G. Heller, R. Neth, G. Schöch, H. Kübler, F. Amelunxen, G. Sander, and A. Parmeggiani: Specific Interactions of Ribosomes in Decoding. Cold Spr. Harb. Symp. quant. Biol. 31, 25—38 (1966).

7. —, and M. Milberg: Mechanisms in protein synthesis V. Evidence for two ribosomal sites from equilibria in binding of Phenylalanyl-tRNA. Biochem. Biophys. Res. Comunun. 29, 593—599 (1967).

8. MATTHAEI, H., G. SANDER, D. SWAN, T. KREUZER, H. CASSIER u. A. PARMEG-
   GIANI: Mechanismen der Proteinsynthese X, über die Reaktionsschritte
   der Polypeptidsynthese an Ribosomen. Naturwissenschaften 55, 281–294
   (1968).
9. NIRENBERG, M. W., and J. H. MATTHAEI: The dependence of cellfree protein
   synthesis in E. coli upon naturally occurring or synthetic polyribonucleotides
   Proc. nat. Acad. Sci. (Wash.) 47, 1588–1602 (1961).
10. —, and P. LEDER: RNA codewords and protein synthesis: The effect of tri-
    nucleotides upon the binding of sRNA to ribosomes. Science 145, 1399
    (1964).
11. TRAUT, R., and R. MONRO: The puromycin reaction and its relation to protein
    synthesis. J. molec. Biol. 10, 63–72 (1964).
12. REVEL, M., and H. H. HIATT: Magnesium requirement for the formation of an
    active messenger-RNA-ribosome-s-RNA complex. J. molec. Biol. 11,
    467–475 (1965).
13. VOIGT, H. P., u. H. MATTHAEI: Mechanismen der Proteinsynthese II. Reinigung
    der Ribosomen von Escherichia coli von Nucleasen und deren Charakteri-
    sierung. Hoppe Seyler's Z. physiol. Chem. 349, 54–64 (1968).
14. — — Mechanismen der Proteinsynthese III. Über einige Aktivitäten hoch-
    gereinigter Ribosomen von Escherichia coli. Hoppe Seyler's Z. physiol.
    Chem. 349, 65–76 (1968).
15. WETTSTEIN, F., and H. NOLL: Binding of transfer RNA to ribosomes engaged in
    protein synthesis: Number and properties of ribosomal binding sites. J.
    molec. Biol. 11, 35–53 (1965).

## Discussion

*Nomura:* You have found two-fold stimulation of Phe-tRNA binding by adding
50 S subunits and have taken this as evidence of there being two tRNA-binding sites
on ribosomes. Have you done similar experiments with other aa-tRNA's? In our
experiments, with some aa-tRNA's (e. g. Lys-tRNA, Tyr-tNRA, Val-tRNA) we
have observed 5 to 10-fold (or higher) stimulation by 50 S subunits.

*Matthaei:* The stimulations of Phe-tRNA-binding by addition of 50 S subunits
reported in the literature vary from almost none to two-fold. These can be explained
only now on the basis of the degrees of Phe-tRNA-saturation reached in these
experiments. We do have the impression, too, that the equilibria for binding of other
adapters may be quite different and, moreover, their values for the different sites
suggested may be in different ratios.

*Nomura:* Have you done similar experiments (30 S compared to 30 S + 50 S)
with random poly-A, U, G?

*Matthaei:* We have not yet compared 30 S versus 30 S + 50 S with poly-A, U, G.
We have seen, however, remarkable differences in the $Mg^{++}$ requirement for binding
F-Met-tRNA dependent on ApUpG, ApUpGpU . . . pU, and poly-A, U, G.

*Nomura:* Have you studied the effect of $Mg^{++}$ concentration on F-Met-tRNA$_F$-
and Met-tRNA$_M$-binding using 30 S subunits?

*Matthaei:* $Mg^{++}$-concentration curves for binding to 30 S-subunits directed by
poly-A, U, G suggest that the optimal binding for F-Met-tRNA to ribosomes at
3–5 mM is due ro their dissociation into 30 S- and 50 S-subunits.

# Modification of DNA-Dependent RNA-Polymerase in $T_4$-infected E. coli

By

G. Walter, W. Seifert, and W. Zillig

Max-Planck-Institut für Biochemie, 8000 München 15, Goethestr. 31

The time course of virus multiplication in E. coli infected with bac-teriophage $T_4$ in a simplified manner may be divided into two phases, one, in which early messengers are transscribed, and, consequently, early proteins are synthesized, and one, in which, after the shutoff of the trans-lation of early genes, DNA is replicated and late genes are transscribed and, consequently, translated into late proteins, before the phage particles are assembled. The problem of the operation of this time clock is of a twofold nature:

1. What is the mechanism of the blockade of the translation of early genes?

2. What is the mechanism of the induction of the transscription and translation of late genes?

This progress report deals with the first of these two questions. Originally we had been interested in the possibility of the existence of bacteriophage-dependent RNA-polymerases in infected cells. Stimuli for the choice of $T_4$-phage for such investigations came from the work of Buchanan and Sköld [1], and later, from that of Hall [2].

It has been shown first by Buchanan and Sköld, that the in vitro activity of DNA-dependent RNA-polymerase in crude extracts, prepared from $T_4$-phage infected E. coli cells, shows a strong decrease within the first minutes after infection, when $T_4$-DNA is used as template. Since this fall is not observed in the presence of chloramphenicol, it may be as-cribed to the production of an inhibitor, in which protein synthesis is involved.

In addition, evidence has been presented by the authors, that the inhibition results from interaction of the inhibitor with the polymerase. A slight increase of activity at later times might be taken as evidence for the appearance of a $T_4$-specific enzyme.

A striking difference between the time curves of activity with $T_4$-DNA and with calf thymus-DNA as templates for RNA-polymerase in $T_4$-phage infected E. coli cells has been demonstrated by Hall. Whereas the curve

for T$_4$-DNA as the template resembles that produced by BUCHANAN and
SKÖLD, the inactivation is much less pronounced, when calf thymus-DNA
is used in the test. In this case, activity even increases again at later
times.

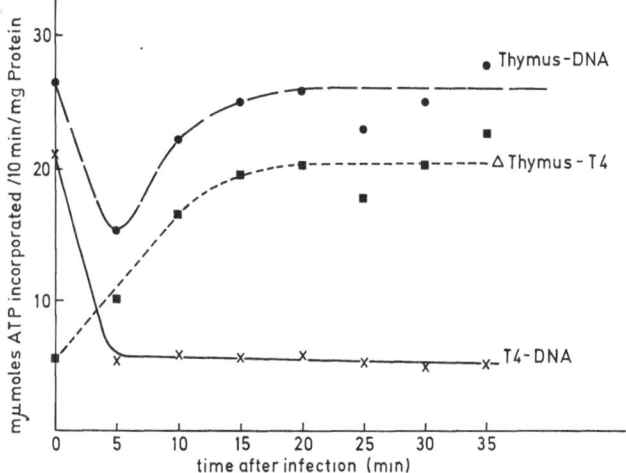

Fig. 1. Template efficiency of T$_4$-DNA and calf thymus-DNA for RNA polymerase
isolated from infected cells at various times after infection

The time course of inactivation of polymerase in the infected cell is
shown in the first figure (Fig. 1). The lower curve is obtained, when T$_4$-
DNA is used as the template, the upper curve, when calf thymus DNA is
used.

What is the reason for the inactivation observed with T$_4$-DNA? And
what is the reason for the reincrease to almost the zero time level with
thymus DNA? Are these observations a reflection of the formation of a
polymerase with changed properties for example as a consequence of substitution of E. coli polymerase by a new and entirely T$_4$-dependent
polymerase, or of a T$_4$-dependent modification of E. coli polymerase or
are they due to the action of a separate inhibitor?

DNA-dependent RNA polymerase was prepared from E. coli CR 63
20 min after infection with T$_4$-am 82, when the activity with thymus DNA
in the crude extract has reached its maximal level after the minimum.
The procedure was exactly the same which is used routinely in our
laboratory for the isolation of pure E. coli polymerase.

The purified enzyme contains less than 5 % impurities. In all steps of
the preparation, that is DEAE chromatography, ammonium sulfate
precipitation and gradient centrifugation in the 24 S (whole molecule) and
13 S (half molecule) form no difference from polymerase isolated from
uninfected cells was observed.

7*

In no step of the purification an inhibitor was removed bringing the properties of the "$T_4$-enzyme" back to those of E. coli polymerase. Throughout the purification the strikingly changed ratio of template activity of thymus to $T_4$- DNA (4.5 as compared to 0.5 for normal E. coli enzyme) stays constant. This is a strong indication, that the polymerase itself is changed, although no difference in physical properties has been demonstrated.

To decide, if the $T_4$ polymerase is an entirely new protein or a modified E. coli enzyme, the immunochemical properties were studied using an antiserum against purified E. coli polymerase.

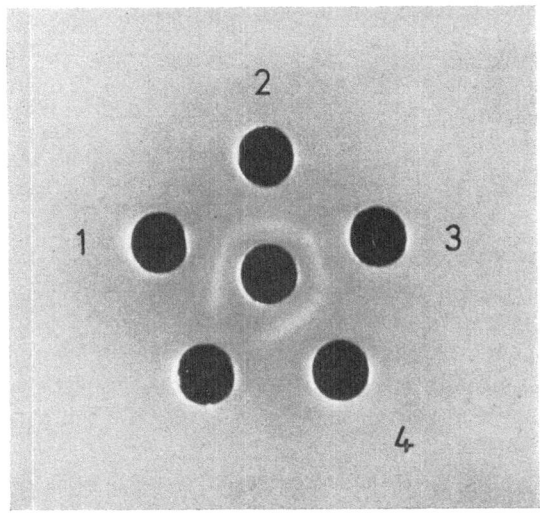

Fig. 2. Comparison of RNA-polymerases from E. coli CR 63, $T_4$ infected E. coli B, E. coli K 12 Hfr and E. coli CR 63 (in this order clockwise) with the immunodiffusion technique of Ouchterlony

No difference from the E. coli enzyme was found with the immundiffusion technique of Ouchterlony (Fig. 2).

With the more specific immunoelectrophoresis technique of Grabar, a main precipitation line is obtained which cannot be distinguished from that produced by the E. coli enzyme. A week second line given by $T_4$ polymerase might or might not be due to impurities in the enzyme preparation and in the antiserum (Fig. 3).

$T_4$ polymerase and E. coli polymerase are both entirely inactivated by preincubation with the antiserum, the same amounts of antibody required for the same amounts of both enzymes which are of equal purity (5 µg of enzyme in DNA saturation, preincubation for 30 min in absence of

template and substrates, serum purified by ammonium sulfate precipitation, nuclease free, concentration 2.1 µg/µl (Fig. 4).

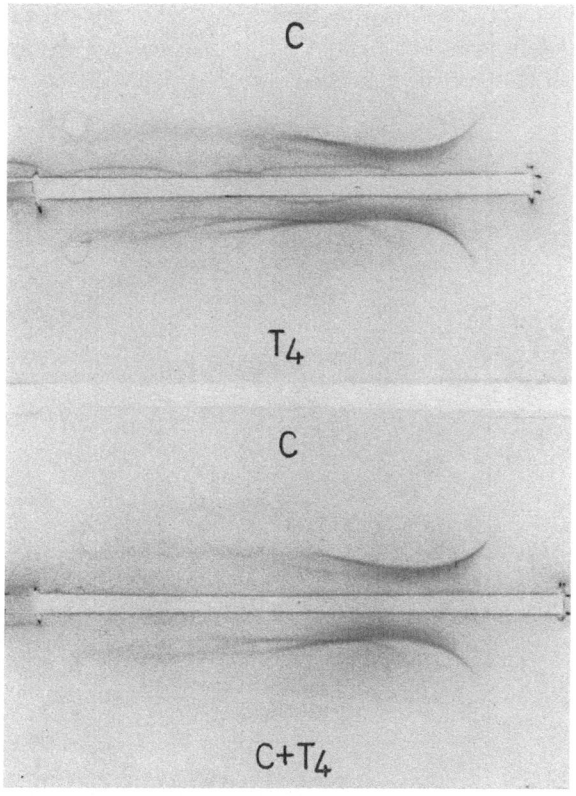

Fig. 3. Immunoelectrophoresis of RNA-polymerases from uninfected and $T_4$ infected E. coli

When thymus DNA is used as the template, higher amounts of antiserum are required than with $T_4$-DNA for the same degree of inactivation.
This could be a consequence of a higher affinity of the enzymes towards
thymus-DNA, which is less influenced by the binding of antibody
molecules.

Thus, with three immunochemical methods $T_4$ polymerase has not
been distinguished from E. coli polymerase. This means, that at least a
large part of the structure of the $T_4$-enzyme is either identical or crossreacting with high efficiency with the E. coli enzyme. The polymerase from
the blue green alga Anacystis nidulans [3] does not crossreact at all with
antiserum against E. coli enzyme. This finding supports the idea, that the

$T_4$-enzyme is a modified E. coli polymerase. What is the nature of this modification?

In 6 M urea at pH 9.5 in the presence of mercaptoethanol E. coli polymerase is dissociated into two main subunits which may be peptide chains and which both sediment with about 3.1 S but are separated into a fast and a slow component (A and B) by polyacrylamide disc electrophoresis under dissociating conditions [4] (Fig. 5).

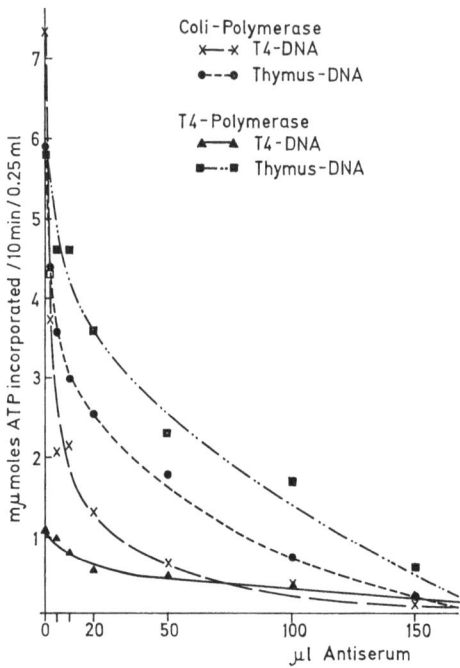

Fig. 4. Inactivation of E. coli- and $T_4$-polymerase by antibodies against E. coli polymerase, using calf thymus- and $T_4$-DNA's as templates

The purified $T_4$-enzyme under these conditions also shows two main bands migrating with velocities very similar to those of A and B. When the E. coli enzyme, prepared from a $K_{12}$Hfr strain, and the $T_4$-enzyme, isolated from infected CR 63, are mixed in equal amounts, two clearly separated faster main bands (A and A') of equal intensity are observed besides one B band. To exclude the possibility that the difference of the mobilities of A' from A is the consequence of a genetic difference of the E. coli strains, purified RNA polymerase was prepared from uninfected CR 63. The mixture of the two E. coli enzymes shows one A band (and B band) only. The mixture of the enzymes from uninfected and infected CR 63 again shows two fast bands A and A'. These observations lead to the conclusion, that the A'-subunit of T$_4$-polymerase is different from the A subunit of the E. coli enzyme and that this difference is a consequence of phage infection. 20 min after infection, the substitution of A by A' is quantitative.

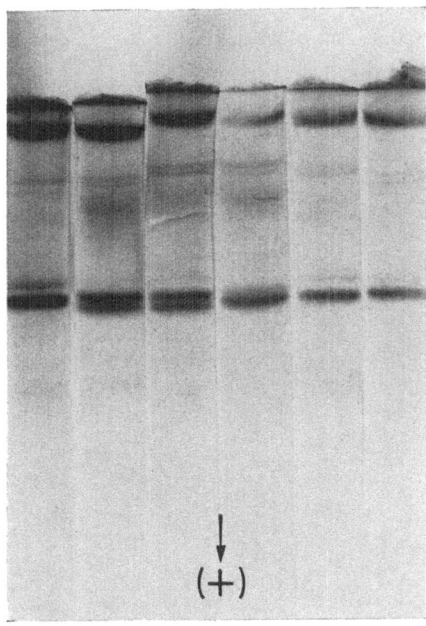

Fig. 5. Polyacrylamide disc electrophoresis in 6 m urea at pH 9.5 of RNA-polymerase from E. coli K 12 Hfr, E. coli CR 63, T$_4$-infected E. coli and of RNA-polymerase mixtures E. coli K 12 Hfr + T$_4$, E. coli CR 63 + T$_4$ and E. coli K 12 Hfr + E. coli CR 63

From the position of a weak front band which may be a minor component of polymerase [4], and the position of a weak satellite band migrating slightly behind the E. coli A subunit in less pure preparations, it appears, that the faster of the two A bands (A') in the mixture stems from T$_4$-polymerase.

Four alternative explanations would account for this finding.

1. A' is an entirely phage genome dependent subunit which has replaced the corresponding E. coli subunit A.

2. A' is the product of covalent binding of a phage dependent substance to the A-subunit of the E. coli enzyme.

3. A' results from A by a loss of part of its structure, for example a piece of the peptide chain.

4. A' is a non covalent complex of the A component of the E. coli enzyme with a phage dependent substance. This complex would have to

be extremely stable, since no dissociation is observed in 6 M urea at pH 9.5, on a DEAE column and at high ionic strength.

Experiments are in progress, which should decide between these possibilities.

The enzymological properties of $T_4$-polymerase are shown in the next figures:

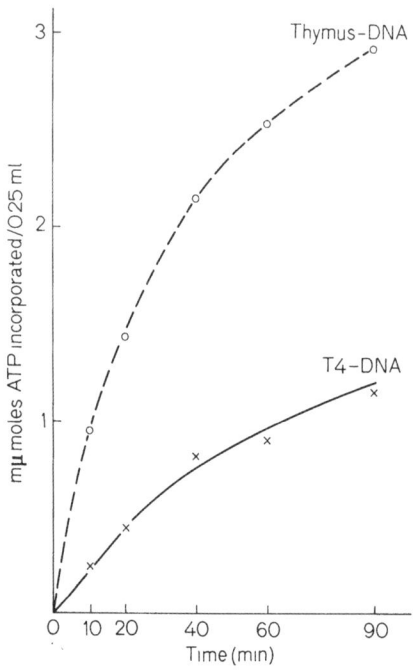

Fig. 6. Kinetics of RNA-synthesis by $T_4$-polymerase using calf thymus-DNA and $T_4$-DNA as templates

Fig. 6 shows the kinetics of RNA synthesis with $T_4$-polymerase using $T_4$-DNA and thymus DNA as templates. The ratio of efficiencies of both templates is time independent. An incubation time of 10 min has therefore been used for the following tests.

Fig. 7 shows saturation curves for E. coli- and $T_4$-polymerase with $T_4$-DNA, calf thymus DNA, and E. coli-DNA. The ordinate gives the amount of RNA synthesized in 10 min with constant enzyme concentration and increasing amounts of DNA which are shown on the abscissa. The amount of DNA required to reach plateau is a measure of the number of binding sites for the enzyme on the DNA and the height of the plateau is a measure of the rate of RNA synthesis, which depends on the rate of single chain growth and on the fraction of active, from total bound, enzyme molecules. In DNA saturation, with $T_4$-DNA, $T_4$-polymerase is 20 fold less efficient than E. coli polymerase. With thymus and E. coli-DNA, this factor is only about 2. The results are summarized in the table (Table 1).

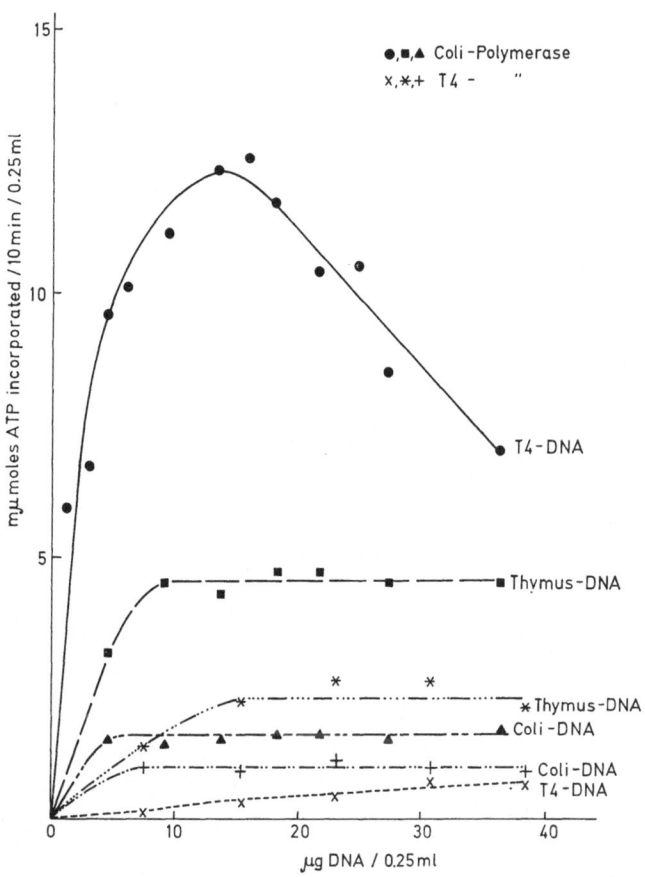

Fig. 7. Saturation of equal amounts of E. coli- and $T_4$-polymerase with $T_4$-, calf thymus- and E. coli-DNA

Table 1. *Ratios of activities of E. coli- and $T_4$-polymerase with $T_4$-, calf thymus- and E. coli-DNA's as templates. 5 μg of enzymes and saturating amounts of DNA's were used*

|  | E. coli-Enzyme | $T_4$-Enzyme | enzyme ration E. coli-/$T_4$- |
|---|---|---|---|
| $T_4$-DNA | 12,2 mμM | 0.6 mμM | 20.3 |
| calf thymus DNA | 4.6 mμM | 2.6 mμM | 1.8 |
| E. coli-DNA | 1.5 mμM | 1.0 mμM | 1.5 |
| DNA ratio $T_4$/calf th. | 2.7 | 0.23 |  |

The last figure (Fig. 8) proposes one possible model for the operation of the time clock in $T_4$ infected E. coli cells, which is in agreement with our results.

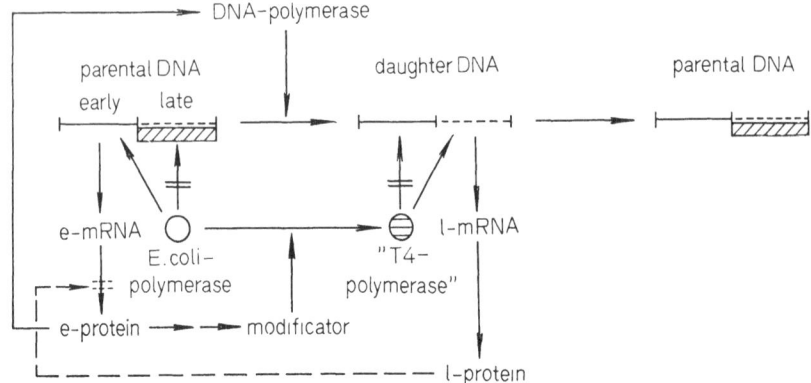

Fig. 8. Tentative model for early-late-regulation in $T_4$ infected E. coli

This would be the first example of the repression of gene transscription by a specific modification of the transscribing enzyme, whereas in the classical repressor model repressors interact with informational nucleic acids (DNA or RNA).

## Acknowledgements

It is a pleasure to thank Dr. Stan Hattman for many stimulating discussions, Dr. E. Fuchs for his help in disc electrophoresis, Dr. P. Palm for the preparation of the antiserum and Miss H. Büger for excellent technical assistance.

## References

1. Sköld, Ola, and J. M. Buchanan: Proc. nat. Acad. Sci. (Wash.) **51**, 553 (1964).
2. Hall, B. D.: Personal communication.
3. Helm, K. v. d., u. W. Zillig: Hoppe Seyler's Z. physiol. Chem. **348**, 902 (1967).
4. Zillig, W., E. Fuchs, et al.: Internat. Symp. on Biochemistry of Ribosomes and mRNA, Reinhardsbrunn, May 1967.
5. —, R. L. Milette, E. Fuchs, G. Walter, P. Palm, and H. Priess: VII. Internat. Congress of Biochemistry, Symp. II, Tokyo, August 1967.

## Discussion

*Bertani:* Do the other T-even phages behave like T 4 in respect to DNA-dependent RNA polymerase activity during the infectious cycle?

*Zillig:* Buchanan has investigated T 4 as we did, Furth has worked with T 2, which appears to behave in the same way. Other types have not been investigated to my knowledge.

*Hofschneider:* Do you already know whether antiserum against Coli enzyme can be exhausted with T 4-enzyme ?

*Zillig:* This has not been done yet.

*Trautner:* Have you attempted to modify *in vitro E. coli* polymerase to the T 4 type enzyme by treating enzyme isolated from uninfected cells with extracts from infected cells ?

*Zillig:* So far we had no success.

# Genetic Control of Translation of Nonsense Triplets

By

A. GAREN

Department of Molecular Biophysics, Yale University,
New Haven, Connecticut U.S.A.

The genetic code, shown in Table 1, designates the relationships between the 64 possible codons present in messenger RNA and the 20 amino acids present in proteins. The triplet nature of this code, which provides a potential surplus of codons for 20 amino acids, requires that the code contain degenerate codons (*i.e.* different triplets coding for the same amino acids) or nonsense triplets (*i.e.* triplets which do not code for any amino acid), or both. It is immediately apparent from a glance at Table 1 that the code exhibits extensive degeneracy, as much as six-fold

Table 1. *The genetic code. Each amino acid listed in the table is coded for by an RNA triplet (codon) which has a nucleotide sequence indicated as follows: The 5′ terminal nucleotide appears in the column on the left, the middle nucleotide appears in the top row, and the 3′ terminal nucleotide appears in column on the right.* The symbols $N1$ *(amber)*, $N2$ *(ochre)* and $N3$ designate the three nonsense triplets, UAG, UAA and UGA

| 1st↓  2nd→ | U | C | A | G | ↓3rd |
|---|---|---|---|---|---|
| U | PHE | SER | TYR | CYS | U |
|   | PHE | SER | TYR | CYS | C |
|   | LEU | SER | N2 (Ochre) | N3 | A |
|   | LEU | SER | N1 (Amber) | TRP | G |
| C | LEU | PRO | HIS | ARG | U |
|   | LEU | PRO | HIS | ARG | C |
|   | LEU | PRO | GLUN | ARG | A |
|   | LEU | PRO | GLUN | ARG | G |
| A | ILEU | THR | ASPN | SER | U |
|   | ILEU | THR | ASPN | SER | C |
|   | ILEU | THR | LYS | ARG | A |
|   | MET | THR | LYS | ARG | G |
| G | VAL | ALA | ASP | GLY | U |
|   | VAL | ALA | ASP | GLY | C |
|   | VAL | ALA | GLU | GLY | A |
|   | VAL | ALA | GLU | GLY | G |

for some amino acids. There also are three triplets, UAG, UAA and UGA, which are designated as nonsense.

## Nonsense Mutations

Protein biosynthesis is a sequential process during which a peptide chain grows inidirectionally, by increments of one amino acid, from the amino-terminal towards the carboxyl-terminal residue [1]. Accordingly, if a nonsense triplet is present at any of the positions in messenger RNA which code for the amino acid residues of a peptide chain, a gap will appear in the chain causing premature termination of chain growth. Nonsense triplets do not normally occur within the coding regions of messenger RNA, but they can be generated from certain codons by mutation. The resulting mutants are called nonsense mutants, as distinguished from missense mutants which result from the transformation of one codon into another codon. Nonsense mutants have been found to occur in certain strains of *E. coli*, as a result of the formation of one of the three triplets UAG, UAA or UGA [2, 3, 4, 5, 27].

## Suppressor Genes for Nonsense Mutations

The coding behavior of nonsense triplets is genetically controlled in bacteria by certain suppressor genes which determine the capacity of the cell to translate the triplets. The properties of bacterial suppressor genes have been studied by the procedure outlined in Fig. 1 for obtaining

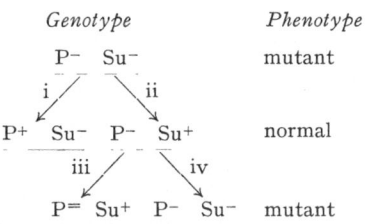

Fig. 1. A procedure for the genetic analysis of suppressor genes for phosphatase nonsense mutations. Experimental details are described in the text and in ref. [6]

suppressor mutants [6]. For these studies, nonsense mutations in the alkaline phosphatase structural gene were used as a marker to detect the suppressor mutants (nonsense mutations in other genes can also be used for this purpose, since the specificity of suppression pertains to the mutation and not to the gene in which the mutation occurs). The procedure starts with a suppressor-negative $Su^-$ strain containing a phosphatase nonsense mutation. The first step is the isolation of revertant strains having a normal phosphatase phenotype. Genetic mapping of the rever-

tants shows that they can occur in either of two ways. One is by a mutations which transform the nonsense triplet into another triplet which is a
codon in the $Su^-$ strain (pathway I in Fig. 1). The second way revertants
occur is by a suppressor mutation which produces a suppressor-positive
$Su^+$ strain capable of translating the nonsense triplet (pathway II). The
revertants obtained in this way comprise the strains needed to analyze the
genetic and biochemical properties of suppressor genes for nonsense mutations. In addition to revertants which suppress phosphatase nonsense
mutations, revertants have been isolated on the basis of their suppressor
activity for nonsense mutations in other genes of *E. coli* and T4 [7−11].

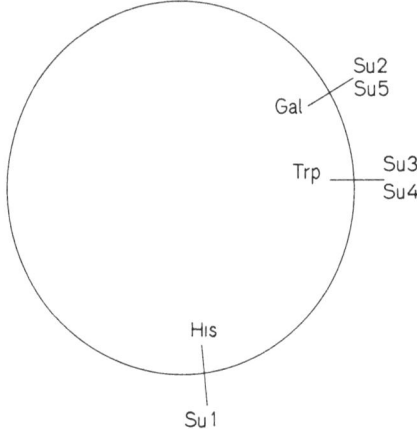

Fig. 2. Genetic map locations of five *E. coli* suppressor genes for nonsense mutations.
The *Su1*, *Su2* and *Su3* suppressor genes [6] correspond to the suppressor genes
designated elsewhere as suI, suII and suIII [20]

The revertants which suppress UAG *(amber)* or UAA *(ochre)* mutations map at several different positions. This article will concentrate on
the three map positions shown in Fig. 2, involving five suppressor genes;
each suppressor gene designates a region of the map controlling a single
suppressor function. The five suppressor genes in Fig. 2 differ in certain
physiological and chemical properties. All five of the genes suppress *amber*
mutations, but only *Su4* and *Su5* also suppress *ochre* mutations (Table 2).
Another difference is in the efficiency of suppression of nonsense mutations,
*i.e.* the extent of restoration of the synthesis of the intact protein molecule.
The efficiencies range from 5 % to about 60 % among the five genes
(Table 2). These values are a measure of the relative rates of translation
of a nonsense triplet in different suppressor strains. A third difference is in
the amino acid residue incorporated into a protein as a result of suppression. The remarkable finding is that four of the genes specify the incorporation of different amino acids (Table 2). Thus, a suppressor gene

controls not only the capacity of a cell to translate a nonsense triplet but also the coding specificity of the triplet. For example, the same *amber* triplet can either be nonsense if all the suppressor genes are $Su^-$, or it can specify any one of the four amino acids serine, glutamine, tyrosine or a basic residue, depending on which of the suppressor genes is $Su^+$ (Fig. 3). There are also other *E. coli* suppressor genes for *amber* and *ochre* nonsense mutations, in addition to the five described in Table 2; it is likely that biochemical studies with these suppressors will reveal new amino acids which can be specified by the nonsense triplets.

| Messenger RNA | Polypeptide | Suppressor Genotype Su 1 Su 2 Su 3 Su 4 Su 5 | | | | |
|---|---|---|---|---|---|---|
| AUGUAC*UGG*GUCGUU | Met. Tyr. Trp. Gly. Val | − | − | − | − | − |
| AUGUAC*UA*GGUCGUU | Met. Tyr. | − | − | − | − | − |
| AUGUAC*UA*GGUCGUU | Met. Tyr. Ser. Gly. Val | + | − | − | − | − |
| AUGUAC*UA*GGUCGUU | Met. Tyr. Gln. Gly. Val | − | + | − | − | − |
| AUGUAC*UA*GGUCGUU | Met. Tyr. Tyr. Gly. Val | − | − | + | − | − |
| AUGUAC*UA*GGUCGUU | Met. Tyr. Tyr. Gly. Val | − | − | − | + | − |
| AUGUAC*UA*GGUCGUU | Met. Tyr. (Lys) Gly. Val | − | − | − | − | + |

Fig. 3. Effects of five suppressor genes on the translation of a messenger RNA containing the *amber* triplet UAG

A suppressor gene for mutations producing the UGA nonsense triplet has recently been reported [12]. This gene has a high efficiency of suppression of about 60 %, but it is inactive against either *amber* or *ochre* mutations. There is no information at present about either the map position of the suppressor gene or the amino acid which is specified by UGA as a result of suppression.

The mutational pathway II in Fig. 1, by which suppressor revertants are obtained, appears to generate a unique change in each suppressor gene, since independently isolated $Su^+$ revertants for the same gene fail to produce $Su^-$ recombinants in pairwise crosses. (This genetic behavior can now be understood in terms of the biochemistry of the suppressor gene product, as discussed below.) The mutational pathway IV in Fig. 1, by which $Su^-$ mutations are induced in an $Su^+$ strain, yields a different result: Studies of the $Su1^+$ gene have shown that $Su^-$ mutations occur at several separate sites within the suppressor gene [6]. The $Su^-$ mutants derived in this way provide material for fine-structure genetic analyses of suppressor genes and, as discussed below, also for biochemical analyses of the suppressor gene product.

## Biochemical Basis of Suppression

The availability of *amber* mutants of the RNA phage [13], combined with the knowledge that the amino acid specified by the *amber* triplet in

Table 2. *Response of nonsense mutations to suppression*. The amino acid inserted by a suppressor gene was determined by analysis of the protein produced as a result of suppression. Experimental details are reported in the following references: Su1+ [29], Su2+ and Su3+ [30], Su4+ [31], Su5+ [27]. The tabulated efficiencies of suppression were estimated by the amount of alkaline phosphatase protein (measured as CRM) produced as a result of suppression of a phosphatase-negative mutant [6, 27]. These efficiencies have also been estimated by the extent of reversal of chain-termination in an *amber* mutant of phage T4 [7, 20]

| Suppressor Gene | Su1+ | | Su2+ | | Su3+ | | Su4+ | | Su5+ | |
| --- | --- | --- | --- | --- | --- | --- | --- | --- | --- | --- |
| Nonsense mutation | Amber | Ochre | Amber | Ochre | Amber | Ochre | Amber | Ochre | Amber | Ochre |
| Amino acid inserted | Serine | 0 | Glutamine | 0 | Tyrosine | 0 | Tyrosine | Tyrosine | (Basic) | (Basic) |
| Efficiency of suppression | 28% | 0 | 14% | 0 | 55% | 0 | 16% | 12% | 5% | 6% |

the Su1+ suppressor strain is serine (see Table 2), provided the methodology for an *in vitro* test of suppression. Successful results were obtained by two groups, by CAPPECCHI and GUSSIN working in J. D. WATSON's laboratory [14] and by a collaborative effort between the laboratories of N. D. ZINDER and A. GAREN [15]. Both groups were able to demonstrate that a transfer RNA (tRNA) for serine was the active suppressor component in an Su1+ strain. The results of the latter group showed that in the absence of the suppressor tRNA a small fragment of the coat protein was formed, and that the addition of tRNA from the Su1+ suppressor strain resulted in the formation of a complete protein molecule with a serine residue replacing a glutamine residue normally present in the molecule. *In vitro* suppressor tests on the tRNA derived from the other four suppressor strains listed in Table 2 suggest that the mechanism of suppression is the same in all of the strains tested: A specific tRNA from each suppressor strain appears to be the active suppressor component [16]. Confirmation of these results has been obtained by a new *in vitro* test, involving the synthesis of the lysozyme specified by phage T4 [17].

The function of a suppressor gene in the formation of a suppressor tRNA has been revealed by recent biochemical studies involving the Su3 gene. This gene maps in the region of the E. coli genome which is transducible by the phage φ80. Transduction by φ80 is of the "restricted" type (analogous to $\lambda$-gal transduction); in this type of transduction, bacterial genes either replace some of the phage genes of the transducing particle, or are added to a complete phage genome. $\phi 80$ particles with transducing activity for the $Su3^+$ gene provide a source of phage DNA, containing the $Su3^+$ gene, which can be used for hybridization tests between DNA and tRNA. The results of such hybridization tests performed in two laboratories [18] showed that $Su3^+$ DNA could hybridize specifically with tyrosyl tRNA. Since tyrosine is the amino acid specified by the *amber* triplet as a result of suppression by the $Su3^+$ gene, this is the expected result if the $Su3^+$ gene is a structural gene for a tyrosyl tRNA.

Purified fractions of two major species of tyrosyl tRNA, called species I and II, were equally effective in the hybridization tests. Thus, as a method for identifying the product of a suppressor gene, hybridization cannot distinguish between the different species of tRNA molecules with the same acceptor activity, presumably because of the close similarity of their base sequences. Furthermore, the tyrosyl tRNA obtained from both an $Su3^+$ and $Su3^-$ strain proved to be equally effective in hybridizing with the DNA of the transducing phage. This is not a surprising result in view of the fact that the two strains differ by a single suppressor mutation, and therefore the two RNA molecules presumably differ by only a single base substitution.

Further biochemical studies of the tRNA specified by a suppressor gene requires purified material. An attempt to purify the seryl tRNA, specified by the $Su1$ gene, was only partially successful and led to the discouraging conclusion that this tRNA was present as a minor species containing less than 10 % of the amount of seryl tRNA in a major species [19]. A similar conclusion was reached in the case of the tyrosyl tRNA specified by the $Su3$ gene [20]. The purification of such minor components presents a formidable experimental obstacle. This obstacle was overcome by an ingenious use of the $\phi 80$ transducing phage carrying the $Su3$ gene [20]. It was found that, in a cell infected with both the transducing phage and a nontransducing $\phi 80$ phage, conditions could be established in which the tyrosyl tRNA specified by the $Su3$ gene becomes the predominant species synthesized by the infected cell, thereby greatly simplifying its purification. With this procedure, enriched samples of tyrosyl tRNA specified by the $Su3^-$ and $Su3^+$ genes (the two genes differing by a single mutation) were prepared, and these were tested for their capacity to bind to ribosomes in the presence of various nucleotide triplets [20]. It was found that the triplets UAU and UAC, which are standard codons for tyrosine, stimulated the binding of the tRNA specified by the $Su3^-$ gene, whereas the *amber* triplet UAG stimulated binding of the tRNA specified by the $Su3^+$ gene, and the *ochre* triplet UAA did not stimulate binding of either of these tRNA species. These results suggest that the two tyrosyl tRNA species differ in their anticodons, the tRNA specified by the $Su3^-$ gene containing the anticodon GUA, and the suppressor tRNA specified by the $Su3^+$ gene containing the anticodon CUA.

| Suppressor gene | Anticodon in tRNA | Matching codon in mRNA | Amino acid specified |
|---|---|---|---|
| Su1⁻ | CGA | UCG | Serine |
| Su1⁺ | CUA | UAG | |
| Su2⁻ | CUG | CAG | Glutamine |
| Su2⁺ | CUA | UAG | |
| Su3⁻ | GUA | UAU, UAC | Tyrosine |
| Su3⁺ | CUA | UAG | |
| Su4⁻ | AUA or GUA | UAU and/or UAC | Tyrosine |
| Su4⁺ | UUA | UAA, UAG | |
| Su5⁻ | UUU | AAA, AAG | Lysine |
| Su5⁺ | UUA | UAA, UAG | |

Fig. 4. A model for the biochemical effect of suppressor mutations. The general assumption is that each suppressor gene is a structural gene for a tRNA species, and that an $Su^+$ mutation alters the anticodon of the tRNA. For further details see the text and ref. [28]

This difference expected in the anticodons of the tyrosyl tRNA specified by the $Su3^-$ and $Su3^+$ genes was demonstrated by comparative base sequence analyses of the purified tRNA. The tRNA specified by the $Su3^-$ gene was found to contain the triplet GUA, which was replaced in the tRNA from the $Su3^+$ gene by the triplet CUA [21]. It is now understandable why, as discussed above, independent $Su3^+$ revertants isolated from an $Su3^-$ strain (see Fig. 4) always arise by the same mutational change at a single genetic site: The mutation must produce a change from G to C in the anticodon of the tyrosyl tRNA, which can only occur by a transversion of a GC base pair to a CG base pair at one site in the suppressor gene.

The impressive biochemical results obtained with the tyrosyl tRNA of the $Su3$ gene supports the general mechanism of suppressor gene function shown in Fig. 4. The basic premise is that for all of the suppressor genes that act on *amber* or *ochre* mutations, the phenotypic difference between the $Su^-$ parental strain and the $Su^+$ revertant strain results from a change in the sequence of the suppressor tRNA. It is further assumed that the anticodon of the tRNA specified by the $Su^-$ gene is complementary to one of the standard codons for the amino acid which the tRNA carries. This would account for the finding that the four amino acids known to be specified by *amber* and *ochre* triplets as a result of suppression (Table 2) are normally coded for by triplets which are related to the two nonsense triplets by a single base substitution. Accordingly, there probably exist other, as yet unidentified, suppressor genes for both *amber* and *ochre* mutations which specify leucyl and glutamyl tRNA species, and a suppressor gene for *amber* mutations which specifies a tryptophanyl tRNA.

The mechanism of suppression in Fig. 4, in conjunction with the "wobble" hypothesis of codon-anticodon specificity proposed by CRICK [22], provides an adequate explanation for the fact that the $Su1^+$, and $Su2^+$ and $Su3^+$ genes suppress only *amber* mutations, while the $Su4^+$ and $Su5^+$ genes suppress both *amber* and *ochre* mutations (Table 2). A CUA anticodon should be capable, according to CRICK's hypothesis, of interacting only with UAG because of the postulated stringency in the GC base pair when $C$ is at the 5' and of the anticodon, in contrast to a UUA anticodon which should be capable of interacting both with UAG and UAA because of the reduced stringency (*i.e.* "wobble") premissible when uracil occupies the 5' end position in the anticodon triplet.

Suppressor mutations which act as shown in Fig. 4 are potentially dangerous to a cell, since these produce alterations in anticodons required for the translation of some of the standard codons. Unless there is available to the cell an additional tRNA species containing an anticodon which is identical (or functionally equivalent) to the anticodon altered in the $Su^+$ strain, a suppressor mutation would be a lethal event. In the case of the tyrosyl tRNA specified by the $Su3^-$ gene, the GUA anticodon is also present in two other species of tyrosyl tRNA which are unaffected by the suppressor mutation and remain available to the $Su3^+$ strain for translating the standard tyrosine codons UAC and UAU [21].

## Fine-structure Analysis of Suppressor tRNA

The genetic and biochemical methodology that has been developed for studying the fine-structure details of suppression is also applicable to an analysis of the relationship between the structure and function of the tRNA which is specified by a suppressor gene. Consider the mutational pathway IV in Fig. 1, by which $Su^-$ mutations are induced in an $Su^+$ gene. The phenotypic effect of an $Su^-$ mutation is the reduction or complete elimination of suppressor activity [6]. Since it is known that a suppressor gene is a structural gene for tRNA, this phenotypic effect must result from base substitutions occurring in the suppressor tRNA which affect its capacity to participate in the translation of a nonsense triplet. There are several ways that mutations might inhibit the suppressor activity of a tRNA. One of the ways is by alteration of the anticodon. The fact that $Su^-$ mutations occur at more than three different genetic sites within a suppressor gene [6] indicates that the anticodon cannot be the only

8*

region of the suppressor tRNA affected by the mutations. Evidently other essential functions of the tRNA molecule, such as its amino acid acceptor and transfer activities, are subjects to inactivation by mutation. Thus, $Su^-$ mutations can provide a family of differentley altered forms of the same species of tRNA, involving base substitutions at various sites in the molecule, which is ideal material for biochemical studies of the structural requirements for the various functions of the molecule.

## Minor Species of tRNA

A striking conclusion from fractionation experiments with suppressor tRNA is that these tRNA species are, at least in the two cases analyzed [19, 20], minor components containing less than 10 % of the tRNA in a major species having the same amino acid acceptor activity. The small amounts of suppressor tRNA may explain why the efficiency of suppression is less than 100 % (Table 2): The rate of translation of a nonsense triplet is probably limited by the amount of suppressor tRNA available.

Identification of a minor species of tRNA is difficult to achieve without a specific assay such as suppression. For example, since the seryl tRNA specified by the $Su1$ gene and the tyrosyl tRNA specified by the $Su3$ gene fractionate almost identically with major species having the same amino acid acceptor activity [19, 21], these components might not have been detected without the availability of the suppressor assay. Probably other minor species of tRNA also exist. Nothing is known about the number of such species or their physiological role in the cell.

## Chain-termination

During the process of protein biosynthesis, the growing peptide chain is covalently linked to a tRNA which is attached (probably non-co-valently) to the messenger-ribosome complex .The finished protein molecule is ultimately released unlinked to tRNA. When premature chain-termination is induced by a nonsense triplet, either *amber* or *ochre*, the protein fragment apparently undergoes the complete reaction sequence involved in normal chain-termination. The evidence now available, all from *in vitro* experiments, indicates that the fragment is not associated with any tRNA [23, 24].

Since virtually nothing is known about the mechanism of normal chain-termination, which is one of the critical steps in protein biosynthesis, attention has focused on premature chain-termination in nonsense mutants as a potential model system for understanding the normal process. The central question is whether any of the nonsense triplets are used as signals for normal chain-termination. There are two reasons why it appears necessary to have certain sequences reserved for this purpose. First, in a polygenic messenger RNA which specifics more than one protein, a terminating signal is needed to prevent the linkage of one protein to another. The chain-initiating triplet AUG, which specifies formyl-methionine [25], is in principle sufficient for this purpose since formyl-methionine cannot form a peptide bond with the preceding amino acid specified by the message. However, when AUG occurs in an internal position in a message rather than at the 5′ terminal position, it acts as a codon for methionine and not for formylmethionine [26]. Therefore, additional information must be associated with AUG to identify it as a chain-initiating formylmethionine codon for the internal genes of a polygenic messenger RNA.

A second reason for requiring a chain-terminating signal is to provide a mechanism for releasing newly synthesized polypeptide chains from the messenger-ribosome complex. There is no information at present about how a release mechanism might operate; possibly additional components, such as special tRNA or protein species, are involved.

The sequence in messenger RNA which normally chain-terminates in E. coli has not been identified. However, there are experiments bearing indirectly on the possible role of nonsense triplets in chain-termination. If nonsense triplets do perform this essential function, suppression of the function could be harmful to the cell. For two of the nonsense triplets, UAG and UGA, suppression can be as efficient as 60 % and still not affect cell growth [6, 11, 20], which argues against a normal chain-terminating role for the two triplets and suggests that they are entirely absent from all messenger RNA. By the same reasoning, UAA is a candidate for this role, since even low levels of suppression of the chain-terminating action of this triplet are inhibitory to cell growth [20, 27]. The present evidence on this question, however, is inconclusive.

### Dispensible and Indispensible Suppressor Genes

It has been shown, as discussed in a preceding section, that the Su3 suppressor gene is a structural gene for a minor species of tyrosyl tRNA. This tRNA species, and also the minor species of seryl tRNA specified by the Su1 gene, appear to be dispensible components, since suppressor mutations which alter the anticodon or affect the activity of the molecules in other (unidentified) ways, are not lethal to the cell. In addition to these dispensible tRNA species, there probably are also indispensible tRNA species which cannot be altered without causing cell death, notably the species required for translation of the standard codons of a cell. However, in a diploid cell containing two copies of the structural gene for an indispensible tRNA, a mutation in one of the genes need not be lethal as long as the cell retains a normal copy of the gene. E. coli cells are usually

haploid and therefore cannot tolerate such mutations, but some of the
episomal strains of E. coli, which have multiple copies of certain regions
of the bacterial chromosome, might contain structural genes for indispensible tRNA species in duplicate. In that case, it should be possible
to obtain viable mutants of the episomal strains with altered forms of
indispensible tRNA species. The problem remains of identifying the
mutants.

Consider the mechanism of suppression outlined in Fig. 4. There are
seven amino acids with codons which are related to the *amber* triplet UAG
by a single base substitution, namely the four amino acids shown in
Fig. 4, serine glutamine, tyrosine and lysine, and three others, glutamic
acid, leucine and tryptophan (see Table 1). The anticodons of the major
tRNA species normally involved in translating these seven codons can be
altered by mutation to yield suppressor tRNA which can translate the
*amber* triplet. If the suppressor mutant is isolated in an appropriate
episomal strain, it should be possible to obtain a heterozygote suppressor
strain (*i.e.*, $Su^+/Su^-$) in which the suppressor gene is a structural gene for
a major species of tRNA. Suppressor strains of this kind can provide
genetic and biochemical information about the structural genes for some
of the major, indispensible tRNA species. Experiments along these lines
are now in progress.

## References

1. DINTZIS, H.: Proc. nat. Acad. Sci. (Wash.) **47**, 247 (1961). — BISHOP, J.,
   J. LEAHY, and R. SCHWEET: Proc. nat. Acad. Sci. (Wash.) **46**, 1030 (1960). —
   GOLDSTEIN, A., and B. J. BROWN: Biochim. biophys. Acta (Amst.) **53**, 438
   (1961).
2. BRENNER, S., A. O. W. STRETTON, and S. KAPLAN: Nature (Lond.) **206**, 994 (1965).
3. WEIGERT, M. G., and A. GAREN: Nature (Lond.) **206**, 992 (1965).
4. —, E. LANKA, and A. GAREN: J. molec. Biol. **23**, 391 (1967).
5. BRENNER, S., L. BARNETT, E. R. KATZ, and F. H. C. CRICK: Nature (Lond.)
   **213**, 449 (1967).
6. GAREN, A., S. GAREN, and R. C. WILHELM: J. molec. Biol. **14**, 167 (1965).
7. KAPLAN, S., A. O. W. STRETTON, and S. BRENNER: J. molec. Biol. **14**, 528 (1965).
8. SCHWARTZ, N. M.: J. Bact. **89**, 712 (1965).
9. ORIAS, E., and T. K. GARTNER: Proc. nat. Acad. Sci. (Wash.) **52**, 859 (1964).
10. EGGERTSSON, G., and E. A. ADELBERG: Genetics **52**, 319 (1965).
11. BECKWITH, J. R.: Biochim. biophys. Acta (Amst.) **76**, 162 (1963).
12. SAMBROOK, J. F., D. P. FAN, and S. BRENNER: Nature (Lond.) **214**, 452 (1967).
13. ZINDER, N. D., and S. COOPER: Virology **23**, 152 (1964).
14. CAPECCHI, M., and G. GUSSIN: Science **149**, 417 (1965).
15. ENGELHARDT, D. L., R. E. WEBSTER, R. C. WILHELM, and N. D. ZINDER:
    Proc. nat. Acad. Sci. (Wash.) **54**, 1791 (1965).
16. WILHELM, R. C.: Cold Spr. Harb. Symp. quant. Biol. **31**, 496 (1966).
17. GESTELAND, R. F., W. SALSER, and A. BOLLE: Proc. nat. Acad. Sci. (Wash.)
    **58**, 2036 (1967).

18. ANDOH, T., and H. OZCKI: Proc. Nat. Acad. Sci. (Wash.) (in press). — LANDY, A., J. N. ABELSON, H. M. GOODMAN, and J. D. SMITH: J. molec. Biol. (in press).
19. —, and A. GAREN: J. molec. Biol. **24**, 129 (1967).
20. SMITH, J. D., J. N. ABELSON, B. F. C. CLARK, H. M. GOODMAN, and S. BRENNER: Cold Spr. Harb. Symp. quant. Biol. **31**, 479 (1966).
21. —, H. N. ABELSON, H. M. GOODMAN, A. LANDY, and S. BRENNER: Symp. Fed. European Biochem. Soc. (1967). In press.
22. CRICK, F. H. C.: J. molec. Biol. **19**, 548 (1966).
23. ZINDER, N. D., D. L. ENGELHARDT, and R. E. WEBSTER: Cold Spr. Harb. Symp. quant. Biol. **31**, 251 (1966).
24. GANOZA, M. C.: Cold Spr. Harb. Symp. quant. Biol. **31**, 273 (1966).
25. MARCKER, K., and F. SANGER: J. molec. Biol. **8**, 835 (1964). — ADAMS, J. M., and M. R. CAPECCHI: Proc. nat. Acad. Sci. (Wash.) **55**, 147 (1966). — WEBSTER, R. E., D. L. ENGELHARDT, and N. D. ZINDER: Proc. nat. Acad. Sci. (Wash.) **53**, 155 (1966).
26. NIRENBERG, M., R. CASKEY, R. MARSHALL, D. BRIMACOMBE, D. KELLOGG, B. DOCTOR, D. HATFIELD, J. LEVIN, F. ROTTMAN, S. PESTKA, M. WILCOX, and F. ANDERSON: Cold Spr. Harb. Symp. quant. Biol. **31**, 11 (1966). — KHORANA, H. BUCHI, N. GHOSH, T. M. GUPTA, J. JACOB, H. KOSSEL, R. MORGAN, S. A. NARANG, E. OHTSUKA, and R. D. WELLS: Cold Spr. Harb. Symp. quant. Biol. **31**, 39 (1966). — MATTHAEI, J. H., H. P. VOIGT, G. HELLER, R. NETH, G. SCHOCH, H. KUBLER, F. AMELUNXEN, G. SANDER, and A. PARMEGGIANI: Cold Spr. Harb. Symp. quant. Biol. **31**, 25 (1966).
27. GALLUCCI, E., and A. GAREN: J. molec. Biol. **15**, 193 (1966).
28. LENGYEL, P.: J. gen. Physiol. **49**, 305 (1966).
29. WEIGERT, M. G., and A. GAREN: J. molec. Biol. **12**, 448 (1965). — NOTANI, G. W., D. L. ENGELHARDT, W. KONIGSBERG, and N. D. ZINDER: J. molec. Biol. **12**, 439 (1965). — STRETTON, A. O. W., and S. BRENNER: J. molec. Biol. **12**, 456 (1965).
30. —, E. LANKA, and A. GAREN: J. molec. Biol. **14**, 522 (1965). — KAPLAN, S., A. O. W. STRETTON, and S. BRENNER: J. molec. Biol. **14**, 528 (1965).
31. — — — J. molec. Biol. **23**, 401 (1967).

# The Recognition of Anomalous Triplet-Codons by Aminoacyl-tRNA

By

D. Grünberger, A. Holý and F. Šorm

Institute of Organic Chemistry and Biochemistry
Czechoslovak Academy of Sciences, Prague

In the last few years many analogues of purine and pyrimidine bases were discovered which are incorporated into nucleic acids. This incorporation results in remarkable changes in the biological activity of the abnormal RNA's. As a consequence of these changes in the genetic material, the protein synthesizing capacity of the cell is affected. For example, it was observed that if the synthesis of TMV occurs in the presence of 5-fluorouracil, more than half of the uracil normally present in the TMV-RNA may be replaced by FU [1]. This substitution leads to mutations, probably because FU is occasionally read as C and not as U.

Since the nature of the genetic code, which provides sixty four possible codons for twenty aminoacids, is now sufficiently well understood [2, 3], it is possible to study the nature of the changes in the genetic material after replacing a nucleotide in the triplet codon by an analogue. Moreover, such a modification of the codon may give us more information of the general process of the recognition of the codons by aminoacyl-tRNA's.

For the assay of the coding properties of the anomalous triplets, the elegant technique of Nirenberg and Leder was used [4]. This method is based on the observation that trinucleotides stimulate the binding of specific aminoacyl-tRNA's to ribosomes and that the formed trinucleotide-ribosome-aminoacyl-tRNA complexes are retained by cellulose nitrate filters. For the synthesis of trinucleotides of defined sequences enzymatic methods were devised. We should like to describe briefly the techniques we have used [5]. The method involves the use of $T_1$ ribonuclease according to this general scheme:

$$N\text{-}2',3'\text{-cyclic phosphate} + YpZ \xrightarrow{T_1} NpYpZ;\ N = G,\ azaG,\ I;\ Y = C,$$
U, A, 5-BrU, 5-IU, $N^3$-Me-5BrU, $N^3$-Me-5JU; Z = as Y and G, azaG, I.

The different dinucleoside phosphates were synthesized either chemically or enzymatically using pancreatic RNase [6, 7].

### Coding Properties of 8-Azaguanosine-Containing Polymers and Trinucleotides

First we would like to consider some of our results with 8-azaguanine which is incorporated into RNA of different organisms where it replaces guanine. It was observed that after addition of azaG to growing cells, especially B. cereus, the analogue was incorporated into mRNA [8–10] and the protein synthesis was inhibited within few minutes [11–13]. It was suggested that azaG alters the normal function of mRNA. To test this hypothesis we prepared copolymers containing uridylic and guanylic, as well uridylic and azaguanylic acid and measured the stimulation of incorporation of different amino acids into proteins in ribosomal system [14] (Fig. 1). Both poly UG and poly UazaG markedly stimulated the incorporation of [14]C-valine but the template activity of poly UazaG was approximately one-half that of poly UG. Although the difference between the template efficiency of poly UazaG and poly UG may reflect differences between the recognition of azaG and G during protein synthesis, other factors such as chain length, secondary structure, base composition, also deserve consideration.

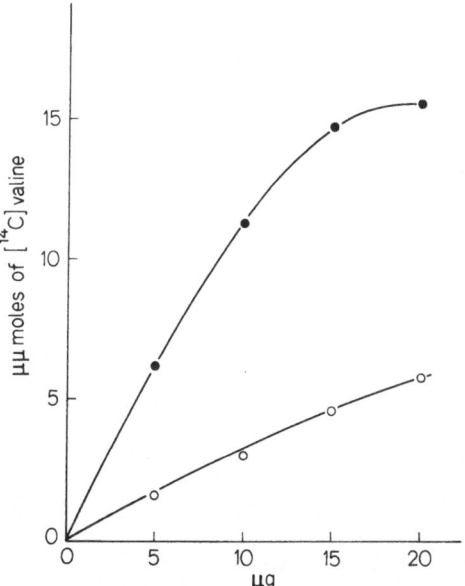

Fig. 1. The effect of poly UG and poly UazaG on [14]C-Valine incorporation into protein in preincubated S-30 E. coli extracts. ●——● poly UG; o——o poly UazaG

For the elimination of these factors which might influence the template activity of the polymers, we have prepared by methods mentioned previously, triribonucleoside diphosphates in which azaG substituted for G in all three possible positions of the triplet [15].

The codon assignment for valine is GpUpU, GpUpA, GpUpC and GpUpG[2]. We have prepared these triplets and the analogues containing azaG instead of G either in the 5'- or in the 3'-position of the codon. Fig. 2 shows the effect of different concentrations of GpUpC and azaGpUpC on the binding of [14]C-Val-tRNA to ribosomes. At equimolar concentrations, GpUpC directed the binding of more Val-tRNA to ribosomes than azaGpUpC. We have obtained very similar results with the other valine codons, namely GpUpU, GpUpA, and azaGpUpU, azaGpUpA, respectively. In all these cases azaG could replace G but these anomalous triplets had a lower template efficiency.

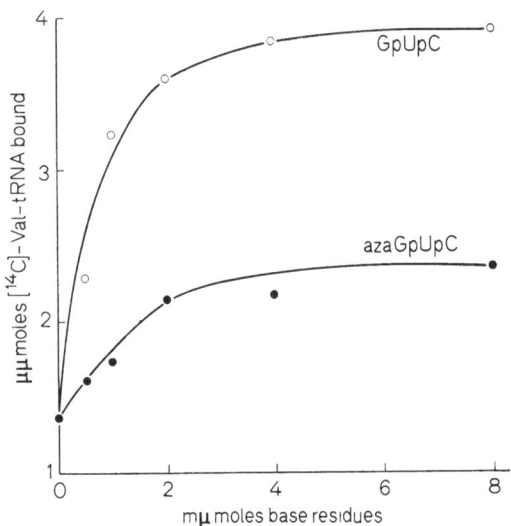

Fig. 2. The effect of GpUpC and azaGpUpC on the binding of [14]C-Val-tRNA to ribosomes. Experimental conditions as described in legend to Table 1

Since GpUpG could be recognized by [14]C-Val-tRNA, it was possible to substitute G by azaG in the 5'- as well in the 3'-position and in both places of this trinucleotide. Fig. 3 shows that [14]C-Val-tRNA responds best to GpUpG, less well to azaGpUpG and GpUpazaG, but not to azaGpUpazaG.

To determine whether the effect of azaG in the codons for valine was specific, we have prepared codons for alanine with general formula GpCpX (X — all four natural nucleosides) and substituted G by azaG in the 5'-position. Fig. 4 shows that [14]C-Ala-tRNA responds to azaGpCpA less well than to GpCpA.

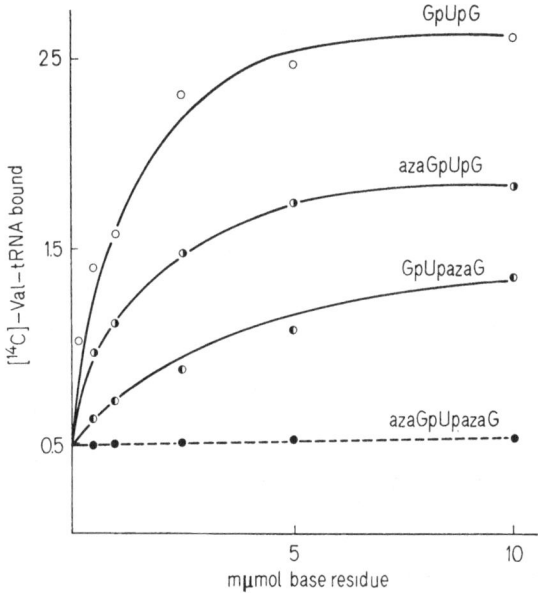

Fig. 3. The effect of substitution of 8-azaguanosine for guanosine in GpUpG codon for valine on the binding of ¹⁴C-Val-tRNA to ribosomes. Experimental conditions as described in legend to Table 1

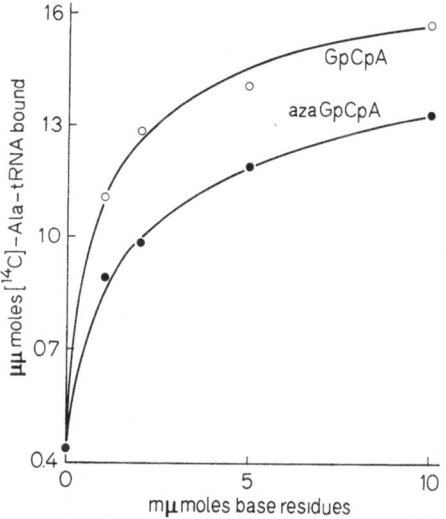

Fig. 4. Stimulation of binding of ¹⁴C-Ala-tRNA to ribosomes by GpCpA and azaGp CpA. Experimental conditions as described in legend to Table 1

Similar results we have obtained with the other codons for alanine. It is of interest that the replacement of G by azaG in alanine codons caused a lower decrease of template activity than in codons for valine.

Other trinucleotides containing guanosine in the 5'-terminal position are the codons for aspartic acid (GpApU and GpApC). Fig. 5 shows that GpApC is recognized by $^{14}$C-Asp-tRNA better than azaGpApC at each concentration tested. Further it was of interest to test whether azaG can

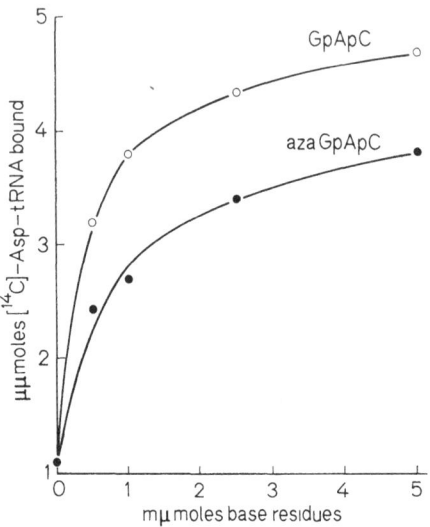

Fig. 5. The effect of GpApC and azaGpApC on the binding of $^{14}$C-Asp-tRNA to ribosomes. Experimental conditions as described in legend to Table 1

replace G in the second position of the triplet. For this purpose we prepared codons for arginine (CpGpU and CpGpC), as well as trinucleotides with azaG in the second position, i.e. CpazaGpU and CpazaGpC. The synthesis of these triplets were carried out from CpG-2',3'-cyclic phosphate resp. CpazaG-2',3'-cyclic phosphate and from the appropriate nucleoside in the presence of T$_1$-ribonuclease. The dinucleoside phosphates were prepared by pancreatic RNA-ase from C-2',3'-cyclic phosphate and G or azaG. The product of this enzymatic reaction i.e. CpG or CpazaG were chemically converted to CpG-, or CpazaG-2',3'-cyclic phosphates and these were used as substrates for T$_1$-ribonuclease in the synthesis of trinucleoside diphosphates.

Fig. 6 shows the effect of different concentrations of CpGpU and CpazaGpU on the binding of $^{14}$C-Arg-tRNA to ribosomes. It follows from these results that azaG can partly substitute for G in the second position of the codons for arginine, too. However, $^{14}$C-Arg-tRNA responds to CpazaGpU less well than to CpGpU. Thus the recognition of azaG in the second position proceeeds with similar efficiency as in the 5'- or 3'-position of the trinucleotides.

All these results have shown that azaG can partly replace G in all three positions of the codons for different amino acids. However, there appeared a possibility that azaG might behave in codons as adenosine or uridine and induce miscoding during the recognition process. To verify

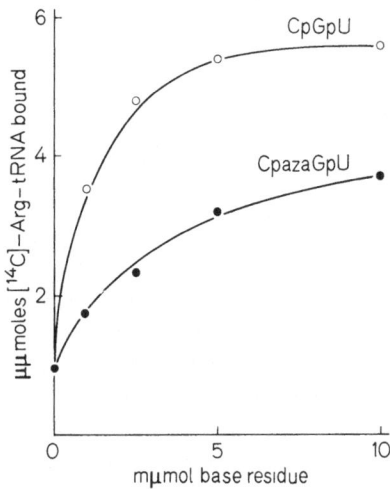

Fig. 6. The effect of CpGpU and CpazaGpU on the binding of ¹⁴C-Arg-tRNA to ribosomes. Experimental conditions as described in legend to Table 1

this assumption we have measured the effect of GpUpU and azaGpUpU on the stimulation of the binding of ¹⁴C-Ile-tRNA (Ile codon ApUpU) and ¹⁴C-Phe-tRNA (Phe codon UpUpU) to ribosomes. The results in Table 1 show that neither GpUpU nor azaGpUpU were recognized by ¹⁴C-Ile- or ¹⁴C-Phe-tRNA.

Table 1. *The effect of GpUpU and azaGpUpU on the binding of ¹⁴C-aminoacyl-tRNA to ribosomes.* The reaction mixture contained in a final volume of 50 $\mu$l : 0.05 M Tris-acetate buffer pH 7.2, 0.1 M ammonium acetate, 0.03 M Mg-acetate, 2.5 $A_{260}$ units of E. coli ribosomes, 0.8 $A_{260}$ unit of tRNA acylated with 19 $\mu\mu$mol ¹⁴C-Val, or 15 $\mu\mu$mol ¹⁴C-Ile or 11 $\mu\mu$mol ¹⁴C-Phe and 20 m$\mu$moles of base residues of tri-nucleotides. Incubation was for 20 min at 24°

| Trinucleotide | $\mu\mu$mol ¹⁴C-Aminoacyl-tRNA bound | | |
|---|---|---|---|
| | ¹⁴C-Val | ¹⁴C-Ile | ¹⁴C-Phe |
| None | 0.91 | 0.89 | 0.80 |
| GpUpU | 2.59 | 0.74 | 0.76 |
| azaGpUpU | 1.93 | 0.83 | 0.69 |

From these results we can conclude that azaG can function in mRNA only like G and not like A or U. However, aminoacyl-tRNA's respond less well to azaG-containing triplets than to respective codons containing G; thus, azaG differs in the coding properties from G only quantitatively and not qualitatively. This seems to be in accord with the fact, that no mutation was observed with the cells growing in the presence of azaG. In addition, it is possible to explain the inhibition of protein synthesis in growing cells in the presence of azaG by the lower recognition of azaG in mRNA. Moreover, the azaGpUpazaG sequence which can occur in mRNA as well, seemed to be totally inactive in the recognition process.

It is not clear why the aminoacyl-tRNA's respond to azaG-containing trinucleotides less well than to the respective codons containing G. It is possible that these differences are due to the different pK values of G and azaG. Substitution in C8 position of guanosine by a nitrogen atom results in the decrease of pK of the enolic function at C6 from 9.2 to 6.7 [16]. It is probable that the increased acidity of this group which is involved in the base pairing, produces a weaker hydrogen bonding with the amino group of the complementary cytidine residue.

### The Effect of Triribonucleoside Diphosphates Containing 5-Bromo- or 5-Iodouridine and their N³-Methyl Derivatives

GRUNBERG-MANAGO and MICHELSON have found that the halogen derivatives of poly U, like poly BrU and poly IU stimulated not only the incorporation of phenylalanine, but also the incorporation of leucine, isoleucine and serine [17]. They suggested that 5-BrU in the codon can be read occasionally as A. Since this is difficult to explain in terms of orthodox base pair hydrogen bonding, we have prepared triplets containing 5-BrU or 5-IU in the second or third position of GpUpU codon

Table 2. *The response of $^{14}C$-Val-tRNA to triribonucleoside diphosphates containing 5-bromo- or 5-iodouridine and their $N^3$-methyl derivatives.* The experimental conditions as described in legend to Tab. 1

| Trinucleotide | $^{14}$C-Val-tRNA bound | | |
|---|---|---|---|
| | $\mu\mu$mol | $\Delta\ \mu\mu$mol | % |
| — | 0.52 | — | — |
| GpUpU | 3.32 | 2.80 | 100 |
| GpBrUpU | 4.50 | 3.98 | 142 |
| GpIUpU | 4.00 | 3.48 | 124 |
| GpUpBrU | 1.96 | 1.44 | 51 |
| GpUpIU | 1.82 | 1.30 | 47 |
| GpUpN³-Me-5BrU | 0.67 | 0.15 | 5 |
| GpUpN³-Me-5IU | 0.68 | 0.16 | 6 |

for valine and measured the codon properties of these anomalous trinucleotides.

Table 2 shows the stimulation of binding of [14]C-Val-tRNA to ribosomes in the presence of different anomalous triplets. In the presence of
GpBrUpU and GpIUpU the binding of [14]C-Val-tRNA to ribosomes is
higher than in the presence of GpUpU which represents one of the normal
codons for valine. In contrast, [14]C-Val-tRNA responds less well to
GpUpBrU and GpUpIU than to GpUpU. By substitution of a methyl
group in $N^3$-position of uridine for the hydrogen atom involved in the
base pair hydrogen bonding, the formed trinucleotides, i.e. GpUp
$N^3$-Me-5BrU and GpUpN$^3$-Me-5IU are totally inactive in this system.
This data demonstrate again that during the recognition process the complementary base pairing according to Watson-Crick model takes place.

The behaviour of BrU and IU in the second and third position of
triplets could be anticipated on the basis of the stability of the binding
between halogen derivatives of uridine and complementary nucleosides
in the anticodon part of tRNA molecule. If we assume that the anticodon
of Val-tRNA contains the IAC sequence[18], then during the recognition
of GpUpU hydrogen bonds between G and C, U and A, U and I arise.
If we substitute uridine in the second position with BrU or IU, then the
binding between BrU and A takes place. Since MASSOULIÉ et al. [19]
have observed that the stability of the complex poly BrU-poly A is
greater than that of poly U-poly A, we suggest that A in the second
position of the anticodon may hydrogen bond more strongly with BrU
or IU in the codon than with U. Therefore, the stimulation of the binding
of [14]C-Val-tRNA to ribosomes might be higher in the presence of
GpBrUpU or GpIUpU than in the presence of GpUpU.

In contrast, the decreased template activity of trinucleotides containing halogen derivatives of uridine in the 3'-position is probably
connected with the lower stability of the binding between inosine and
BrU or IU than between inosine and uridine.

To test the possibility that 5-BrU can behave in codon as adenosine
as it was suggested by GRUNBERG-MANAGO and MICHELSON [12] we have
measured the binding of [14]C-Asp-tRNA to ribosomes in the presence of
GpApU (codon for Asp), GpBrUpU and GpIUpU. If 5-BrU behaves as
adenosine, then GpBrUpU should stimulate the binding of [14]C-Asp-tRNA
to ribosomes, too. However, Table 3 shows that [14]C-Asp-tRNA responses
only to GpApU but neither to GpBrUpU nor GpIUpU.

Thus these data do not support the hypothesis that 5-BrU can
behave in the codon as adenosine. We suggest that 5-BrU or 5-IU can
replace only uridine in the codon. The template efficiency of these
anomalous triplets depends on the stability of the binding between the
halogen-derivative and the complementary base in the anticodon.

Table 3. *The effect of GpApU, GpBrUpU and GpIUpU on the binding of* [14]*C-Asp-tRNA to ribosomes.* The experimental conditions as described in legend to Table 1

| Triplet | [14]C-Asp-tRNA bound | |
|---|---|---|
| | $\mu\mu$mol | $\Delta$ $\mu\mu$mol |
| — | 1.46 | — |
| GpApU | 5.26 | 3.80 |
| GpBrUpU | 1.35 | − 0.11 |
| GpIUpU | 1.39 | − 0.09 |

## The Recognition of Inosine-Containing Trinucleotides by Aminoacyl-tRNA's

The similarity of G and I with regard to hydrogen bonding [20] led to the assumption that inosine could substitute for guanosine in the genetic code. BASILIO et al. [21] have observed that randomly ordered poly UI stimulated the incorporation of the same amino acids into proteins as poly UG, though less efficiently. Moreover, inosine was discovered in different tRNA molecules as a possible part of the supposed anticodon [18, 22]. CRICK [23] has proposed models in which inosine in the anticodon of tRNA is assumed to be hydrogen bonded with A, C or U in mRNA codon.

For the elucidation of the possible function of inosine in codon-anticodon interaction we have substituted inosine for guanosine in all three positions of the codons for several amino acids and tested their effect on the binding of [14]C-aminoacyl-tRNA to ribosomes.

Table 4. *The effect of substitution of inosine for guanosine in valine codons on the binding of* [14]*C-Val-tRNA to ribosomes.* The experimental conditions as described in legend to Table 1

| Trinucleotide | [14]C-Val-tRNA bound | |
|---|---|---|
| | $\mu\mu$moles | $\Delta$ $\mu\mu$moles |
| None | 0.64 | — |
| GpUpU | 4.27 | 3.63 |
| IpUpU | 0.65 | 0.01 |
| GpUpC | 2.57 | 1.93 |
| IpUpC | 0.65 | 0.01 |
| GpUpA | 4.70 | 4.06 |
| IpUpA | 0.67 | 0.03 |

The effect of the substitution of inosine in the 5'-position of valine codons on the binding of [14]C-Val-tRNA to ribosomes is shown in Table 4. As expected, GpUpU, GpUpC and GpUpA were recognized by [14]C-Val-tRNA well. On the other hand, none of the inosine-containing tri-nucleotides had any template activity under the same experimental conditions.

Since [14]C-Val-tRNA responses to GpUpG as well, it was possible to replace G by inosine in the 5'-, or in the 3'-position and in both positions of this codon. Fig. 7 shows that [14]C-Val-tRNA responds well only to GpUpG and GpUpI but not to IpUpG and IpUpI. We have obtained similar results with codons for alanine. The results presented in Fig. 8 show that [14]C-Ala-tRNA responds best to GpCpG and GpCpI, less well to IpCpG, but not to IpCpI.

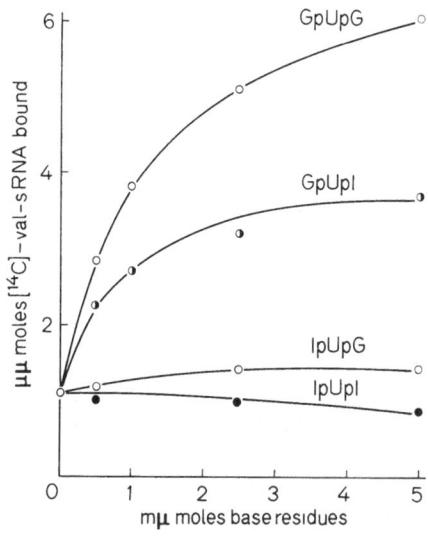

Fig. 7. The effect of substitution of inosine for guanosine in GpUpG codon for valine on the binding of [14]C-Val-tRNA to ribosomes. Experimental conditions as described in legend to Table 1

In order to determine whether inosine can replace guanosine in the second position of a triplet, we have compared the response of [14]C-Arg-tRNA to CpGpU and CpIpU. Fig. 9 illustrates the effect of different concentrations of CpGpU and CpIpU on the binding of [14]C-Arg-tRNA to ribosomes. The stimulation of the binding was dependent on the concentration of the triplet. However, at equimolar concentrations CpGpU directed the binding of [14]C-Arg-tRNA to ribosomes much more than CpIpU. These results led to the conclusion that substitution of inosine for guanosine in the second position of the codon for arginine has a similar effect as the replacement of G in the 5'-position of the codons.

We can conclude from our results that the recognition of inosine-containing trinucleotides by different aminoacyl-tRNA's depends on the position of G—I substitution. Aminoacyl-tRNA's respond to triplets containing inosine in the 5'-or second position only slightly. In contrast,

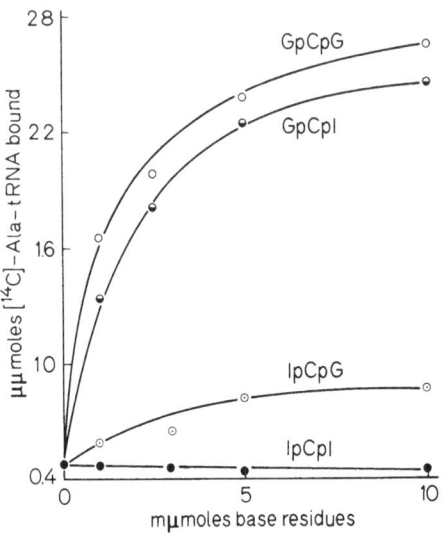

Fig. 8. The effect of substitution of inosine for guanosine in GpCpG codon for alanine on the binding of $^{14}$C-Ala-tRNA to ribosomes. Experimental conditions as described in legend to Table 1

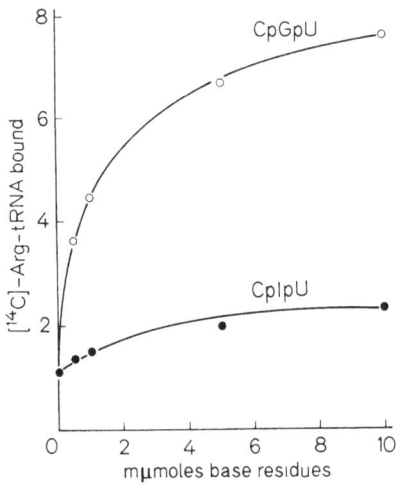

Fig. 9. Stimulation ob binding of $^{14}$C-Arg-tRNA to ribosomes by CpGpU and CpIpU. Experimental conditions as described in legend to Table 1

no marked difference was observed in the response of Val- and Ala-tRNA's to trinucleotides containing inosine in the 3'-terminal position. Recently, SEKIYA et al. [24] also reported that GpApI stimulated the binding of ¹⁴C-Glu-tRNA to ribosomes.

The different behaviour of inosine in various positions of triplet could be a reflection of the fact that the stability of inosine-cytidine pair with only two hydrogen bonds is lower than that of the guanosine-cytidine pair with three hydrogen bonds. Whereas the reading of the codons in the 5'- and second positions proceeds with very precise mechanism, the 3'-wobble position related to degeneration of the codon, does not show such a high degree of specifity.

## References

1. KRAMER, G., H. G. WITTMANN, and H. SCHUSTER: Z. Naturforsch. 19 b, 46 (1964).
2. NIRENBERG, M., P. LEDER, M. BERNFIELD, R. BRIMACOMBE, J. TRUPIN, F. ROTTMAN, and C. O'NEAL: Proc. nat. Acad. Sci. (Wash.) 53, 1161 (1965).
3. SÖLL, D., E. OHTSUKA, D. S. JONES, R. LOHRMANN, H. HAYATSU, S. NISHI-MURA, and H. G. KHORANA: Proc. nat. Acad. Sci. (Wash.) 54, 1378 (1965).
4. NIRENBERG, M., and P. LEDER: Science 145, 1399 (1964).
5. GRÜNBERGER, D., A. HOLÝ, and F. ŠORM: Coll. Czech. Chem. Commun. 33, 286 (1968).
6. HOLÝ, A., and J. SMRT: Coll. Czech. Chem. Commun. 31, 1528 (1966).
7. ŽEMLIČKA, J., S. CHLÁDEK, A. HOLÝ, and J. SMRT: Coll. Czech. Chem. Commun. 31, 3198 (1966).
8. CHANTRENNE, H.: J. cell. comp. Physiol. Suppl. 1, 64, 149 (1964).
9. GRÜNBERGER, D., R. N. MASLOVA, and F. ŠORM: Coll. Czech. Chem. Commun. 29, 152 (1964).
10. LEVIN, D. H.: Biochemistry 5, 1618 (1966).
11. MATTHEWS, R. E. F.: Pharmacol. Rev. 10, 359 (1954).
12. MANDEL, H. G.: Arch. Biochem. 76, 230 (1958).
13. GRÜNBERGER, D., and F. ŠORM: Coll. Czech. Chem. Commun. 28, 1044 (1963).
14. —, C. O'NEAL, and M. NIRENBERG: Biochim. biophys. Acta (Amst.) 119, 581 (1966).
15. —, L. MEISSNER, A. HOLÝ, and F. ŠORM: Biochim. biophys. Acta 119, 432 (1966).
16. LEVIN, D. H., and M. LITT: J. molec. Biol. 14, 506 (1965).
17. GRUNBERG-MANAGO, M., and A. M. MICHELSON: Biochim. biophys. Acta (Amst.) 80, 431 (1964).
18. BAYEV, A. A., T. V. VENKSTERN, A. D. MIRZABEKOV, A. I. KRUTILINA, V. D. AXELROD, L. LI, and V. A. ENGELGARDT: Proc. 3rd Symp. Fed. Europ. Biochem. Soc., p. 287 (1967).
19. MASSOULIÉ, J., A. M. MICHELSON, and F. POCHON: Biochim. biophys. Acta (Amst.) 114, 16 (1966).
20. DAVIS, D. R., and A. RICH: J. Amer. Chem. Soc. 80, 1003 (1958).
21. BASILIO, C., A. J. WAHBA, P. LENGYEL, J. F. SPEYER, and S. OCHOA: Proc. nat. Acad. Sci. (Wash.) 48, 613 (1962).

22. HOLLEY, R. W., J. APGAR, A. EVERETT, J. T. MADISON, M. MARQUISEE, S. H. MERRILL, J. P. PENSWICK, and A. ZAMIR: Science 147, 1462 (1965).
23. CRICK, F. H. C.: J. molec. Biol. 19, 548 (1966).
24. SEKIYA, T., M. YOSHIDA, and T. UKITA: Biochim. biophys. Acta (Amst.) 149, 610 (1967).

## Discussion

*Matthaei:* For a while least, we should synthesize the triplets in labeled form to make sure whether their binding to the ribosome does not influence their apparent activity in coding the binding of amino acyl-tRNA.

We found IpUpC in IpUpC ... pC $\sim_{50}$ not to code valine in the binding assay, whereas "I" rather actively replaces "G" in amino acid incorporation coded by random polynucleotides.

A comparison between efficiencies of codons that could form a minimum or a maximum of hydrogen bonds with base complementary anticodons does not indicate a clearly less favourable equilibrium for binding of amino acid-adapters which could form 6 rather than 9 hydrogen bonds. Thus the poor efficiency of IXY compared to GXY would not seem to result from the formation of two instead of three hydrogen bonds.

# Structure and Cellular Function of
# Aminoacyl Ribonucleic Acid Synthetases

By

F. C. NEIDHARDT

Department of Biological Sciences Purdue University
Lafayette, Indiana

Temperature-sensitive *(ts)* mutants have several uses in molecular and cellular biology. First of all, since phenotypic expression of such mutations is conditional, they permit identification of the genetic elements concerned with indispensable enzymes, thereby permitting progress toward the goal of a total definition of the genome of one cell. Careful study of such mutants can sometimes also reveal ancillary cellular roles for indispensable enzymes.

Secondly, *ts* mutations potentially provide a tool for exploring the structure, enzymatic activity, and allosteric control of any cellular protein, whether indispensable or not (cf. [*13*]). There are also hints that such mutants may provide a means to study late steps in the ontogeny of biologically active protein molecules.

Several of my colleagues and associates during the past five years have been studying aminoacyl ribonucleic acid (RNA) synthetases by means of *ts* mutants. In this report I would like to review our principal findings to illustrate at least some uses of *ts* mutants. All of this work was done with *Escherichia coli*, and most of it concerns the synthetases for phenylalanine and valine.

## Frequency of *ts* Mutants

*Ts* mutants in the synthetases for valine and phenylalanine are readily isolated without deliberate selection. Their frequency is high relative both to the frequency of all *ts* mutants and to the frequency of *ts* mutations in other aminoacyl RNA synthetases.

We have summarized in Table 1 the record of our laboratory in isolating such mutants. We have not attempted precise quantitation (which would be possible only if we eliminated steps of liquid cultivation in the isolation procedure) but some idea of the abundance of valyl and phenylalanyl RNA synthetase mutants can still be given. When a culture of *E*.

*coli* (any of five strains used) is exposed to ethyl methane sulfonate, and then subjected to penicillin counterselection favoring all variants that can grow at 30° but cannot grow in rich medium at 40°, a significant fraction of these mutants have altered valyl or phenylalanyl RNA synthetases

Table 1. *Ts mutants from E. coli strains*

| Wild strain | *ts* mutants |
| --- | --- |
| Strain NP3 (K 10): | *phe*S 5 |
|  | *phe*S 6 |
|  | *phe*S 7 |
| NP5 (C): | *phe*S 8 |
| NP2 (KB): | *val*S 1 |
|  | *val*S 2 |
| K 12 (Yanofsky): | *val*S 3 |
|  | *val*S 4 |
|  | *val*S 5 |

[10]. Furthermore, there is considerable strain specificity. Every *ts* mutant of *E. coli* NP2 we have ever isolated that has an altered aminoacyl RNA synthetase is altered in the valine enzyme, and such mutants may account for up to 5 % of the total *ts* mutants surveyed in any one experiment. A K 12 strain obtained from C. Yanofsky yields valyl RNA synthetase mutants even more readily (M. Tingle, unpublished observations). On the other hand, with *E. coli* strains NP3 and NP5 the same procedures have always yielded phenylalanyl RNA synthetase mutants in high frequency, and never a valine mutant [10]. In Paris, Yanif, Jacob, and Gros [20], using a different mutagen, N-methyl N-nitroso guanidine, have similarly found it easy to isolate *ts* valyl RNA synthetase mutants in large numbers; their only other *ts* synthetase mutant is one altered in the alanine enzyme.

Simple changes in the isolation procedure have not eliminated the recurring isolation of valyl or phenylalanyl RNA synthetase mutants, and the isolation of mutants for other synthetases has so far been achieved only by more selective means, such as the use of amino acid analogs (reviewed in [10]).

The failure to isolate *ts* mutants for other synthetases in good yield by non-selective means is not surprising; in our procedure the odds against any one *ts* mutant being altered in a particular synthetase are perhaps 500:1, since only *ts* mutants in nutritionally bypassable enzymes are eliminated. The surprising finding is the high frequency of valyl and phenylalanyl mutants.

We suspect that this behavior is related to the quaternary structure of these enzymes, as discussed later, rather than a reflection of mutational hotspots.

## Genetic Mapping

The structural genes for aminoacyl RNA synthetases appear not to be contiguous with each other or, for the most part, with genes for their cognate biosynthetic enzymes.

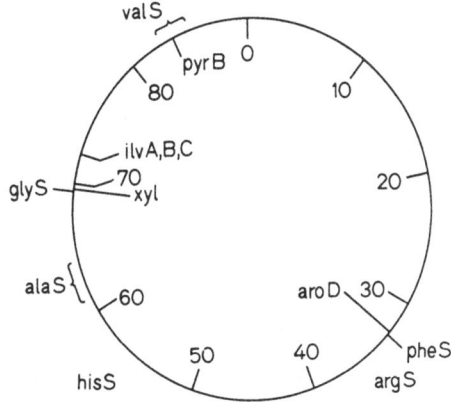

Fig. 1. Location of the structural genes (S) for some aminoacyl RNA synthetase in *E. coli*. The numbers on the map are from the TAYLOR-THOMAN [*18a*] determinations of marker entry times. References to synthetase mapping is as follows: *alaS*, 20; *argS*, HIRSHFIELD and MAAS, unpublished; *glyS*, 2; *hisS*, 16 *(Salmonella)*; *pheS*, 3; *valS*, 7, 20

In Fig. 1 is shown the chromosomal site of several structural genes for aminoacyl RNA synthetases. They are quite evenly distributed on this map. Of the six synthetase genes that have been mapped, only one is known to cotransduce with the gene for a biosynthetic enzyme for that amino acid; *pheS* cotransduces 33 % of the time with *aroD* (controlling one of the enzymes leading to shikimic acid in the pathway common to the aromatic amino acids) [*3*]. The genes of the aromatic pathway, however, are not clustered into one operon, but are scattered over the map. The proximity of *pheS* and *aroD* seems fortuitous.

## Stability of *ts* Enzymes

All *ts* valyl and phenylalanyl RNA synthetases we have studied become rapidly inactivated in cell free extracts.

Table 2 summarizes some early measurements made on two mutant enzymes. The cells had been grown at 30°, and extracts were prepared and kept at 0°. The enzyme measurements were made several hours after preparation of the extracts. It is clear that in each case the mutant synthetase was not measurable by the attachment assay run at 30° even through each enzyme had presumably functioned at a normal rate *in vivo* at that temperature. Some of the Yanif et al. [20] *valS* mutants behave similarly, but others they isolated are considerably more stable *in vitro* than are ours.

Table 2. *Synthetase activities of extracts of normal and Ts mutant strains* [a]

| Strain | Attachment activity [b] | Activation activity [c] |
|---|---|---|
| | valyl RNA synthetase | |
| Wild (NP2) | 0.063 | 3.51 |
| Mutant (NP29) | <0.001 | 1.38 |
| | phenylalanyl RNA synthetase | |
| Wild (NP3) | 0.030 | 1.61 |
| Mutant (NP37) | <0.001 | 0.90 |

[a] All assays were performed at 30°. Taken from Eidlic and Neidhardt [8].

[b] Specific activity, expressed as micromoles of aminoacyl tRNA formed per hour per mg of protein.

[c] Specific activity, expressed as micromoles of pyrophosphate exchanged per hour per mg of protein.

We have not been successful in stabilizing these *ts* enzymes, but recently we have been able to find enzyme activity in extracts immediately after cell rupture by omitting dialysis and other preparative steps that usually delay the start of the assay.

For both valyl and phenylalanyl *ts* mutants, the activation activity of extracts is retained longer than the attachment activity. Part of this residual activation activity may be the result of the action of other synthetases.

## Transfer RNA Charging in ts Mutants

The *in vivo* level of charging of $tRNA_{val}$ but not of $tRNA_{phe}$ decreases upon exposure of *ts* mutants to the restrictive temperature.

During exponential growth at 30° in a minimal medium supplemented with a complete mixture of amino acids, a double mutant having a block in valine biosynthesis as well as a *ts* valyl RNA synthetase maintains a steady state level of charged $tRNA_{val}$ of approximately 86 % (Table 3). Either withdrawal of valine or warming to 40° causes a drop in the charging of $tRNA_{val}$ to 7 % or 23 %, respectively [7]. This behavior indicates that at 40° the *valS^ts* mutant ceases protein synthesis as a result of an impairment in the charging of tRNA$_{val}$, and is consistent with the *in vitro* behavior of the damaged enzyme.

Table 3. *Effect of temperature and amino acid starvation on valyl tRNA charging*[a]

| Strain | Medium | Temp. ° C | Valyl tRNA charging[b] % |
|---|---|---|---|
| Wild | Minimal + aa[c] | 30 | 85 |
| Wild | Minimal + aa | 40 | 76 |
| *valS^ts*, IV⁻ | Minimal + aa | 30 | 86 |
| *valS^ts*, IV⁻ | Minimal + aa − val | 30[d] | 7 |
| *valS^ts*, IV⁻ | Minimal + aa | 40[e] | 23 |

[a] From BÖCK, FAIMAN, and NEIDHARDT [7].
[b] % of total tRNA$_{val}$ resistant to periodate oxidation.
[c] aa = mixture of 20 L-amino acids, 50 μg/ml each.
[d] Incubation for 30 min in valine-deficient medium.
[e] Incubation for 30 min at 40°.

The *pheS^ts* mutants, in contrast, show a paradoxical behavior. In Table 4 it can be seen that the tRNA$_{phe}$ is approximately 94 % charged in a *pheS^ts* mutant (that also carries a block in phenylalanine biosynthesis) during growth at 30° in a phenylalanine-containing medium.

Table 4. *Effect of temperature and amino acid starvation on phenylalanyl tRNA charging*[a]

| Strain | Medium | Temp. ° C | Phe tRNA charging[b] % |
|---|---|---|---|
| *pheS^ts* phe⁻ | Minimal + phe | 30 | 94 |
| *pheS^ts* phe⁻ | Minimal + phe | 40[c] | 97 |
| *pheS^ts* phe⁻ | Minimal | 30[d] | 4 |
| *pheS^ts* phe⁻ | Minimal | 40[e] | 0 |

[a] From BÖCK and NEIDHARDT (unpublished experiments).
[b] % of total tRNA$_{phe}$ resistant to periodate oxidation.
[c] Incubation at 40° for 1 h.
[d] Starvation for phenylalanine for 1 h at 30°.
[e] Starvation for phenylalanine for 1 h at 40°.

Removal of phenylalanine reduces this level quickly to 4 %. Warming the culture to 40°, on the other hand, allows growth to about 5 % of normal, but the charging level remains high (cited in [10]).

We have suggested two possible explanations for this behavior [10]. The first starts with the assumption that there is at least one minor species of tRNA$_{phe}$ that is necessary, along with the major species, for protein synthesis in E. coli. The pheS$^{ts}$ enzyme is then assumed to charge both species at near normal rates in the cell at 30°, but at 40° to be virtually unable to charge the minor species. The limiting amount of charged minor species slows down protein synthesis drastically, resulting in an accumulation of the acylated form of the major species. This suggestion has received indirect support from recent evidence [4] that there may be at least one minor species of tRNA$_{phe}$ in E. coli.

Alternatively one may view the failure of the charging level of tRNA$_{phe}$ to drop when protein synthesis is inhibited at 40° as evidence that the phenylalanyl RNA synthetase must perform at least one function in addition to charging tRNA$_{phe}$ with phenylalanine. This additional function, which would be blocked in the mutants at 40°, might be the transport of the charged tRNA$_{phe}$ to the ribosomal site of protein synthesis. (This function might be only a formal one and the role of the enzyme might be merely the release of the charged tRNA$_{phe}$ from its surface.) Or this additional function might involve the release of tRNA from the ribosome after peptide bond formation.

We have no evidence that permits a choice between these explanations. The recent genetic work of Dr. BÖCK, however, reinforces our contention that the growth lesion in our pheS$^{ts}$ mutant is really a structurally damaged phenylalanyl RNA synthetase despite the in vivo charging data. He has found that the pheS$^{ts}$ markers are tightly linked (97 % cotransduced) and probably allelic with, markers that modify the amino acid specificity of the phenylalanyl RNA synthetase and do result in reduced charging in vivo [3].

Some other findings of his that we shall discuss later on enhance our interest in the pheS$^{ts}$ charging paradox.

## RNA Regulation

Both the valyl and phenylalanyl RNA synthetases must be functional to permit RNA synthesis in RC$^{st}$ (stringent) cells; neither are necessary to permit RNA synthesis in RC$^{rel}$ (relaxed) cells.

All valS$^{ts}$ mutants have been isolated in stringent cells (those having normal amino acid control of RNA synthesis), and all cease RNA accumulation upon a shift to 40° (Fig. 2, upper left). From bacterial mating experiments between valS stringent cells and relaxed cells with a normal synthetase, some recombinants were isolated that were both valS$^{ts}$ and relaxed. These recombinants overproduce RNA at 40° (Fig. 2, upper right) Whether combined with relaxed or stringent control alleles, valS cells show a drop in tRNA$_{val}$ charging [7].

Most, but not all *ts pheS* mutants have been isolated in relaxed cells, and they overproduce RNA at 40° (Fig. 2, lower right). Recombinants that are both *pheS$^{ts}$* and stringent were isolated, and these display the stringent pattern upon a shift to 40° (Fig. 2, lower left). In addition, the one *pheS$^{ts}$* mutant isolated in a stringent strain ceased RNA accumulation at the restrictive temperature [6]. As in the *valS$^{ts}$* mutants, the tRNA charging pattern is unaffected by the RC allele; whether relaxed or stringent, *pheS$^{ts}$* mutants show the paradoxical increase in tRNA$_{phe}$ charging [10].

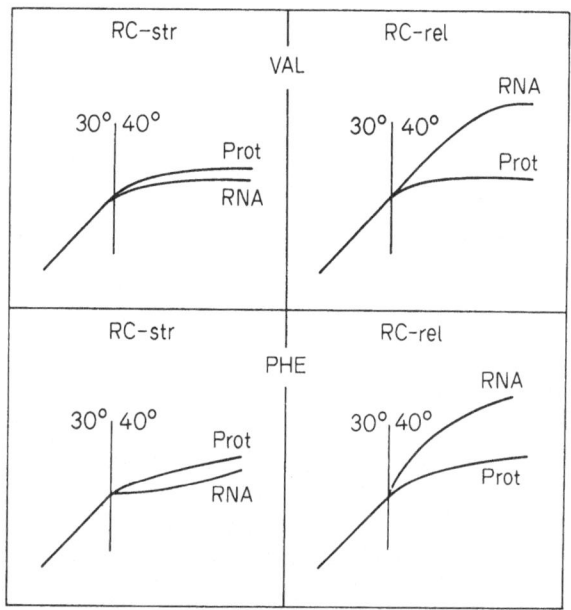

Fig. 2. Patterns of RNA and protein synthesis at 40° in *pheS$^{ts}$* and *valS$^{ts}$* mutants having either relaxed or stringent control. The abscissa is time, and the ordinate is the logarithm of the amount of each component per ml relative to the time of the temperature shift. From EIDLIC [6] and EIDLIC and NEIDHARDT [8]

Work is currently underway to repeat these experiments in strains having the above patterns of *ts* and RC alleles but which are otherwise isogenic.

## Biosynthetic Enzyme Repression

At growth restricting temperatures *valS$^{ts}$* mutants derepress valine-controlled enzymes; but *pheS$^{ts}$* mutants do not derepress phenylalanine-controlled enzymes.

At 35.5° growth of NP29, a *valS$^{ts}$* mutant, occurs at about one-third the rate of a wild strain. One hour of growth at such a temperature, even in a medium containing L-valine, is sufficient to effect over a ten-fold derepression of acetolactate synthetase; growth of the wild strain at this temperature causes no such effect (Table 5). Other controls that have been run have indicated that the temperature-induced derepression of $valS^{ts}$ mutants is restricted to valine-controlled enzymes and requires protein synthesis [9].

Table 5. *Effect of temperature on biosynthetic enzyme repression in vals[ts] mutants*[a]

| Strain | Temp. °C | Acetolactate synthetase units/mg protein |
|---|---|---|
| Wild | 28 | 0.42 |
|  | 35.5 | 0.59 |
| $valS^{ts}$ | 28 | 0.36 |
|  | 35.5 | 5.00 |

[a] From EIDLIC [6]. All cultures were grown in glucose minimal medium supplemented with 100 $\mu$g/ml of L-valine, 50 $\mu$g/ml each of L-isoleucine and L-leucine, and 1 $\mu$g/ml of pantothenic acid. The values at 28° are steady state specific activities; the values at 35.5° are specific activities one h after a shift from 28° to 35.5°.

This result unequivocally implicates valyl RNA synthetase in valine-mediated repression. The result has been confirmed by YANIF and GROS [19] with similar mutants, and is consistent with the recent work of FREUNDLICH [9a] with valine analogs. Furthermore there is a parallel in the evidence of ROTH and AMES [15] that histidyl RNA synthetase is involved in repression of the histidine operon. The latter finding is of particular interest to us for it indicates that synthetase-mediated repression is not restricted to multivalent control systems (such as the leucine-valine-isoleucine-pantothenate pathway control).

Implication of tRNA itself in repression does not follow from our $valS^{ts}$ results. Strictly speaking the behavior of these mutants says only that some thermally-induced modification of valyl tRNA synthetase that reduces its ability to charge $tRNA_{val}$ for protein synthesis, reduces its ability to make whatever derivative of valine is the active repressor. This derivative might be valyl-$tRNA_{val}$, it might be a similar molecule, or it might be a valine peptide, or a complex involving the synthetase itself. The fact that the overall charging level of $tRNA_{val}$ drops in $valS^{ts}$ mutants at growth restricting temperatures does not prove that it is $tRNA_{val}$ charging that is important in repression, but it is consistent with such a notion. The evidence of SILBERT et al. [17] about $tRNA_{his}$ mutants has been taken to implicate this molecule in histidine-mediated repression.

The behavior of cells with a temperature-sensitive phenylalanyl RNA synthetase is totally unlike the valine mutants. There is no derepression of 3-deoxy-D-arabinoheptulosonic acid 7-phosphate (DAHP) synthetase at growth-restricting temperatures (Table 6). This result is consistent with the work of RAVEL, WHITE, and SHIVE [14] with tyrosine analogs, and suggests that repression of enzymes in the aromatic amino acid pathway does not require aminoacyl RNA synthetase participation (a conclusion that appears to be true for the arginine pathway as well).

Table 6. *Effect of temperature on biosynthetic enzyme repression in pheS$^{ts}$ mutants*[a]

| Strain | Medium | Temp. ° C | DAHP synthetase units/mg protein |
|---|---|---|---|
| Wild | minimal | 30 | 9.7 |
| | minimal | 36.5 | 11.0 |
| | minimal + phe | 30 | 2.6 |
| | minimal + phe | 36.5 | 2.6 |
| pheS$^{ts}$ | minimal | 30 | 3.5 |
| pheS$^{ts}$ | minimal | 36.5 | 3.2 |
| pheS$^{ts}$ | minimal + phe | 30 | 11.0 |
| pheS$^{ts}$ | minimal + phe | 36.5 | 11.0 |

[a] From EIDLIC [6]. Both wild and *ts* cells were $RC^{st}$. L-Phenylalanine, where added, was at 50 $\mu$g/ml. The enzyme levels at 30° are steady state specific activities; the levels at 36.5° are specific activities after one h of growth at 35.5°.

Nevertheless the story is not yet complete, and one need only turn to the apparently complete tRNA$_{phe}$ charging at growth restricting temperatures in *pheS$^{ts}$* mutants to appreciate some of the difficulties one encounters in proving that a particular synthetase is not involved in repression. It will be very difficult to establish that a certain synthetase does not mediate repression if one accepts the possibility that this role may not always involve the tRNA species operative in protein synthesis. The demonstration will have to come, I suppose, through evidence of continued repression in cells possessing a synthetase incapable of binding its amino acid substrate under particular conditions.

Work in our laboratory with *valS$^{ts}$* mutants that have either relaxed or stringent control over RNA synthesis has turned up one additional fact of some interest: biosynthetic enzyme derepression, just as induced enzyme formation, is severely impaired in relaxed cells. When the experiment described in Table 5 is performed with a *val$^{ts}$ $RC^{rel}$* cell, no derepression of acetolactate synthetase occurs at growth restricting temperatures [6].

## Nature of the *pheS^{ts}* Lesion and Structure of its Synthetase

Temperature inactivation of phenylalanyl RNA synthetase in *pheS^{ts}* mutants appears to be a consequence of dissociation of the enzyme into subunits with impaired catalytic activity.

Several determinations have been made of the molecular weight of *E. coli* phenylalanyl RNA synthetase (e.g., [18]), and these determinations all indicate a value of approximately 180,000 making this enzyme close to double the size of *E. coli* synthetases for other amino acids (reviewed in [12]). Dr. Böck [1] has determined that the phenylalanine enzyme in NP3 has a MW of approximately 180,000 by sucrose density gradient centrifugation and 210–220,000 by gel filtration. Immunological analysis of the mutant extracts by means of antiserum prepared against purified normal phenylalanyl RNA synthetase shows no cross reacting material (CRM) at the position where the native enzyme sediments (or elutes), but increased CRM in fractions expected to contain molecules of MW 100 to 110,000 (gel filtration). These findings suggested to Dr. Böck that the normal enzyme is composed of subunits that are less tightly associated in *pheS^{ts}* mutants. His guess has recently received additional support from the demonstration that the normal enzyme can be dissociated into subunits of MW 100,000 by means of urea or guanidine hydrochloride (A. Böck, ms in preparation). These subunits, like the residual enzyme protein in *pheS^{ts}* mutants have immunologic activity, but probably little or no enzymatic activity *in vitro*.

It is interesting to combine this information with the *in vivo* behavior of the *ts* phenylalanyl RNA synthetase. If our first explanation of the high level of tRNA_{phe} charging in *pheS^{ts}* cells at 40° is correct, then it appears that dissociation of this enzyme may differentially affect the charging of a minor tRNA_{phe} species (or, alternatively, this species, though minor in amount, is the major one used in coli messages). On the other hand, if the second explanation is correct it suggests that dissociation of this enzyme interferes with some additional function of the synthetase.

## Effect of Phage

Another facet of aminoacyl RNA synthetase structure has come to light as a result of a study of phage-infected cells; apparently the valyl RNA synthetase can exist in more than one molecular form.

We have explored the use of the host cell's translating machinery in making viral protein. Our study began with phenylalanyl RNA synthetase because the existence of both *ts* and *Km* mutants provided an excellent opportunity to test the *in vivo* function of the host enzyme in phage protein synthesis [5]. First we compared phage production and the pattern of amino acid incorporation into phage protein in cells with a normal phenylalanyl RNA synthetase with that in cells with a modified enzyme that cannot incorporate $p$-fluorophenylalanine into bacterial protein. From this study it could be established that under normal conditions the host phenylalanyl RNA synthetase is responsible for the bulk of phenylalanine incorporation into phage T4 protein, because $p$-fluorophenylalanine is incorporated into T4 phage protein only if the host enzyme can attach this analog to tRNA. Measurements of viable phage production and of *in vitro* enzymatic activities corroborated this conclusion.

The next question was whether the host enzyme was really necessary for phage protein synthesis, and this point was settled by the use of a $pheS^{ts}$ mutant [5]. It could be demonstrated that viable phage production was virtually eliminated by incubation of T4-infected $pheS^{ts}$ cells at 40°. Synthesis of T4 protein is just as susceptible to temperature inhibition as is host protein synthesis.

It can therefore be said that the host enzyme must function in T4 protein synthesis and that it is responsible for at least the major if not all the incorporation of phenylalanine. Hydroxylapatite chromatography of infected and uninfected extracts revealed no gross change in the molecular form of the phenylalanyl RNA synthetase [5].

In contrast, the first T4 experiment [11] performed with a $valS^{ts}$ mutant revealed the surprising fact that after 5 min of infection T4 protein synthesis in a $valS^{ts}$ host was accelerated by a shift to 40°. This result was followed by the ready demonstration of a heat-stable valyl RNA synthetase appearing in extracts of infected $valS^{ts}$ cells. The new activity was low (not more than 10 % of the activity of a wild type uninfected cell) and was associated with a molecule eluting much later from hydroxylapatite columns than the normal *E. coli* enzyme. Subsequent work has revealed that it has a MW nearly double that of the normal enzyme as judged by sucrose density gradient centrifugation.

Recent studies have been aimed at learning the origin of the T4-induced valyl RNA synthetase activity. By sucrose density gradient centrifugation we have found that T4 infection of wild type *E. coli* results in the gradual loss (complete by 20 min) of the 90−100,000 MW valyl RNA synthetase activity and the gradual appearance (also complete by 20 min) of a new activity sedimenting as though it has a MW of 170 to 180,000. There is a constant proportionality between the loss of the light enzyme and the appearance of the heavy enzyme throughout this period; approximately one unit of heavy enzyme is gained for every 3 units of light enzyme lost (CHRISPEELS, BOYD, WILLIAMS, and NEIDHARDT, in preparation). This constant conversion factor suggests that the host enzyme is being converted during the early phase of virus infection into a new molecular form.

A density-shift experiment has demonstrated that this suggestion is correct. The experiment is outlined in Fig. 3, and consisted essentially of a pre-labeling of polypeptide chains made before infection with deuterium (from $D_2O$), followed by infection in $H_2O$-medium. The results of equilibrium centrifugation in $CsCl_2$ indicated that the new valyl RNA synthetase activity appearing during T4 infection consists of polypeptide chains made before infection.

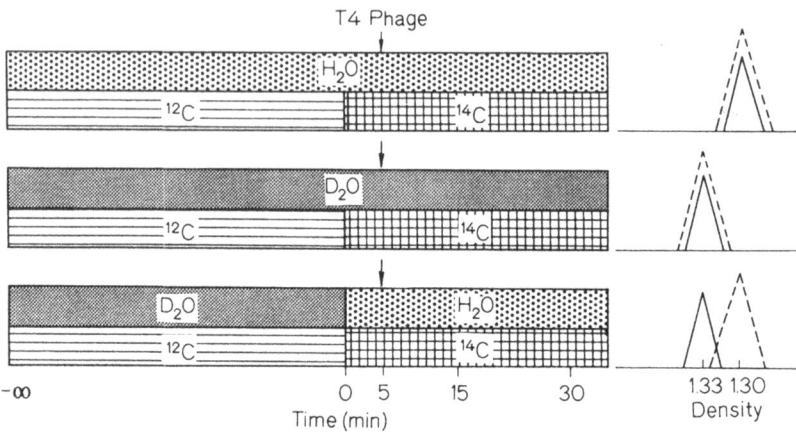

Fig. 3. Summary of an experiment designed to test the origin of the phage induced valyl RNA synthetase. The strain employed was a $valS^{ts}$ mutant. In one part of the experiment (upper third of figure) cells were grown in $H_2O$ medium and then infected with T4 5 min after placing them in a $^{14}C$-glucose medium. An extract made from cells 30 min after infection was analyzed by $CsCl_2$ equilibrium centrifugation and the banding of valyl RNA synthetase activity (which must be phage induced since the host enzyme is inactivated in extracts) and radioactivity (mostly protein made after infection) was determined. The two peaks overlapped at a density of approximately 1.30 (– – – – enzyme activity; ———— radioactivity). The middle third of the figure shows an analogous control run completely in 80% $D_2O$-medium; the enzyme and the phage protein band at a density of approximately 1.33. The lower third of the figure shows the critical experiment in which the $D_2O$-medium is replaced by $H_2O$ 5 min prior to infection. From CHRISPEELS, BOYD, WILLIAMS, and NEIDHARDT (unpublished experiments)

This result is reminiscent of the discovery that phenylalanyl RNA synthetase is composed of subunits, for the change in the valine enzyme during phage infection could be due to a doubling in MW, and may be the result of a dimerization (though this point has not yet been established). The phenylalanine enzyme dissociates into subunits upon thermal inactivation of $pheS^{ts}$ mutants and protein synthesis halts despite the charging of the major species of $tRNA_{phe}$. Does aggregation of the valine enzyme during T4 phage infection change its role in protein synthesis? This is the subject of our current work.

# Summary

The structural genes (*valS* and *pheS*) for valyl and phenylalanyl RNA synthetase in *E. coli* are not contiguous; they readily mutate to yield temperature-sensitive *(ts)* enzymes. The *ts* enzymes are generally unstable in cell free extracts. Both the valine and phenylalanine enzymes must be functional to permit RNA synthesis in stringent cells, but not in relaxed cells. The valine enzyme has been implicated in repression of biosynthetic enzymes, but the phenylalanine enzyme has not been. At growth restricting temperatures the overall charging level of tRNA drops in *valS^{ts}* mutants, but that of tRNA$_{phe}$ does not in *pheS^{ts}* mutants. The phenylalanine enzyme appears to be composed of subunits that are found dissociated in extracts of *pheS^{ts}* mutants. The phenylalanine enzyme is responsible for the incorporation of most if not all phenylalanine residues into T4 protein and is an indispensable enzyme for this process. The valine enzyme is converted by T4 infection into a form with a 50% greater *s* and, in the case of *valS^{ts}* mutants, an increased resistance to heat inactivation.

# Acknowledgements

The investigations reported from the author's laboratory were supported by grant GB 2094 from the National Science Foundation and by Public Health Service grant GM 08437 from the Institute of General Medical Sciences. Many colleagues have participated in this work; particular mention should be made of the contributions of Drs. A. Böck, M. Chrispeels, C. Earhart, L. Faiman, G. Nass, and M. Tingle.

# References

1. Böck, A.: Europ. J. Biochem. (in press).
2. —, and F. C. Neidhardt: Z. Vererbungsl. 98, 187—192 (1966).
3. — — Science 157, 78—79 (1967).
4. Dunn, T. F., and F. R. Leach: J. biol. Chem. 242, 2693—2699 (1967).
5. Earhart, C. F., and F. C. Neidhardt: Virology (1967) (in press).
6. Eidlic, L.: Ph. D. Thesis, Purdue University 1965.
7. Böck, A., L. E. Faiman, and F. C. Neidhardt: J. Bact. 92, 1076—1082 (1966).
8. Eidlic, L., and F. C. Neidhardt: J. Bact. 89, 706—711 (1965).
9. — — Proc. nat. Acad. Sci. (Wash.) 53, 539—543 (1965).
9a. Freundlich, M.: Science 157, 823—825 (1967).
10. Neidhardt, F. C.: Bact. Rev. 30, 701—719 (1966).
11. —, and C. F. Earhart: Cold Spr. Harb. Symp. quant. Biol. 31, 557—563 (1966).
12. Novelli, G. D.: Ann. Rev. Biochem. 36, 449—484 (1967).
13. O'Donovan, G. A., and J. L. Ingraham: Proc. nat. Acad. Sci. (Wash.) 54, 451—457 (1965).
14. Ravel, J. M., M. N. White, and W. Shive: Biochem. biophys. Res. Commun. 20, 352—359 (1965).
15. Roth, J. R., and B. N. Ames: J. molec. Biol. 22, 325—334 (1966).
16. —, D. N. Anton, and P. E. Hartman: J. molec. Biol. 22, 305—323 (1966).
17. Silbert, D. F., G. R. Fink, and B. N. Ames: J. molec. Biol. 22, 335—347 (1966).
18. Stulberg, M. P.: J. biol. Chem. 242, 1060—1064 (1967).
18a. Taylor, A. I., and M. S. Thoman: Genetics 50, 659—677 (1964).
19. Yanif, M., and F. Gros: Proc. Fed. Europ. Biochem. Soc., 3rd, (1966) (in press).
20. —, F. Jacob, and F. Gros: Bull. Soc. Chim. biol. (Paris) 47, 1609—1656 (1965).

## Discussion

*Zillig:* Does the modified enzyme contain any newly found material, for example as evidenced by $^{14}$C-incorporation?

*Neidhardt:* This is a good question because the result of the density shift experiment cannot tell us whether a small addition of phage-made material has occurred during the modification. To answer the question we must purify the modified enzyme to a high degree and then examine it for radioactivity; this work is in progress.

*Fincham:* What do you think is the nature of the modification of valyl-RNA synthetase in T4 infected cells?

*Neidhardt:* At the present time we know only that the modified enzyme is composed chiefly of parts made before infection, that it does not readily return to its original form, that it is stable both *in vivo* and *in vitro* even in a *valS*$^{u}$ host, and that the modification is completely blocked by chloramphenicol (100 $\mu$g/ml). Speculation as to the chemical nature of the modification will not serve too much purpose, since purification and analysis of the modified enzyme should settle the question very soon.

# The Formation of Phenylalanyl-tRNA-synthetase Crossreacting Material in Temperature sensitive Mutants of E. coli

By

GISELA NASS

Max Planck-Institut für Molekulare Genetik, Berlin, Germany

The physiological role of the aminoacyl-tRNA-synthetases has been presented by F. C. NEIDHARDT (1967). In his laboratory, as already mentioned, several temperature sensitive mutants of E. coli have been isolated which hardly grow at 37° but grow almost like the wildtype cells at 25° and which show no enzymatic activity of the phenylalanyl-tRNA-synthetase in vitro (cf. NEIDHARDT, 1966).

No measurable enzymatic activity in vitro can be due to at least two reasons: either enzymatically inactive protein is still present in the mutants or there is a regulatory mutation which prevents enzyme formation.

Antibodies against enzymes are tools to investigate these two possibilities since the antibodies will bind to enzymatically inactive protein, but no antibody binding will occur if a regulatory mutation inhibits enzyme synthesis.

Antibodies against the phenylalanyl-tRNA-synthetase of E. coli have been previously prepared; they react with the enzyme by neutralizing its activity (FANGMANN, NASS, and NEIDHARDT, 1965). The first slide (Fig. 1) shows the neutralization curve for the phenylalanyl-tRNA-synthetase of E. coli. This curve is obtained when a constant amount of crude extract is treated with increasing amounts of antibody. The remaining enzyme activity is plotted against the amount antiserum used. The enzymatic activity in this experiment and all the following ones is measured by the attachment of the aminoacid to tRNA (BERG et al., 1961; NASS and STÖFFLER, 1967).

A summary of some of the results reported here has been described earlier (NASS, 1967a).

The presence of enzymatically inactive protein, the socalled crossreacting material (CRM), is determined in a way similar to a method

10*

described by SUSKIND (1957): A constant amount of phenylalanyl-tRNA-synthetase of the wildtype cells is incubated with an amount of antiserum which neutralizes the wildtype enzyme almost completely.

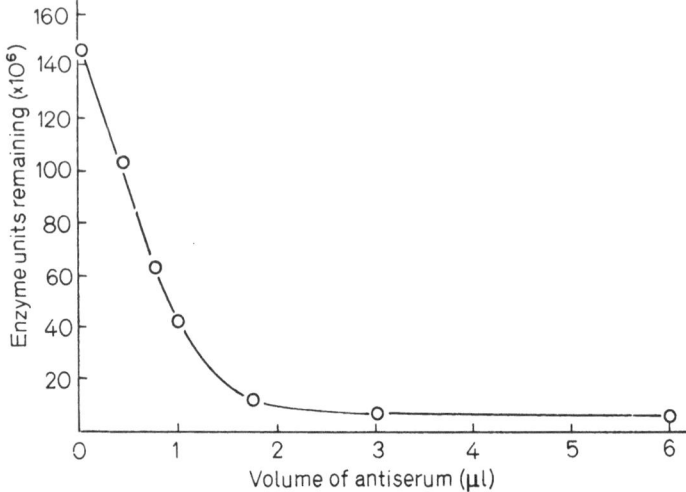

Fig. 1. *Neutralization curve for the phenylalanyl-tRNA-synthetase of E. coli KB.* A constant amount of crude extract of E. coli KB (0.5 ml, 20 μg protein) was incubated for 20 min at 37° with increasing amounts of antiserum. 0.1 ml of the incubation mixtures was used as enzyme source for the enzyme assay; the preparation of the crude extract and the enzyme assay was performed as previously described (BERG et al., 1961; FANGMANN et al. 1965; NASS, 1967b)

Preabsorbing the antiserum with phenylalanyl-tRNA-synthetase CRM will, however, lead to binding of antibodies to the CRM and this means less antibody is available for neutralizing the wildtype enzyme. Or in other words: enzyme acitivity will be released by the presence of CRM. The more CRM there is, the more antibody is bound and the more enzyme activity is released. This allows a quantitative measurement of the CRM. The second slide (Fig. 2) shows that preabsorbing a constant amount of antiserum with increasing amounts of CRM, expressed in μg protein per ml crude extract, leads to an increasing release of enzymatic activity of the wildtype enzyme. So far the test conditions. The question asked was: is there phenylalanyl-tRNA-synthetase CRM in the temperature-sensitive mutants which show no enzymatically active phenylalanyl-tRNA-synthetase in vitro? Three mutants, isolated by NEIDHARDT and his coworkers (cf. NEIDHARDT, 1966) were assayed and all of them showed the presence of CRM in an amount comparable to the wildtype enzyme, when grown at none-growthrestrictive temperature, indicating a mutation in the structural gene of this enzyme.

The next question asked was the following: If the temperature-sensitive mutants are grown during growthrestrictive temperatures is there any variation in the amount of phenylalanyl-tRNA-synthetase

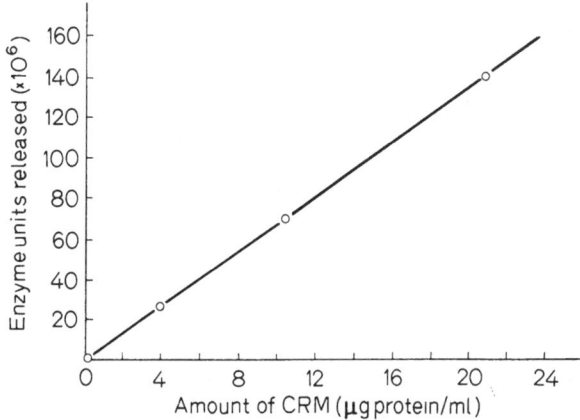

Fig. 2. *Quantitative determination of phenylalanyl-tRNA-synthetase-CRM in crude extracts of E. coli IV-4.* Increasing amounts of crude extract of E. coli IV-4 (0.5 ml containing $10-100$ µg protein) were incubated with a constant amount of antiserum (1.5 µl) for 20 min at 37°. Subsequently the incubation mixtures were chilled in an ice bath and to each testtube a constant amount of wildtype crude extract was added (0.5 ml of E. coli KB crude extract containing 20 µg protein). The incubation mixtures were incubated for further 20 min at 37° and then chilled in an ice bath. The phenylalanyl-tRNA-synthetase activity in the incubation mixtures was determined by using 0.1 ml of each for the enzymes assay; for preparation of the crude extract and performing the enzyme assay see Fig. 1; the mutant cells E. coli IV-4 were grown at a none-growthrestrictive temperature of 25°

CRM present in the cells? Crude extracts were prepared from the mutant IV-4 grown during none-growthrestrictive temperatures and during growth-restrictive temperatures. The third slide (Tab. 1) shows

Table 1. *Growth rate of E. coli KB and IV-4 during various growth temperatures.* Cultures of E. coli KB and IV-4 were grown at 25° for at least two generation times. Subsequently the cultures were shifted to a growth temperature of 32°. After growth for two generation times at this temperature the cultures were shifted to 38° for further growth. Rich growth media were used; the media composition and determination of the growthrate constants were performed as previously described (NASS, 1967 b)

| Temperature | K (gen./h) | |
|---|---|---|
| °C | Wild | Mutant |
| 25 | 0.83 | 0.82 |
| 32 | 1.18 | 0.58 |
| 38 | 1.33 | 0.39 |

that at 25° the growth rate expressed in k-values, generations per hour, of this mutant and of the wildtype are almost identical. At 32° a growth restriction of the mutant is observed, which is more pronounced at 38°. The amount of CRM present in the mutant grown at these temperatures was determined (slide 4 = Fig. 3): when grown at 25°, the none growth restictive temperature, a certain amount of CRM per μg protein extract of the cells was present. The shift of the cells to 32° resulted in an increase of the CRM, since less crude extract protein was necessary to release the same amount of enzyme units. And at 38° there was about three times as much CRM per μg protein in the cells as when they were grown at 25°.

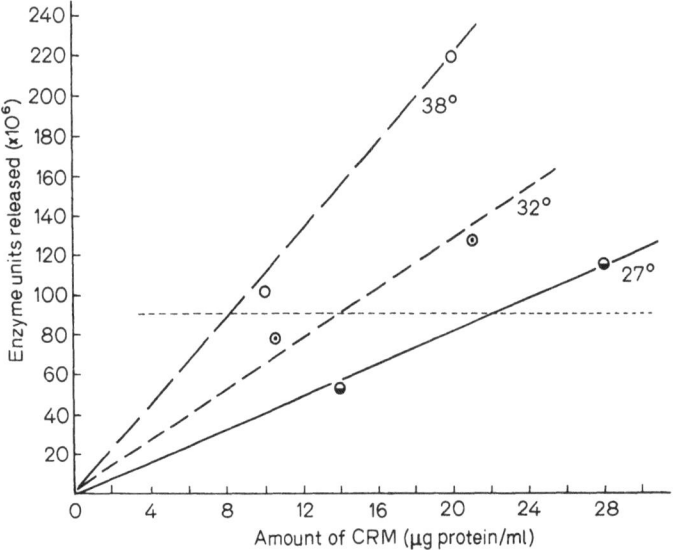

Fig. 3. *Quantitative determination of phenylalanyl-tRNA-synthetase-CRM in E. coli IV-4 grown at various temperatures.* Cultures of E. coli IV-4 were grown at 25° for at least two generation times. Subsequently half of the cells were harvested, the other half was shifted to a growth temperature of 32° and growth was continued for further two generations; then the culture was again divided into two parts, one harvested, the other grown for more than two generation times at 38° before harvesting the cells. E. coli KB, the wildtype was grown in rich media at 37°. The media composition, preparation of crude extracts and the determination of phenyl-alanyl-tRNA-synthetase-CRM are described in Fig. 2

This increase of the CRM in the mutant could have been due to either the synthesis of more enzyme protein or to the derival of CRM from precursors which do not bind the antibodies. Therefore chloramphenicol was added when the growth temperature of the mutant was shifted from 25° to the growthrestrictive temperature. The fifth slide (Fig. 4) shows that the addition of chloramphenicol prevented the increase of the phenylalanyl-tRNA-synthetase CRM after the temperature shift. In the control where no chloramphenicol was added the described increase in the amount of CRM per µg protein occurred. This means that protein synthesis is necessary for the increase in CRM to occur.

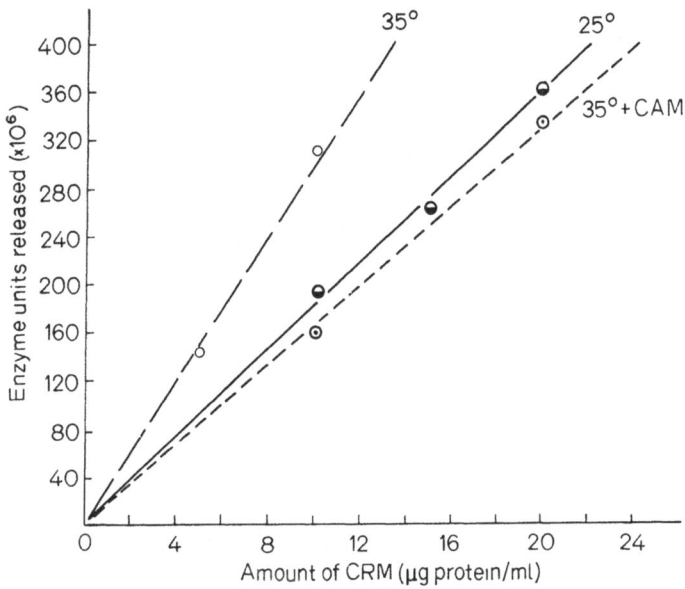

Fig. 4. *Effect of chloramphenicol on the formation of phenylalanyl-tRNA-synthetase CRM in E. coli IV-4.* The experimental details are the same as described in Fig. 3. However, the cells were grown at only two growth temperatures; and when shifting the cells from 25° to 35° the cultures were divided into three parts: one part harvested, to another part chloramphenicol was added before growth of the cells at 35° was started (100 µg/ml final concentration)

The same experiments were performed using two other phenylalanyl-tRNA-synthetase mutants (C-12, V-5). In these two mutants no increase of the CRM was observed after shifting the cells from nongrowth-restrictive (25°) to growthrestrictive temperatures (35°).

Meanwhile, using the antiserum, sucrose gradient and gelfiltration-techniques, A. Böck (1967) has shown that the molecular weight of the CRM in the temperature sensitive mutants is about half the one of the wildtype phenylalanyl-tRNA-synthetase; this is the first evidence that some aminoacyl-tRNA-synthetases might consist of subunits. Berg and his coworkers indicated for the purified isoleucyl- and tyrosyl-tRNA-synthetase of E. coli the existence of only one polypeptide chain and they determined the molecular weights of these enzymes, being

112,000 and 95,000, respectively (Baldwin and Berg, 1966; Calendar and Berg, 1966); values similar to the one found for the crossreacting material of phenylalanyl-tRNA-synthetase (Böck, 1967). From these results the possibility can be deduced that only aminoacyl-tRNA-synthetases with a high molecular weight like the wildtype phenylalanyl-tRNA-synthetase (Stulberg, 1967) of E. coli might consist of subunits.

Therefore the molecular weight of 18 aminoacyl-tRNA-synthetases in E. coli has been determined by means of gelfiltration on Sephadex G-200 (Nass and Stöffler, 1967). The sixth slide (Fig. 5) shows that the

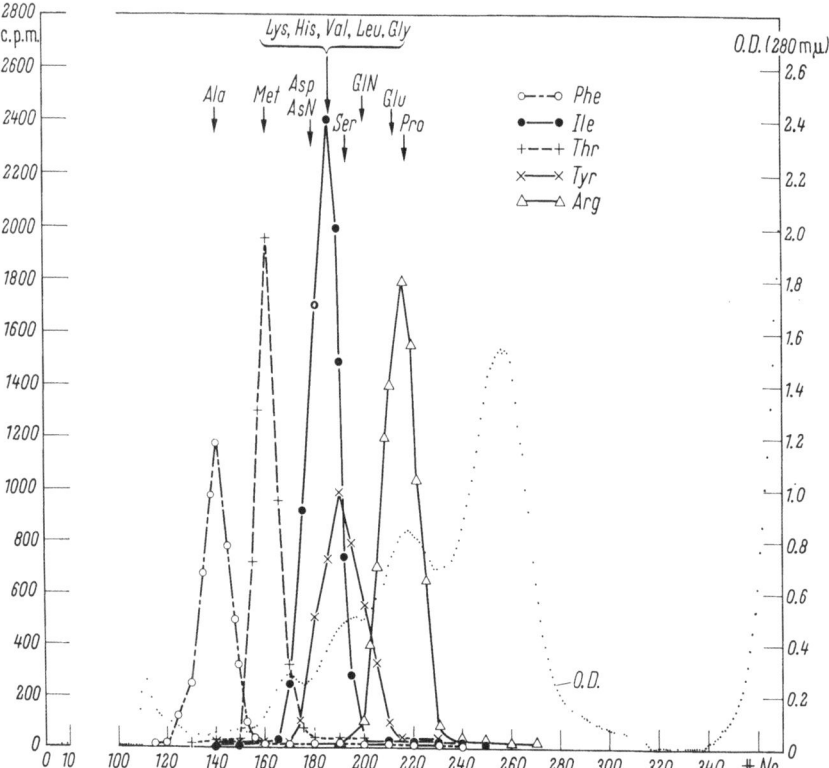

Fig. 5. *Molecular weight distribution of 18 aminoacyl-tRNA-synthetases of E. coli B.* E. coli B was purchased from Fallek Products, New York (USA). Crude ribosome free extract of these cells prepared according to Traub and Zillig, 1966 was layered onto a Sephadex G-200 column (4.5 × 200 cm, equilibrated with 0.01 molar potassium phosphate buffer, pH 7.3, containing 0.005 molar β-mercaptoethanol); the aminoacyl-tRNA-synthetases were eluted with the same buffer and the activity of various of these enzymes was determined as previously described (Berg et al., 1961; Nass and Stöffler, 1967). The molecular weights were estimated according to the position of the maximum activity of these enzymes compared with the elution pattern of the aminoacyl-tRNA-synthetases of which the molecular weights are known from ultracentrifugations analysis (cf. Nass and Stöffler, 1967)

molecular weight of the aminoacyl-tRNA-synthetases varies between 180,000 as for the phenylalanyl- and alanyl-tRNA-synthetase and about 75,000 for the arginyl-, prolyl-, and glutamyl-tRNA-synthetase. Ultracentrifugation analyzes of the purified arginyl-tRNA-synthetase resulted in a molecular weight of 74,000, confirming the value estimated from gelfiltration (STÖFFLER and NASS, unpublished results).

## In Summary

Three temperature sensitive mutants isolated by F. C. NEIDHARDT and his coworkers (cf. NEIDHARDT, 1966) which show no measurable phenylalanyl-tRNA-synthetase activity in vitro are investigated for the presence of CRM. Enzyme protein is found which binds corresponding antibodies; this indicates that the mutations are in a structural gene for the phenylalanyl-tRNA-synthetase and not in a regulatory gene for this enzyme in the investigated mutants. In one of the three mutants (IV-4) the CRM is increased about three times when the cells are shifted to growthrestrictive temperatures. Protein synthesis is necessary for this increase of the CRM to occur as shown by means of chloramphenicol. Investigations by A. BÖCK (1967) about the reduced molecular weight of the phenylalanyl-tRNA-synthetase CRM in one of the mutants and experiments about the distribution of the molecular weight of 18 aminoacyl-tRNA-synthetases in E. coli wildtype cells (G. NASS and G. STÖFFLER, 1967) are mentioned. The distribution of the molecular weights of these enzymes might lead to looking for subunits in aminoacyl-tRNA-synthetases with high molecular weights as the alanyl-, methionyl- and threonyl-tRNA-synthetases and it indicates different structures of the aminoacyl-tRNA-synthetases though their substrates and functions are very similar.

## References

BALDWIN, A. N., and P. BERG: Purification and properties of Isoleucyl ribonucleic acid synthetase from Escherichia coli. J. biol. Chem. **241**, 881 (1966).

BERG, P., F. H. BERGMANN, E. J. OFENGAND, and M. DIECKMANN: The enzymic synthesis of amino acyl derivatives of ribonuleic acid I. The mechanism of leucyl-, valyl-, isoleucyl-, and methionyl ribonucleic acid formation. J. biol. Chem. **236**, 1726 (1961).

BÖCK, A.: Studies on mutant phenylalanyl RNA synthetases of Escherichia coli. Europ. J. Biochem. **2**, 165 (1967).

CALENDAR, R., and P. BERG: Purification and physical characterization of tyrosyl ribonucleic acid synthetases from Escherichia coli and Bacillus subtilis. Biochemistry **5**, 1681 (1966).

FANGMAN, W. L., G. NASS, and F. C. NEIDHARDT: Immunological and chemical studies of phenylalanyl sRNA synthetase from Escherichia coli. J. molec. Biol. **13**, 202 (1965).

NASS, G.: Phenylalanyl-tRNA-synthetase. Cross-reacting material in temperatursensitive mutants of Escherichia coli. Bact. Proc. **1967**a, 125.

— Regulation of histidine biosynthetic enzymes in a mutant of Escherichia coli with an altered histidyl-tRNA synthetase. Molec. and Gen. Genetics **100**, 216 (1967b).

—, and G. STÖFFLER: Molecular weight distribution of the aminoacyl-tRNA-synthetases of Escherichia coli by gel filtration. Molec. and Gen. Genetics **100**, 378 (1967).

NEIDHARDT, F. C.: this symposium.
— Roles of amino acid activating enzymes in cellular physiology. Bact. Review 30, 701 (1966).
STULBERG, M. P.: The isolation and properties of phenylalanyl ribonucleic acid synthetase from Escherichia coli B. J. biol. Chem. 242, 1060 (1967).
SUSKIND, S. R.: Properties of a protein antigenically related to tryptophan synthetase in Neurospora crassa. J. Bact. 74, 308 (1957).
TRAUB, P., and W. ZILLIG: Untersuchungen zur Biosynthese der Proteine VI. Eine neue Methode zur Darstellung eines zellfreien Systems aus Escherichia coli und deren Eigenschaften in der nucleinsäureabhängigen Proteinsynthese. Hoppe-Seyler's Z. physiol. Chem. 343, 246 (1966).

## Discussion

*Neidhardt:* The data you have shown indicate that alanine and phenylalanine have the two largest RNA synthetases, and these are two of the only three synthetases for which temperature sensitive *ts* mutants have been isolated. Dr. BÖCK has shown that the phenylalanine mutants have a CRM approximately one half the size of the normal enzyme. Furthermore I have mentioned the finding in our laboratory of a phage-induced conversion of the valine enzyme into a form apparently twice the normal size. Do you think the higher frequency of *ts* mutants for alanine, valine and phenylalanine than for other synthetases might be the result of a particularly fragile quaternary structure, to begin with?

*Nass:* That's how it looks.

# Host-Controlled Modification of DNA

By

WERNER ARBER

Laboratoire de Biophysique, Université de Genève (Suisse)

The study of host-specific modification and restriction of DNA has its root in a series of independent observations, made some 15 years ago by several groups of authors who worked with various bacteriophages (see ARBER, 1965). Let us recall the results of the experiments made by BERTANI and WEIGLE (1953) with bacteriophage $\lambda$. This phage is usually grown on strains of *E. coli* K 12, and we call it then $\lambda$.K. Phage $\lambda$.K also grows with full efficiency of plating on *E. coli* strain C, but if one tries to infect again strain K 12 with $\lambda$.C, only a low proportion, or some $4 \times 10^{-4}$ of the phage give rise to the production of a progeny; the others are said to be restricted by K 12. Phage isolated from the rare plaques grown on K 12 show to be modified $\lambda$.K since they grow again with full efficiency on K 12. They cannot be explained as being host range mutants, since the ability to grow on K 12 is not maintained upon consecutive multiplication of the phage on strain C.

In the last 7 years a series of experiments were undertaken in our and other laboratories that were designed to explore the molecular mechanisms of host-specific restriction and modification of phage $\lambda$. Since these experiments were reviewed recently (ARBER, 1965, 1968) I would like to summarize here only some of the most relevant conclusions. It was found that it is the DNA molecule of phage $\lambda$ on which host-specific modification and restriction are exerted. It also soon became obvious that both restriction and modification are under the control of bacterial genes located either on the bacterial chromosome, as is the case for the genes involved in K- and B-host specificity, or else on plasmids such as certain resistance transfer factors or even prophage P 1. But the phage $\lambda$ itself, with which the system was explored, plays an absolutely passive role in providing only its DNA as a convenient substrate in the restriction and modification reactions. Many other phages undergo the same kind of host-specific modification and restriction as $\lambda$. Furthermore, bacterial DNA itself is also submitted to these reactions, an effect which becomes important in any type of exchange of genetic information between two bacterial strains of different specificity types. In bacterial conjugation,

for example, female cells of strain K 12 will accept donated DNA only
from males which provide the DNA with K-specific modification. Any
other DNA is submitted to restriction which seems to give rise to destruction of most of the DNA molecules penetrating into the acceptor cells.
These observations may reasonably explain the biological function of
host-specific restriction and modification, which could consist in the
provision of bacterial strains with a relatively efficient system of self-
protection against infection by any kind of DNA except the type
corresponding to their own specificity. This barrier would prevent most
of undesirable virus and other plasmid infections as well as exchange
of bacterial DNA between unrelated strains.

But let us come back to the molecular basis of the restriction and
modification reactions. We know that restriction of phage λ leads to a
quite rapid and rather extensive degradation of the restricted DNA
molecule. It is most probable that the observed acid soluble breakdown
products are a consequence of more than one reactions occurring in series
and that the initial restriction event consists in a less drastic reaction,
perhaps such as a cutting of the DNA molecule.

The mechanism of modification is better characterized than that of
restriction. DNA molecules are provided with host-specific modification
in a process independent of DNA replication, and also independent of
whether the same DNA molecule already carries another type of host-
specific modification. We also know that host specificity remains stably
associated with the DNA. This association is ruptured neither upon
chemical purification, nor during replication of the DNA, in which
parental DNA strands maintain their parental host specificity even if the
host cell is not equipped to produce this same type of modification. *In vivo*
modification is dependent on the presence of methionine, but not on that
of several other amino acids tested.

All these observations could best be explained if host-specific modification consisted in the methylation of certain bases of the DNA. This
seems indeed to be the case, although it was first difficult to prove
because *E. coli* DNA and the DNA of phage λ are methylated to a level
that is more than 20 fold higher than the one recognized now as responsible
for host-specific modification. The biological functions of this majority
type of methylation still remain unknown. We were only able to track
methylation specific to *E. coli* strains carrying the genetic markers for
B-specific modification when we started to work with phage fd (HOFF-
MANN-BERLING, MARVIN, and DÜRWALD, 1963). This phage has a single
stranded DNA molecule of only about 6000 nucleotides. It undergoes
host-specific restriction and modification in B strains. In using several
hybrid strains between K 12 and B we could show that the same bacterial
genes are responsible for B-specific restriction and modification of fd and

of λ (ARBER, 1966). In fact our most frequently used B strain 2027 is such a hybrid containing a K 12 chromosome in which the region with the markers for threonine biosynthesis and for B-specific restriction and modification have been transduced from strain B as a donor. The other bacterial strain used in the experiments which I am going to describe is a nonrestricting and nonmodifying mutant of *E. coli* K 12, strain 993, called also 0 for its inability to provide host specificity to any DNA.

In collaboration with Dr. J. D. SMITH, we measured the methylation of phage fd grown either on the B or on the 0 strains, both of which are methionine auxotrophs and to which we gave methionine (methyl-C 14) as methyl donor during phage growth. The phage DNA was extracted, hydrolysed and the radioactive bases identified by chromatographic or electrophoretic techniques. It was then found that fd DNA is much less methylated than *E. coli* or phage λ DNA (SMITH and ARBER, unpublished results): only about one to one and a half 6-methylamino-purines (6MAP) could be detected per DNA molecule of fd. 0 and certainly less than one 5-methylcytosine, if any. Secondly fd. B DNA definitely showed more methylation than fd. 0: about 3 6MAP per DNA molecule. Our interpretation for these results are that the level of methylation encountered in fd. 0 represents a type of methylation the function of which we do not know. Since the 0 and B hosts used are to a large extent isogenic, it is reasonable to assume that the same level of background methylation is imprinted on fd. B DNA as on fd. 0 DNA. The difference of some $1^{1}/_{2}$ or 2 extra 6MAP carried by fd. B DNA could then indeed be caused by B-specific modification. On the other hand, one must postulate that in a phage population grown on B all the DNA molecules are fully modified as none of them encounter restriction upon infection of B. It could then appear that the fd DNA molecule has two, or perhaps only one B-specific sites on which modification can act.

One of the most appealing hypothesis to explain the molecular structure of a specificity site is the assumption that a very particular sequence of bases determines which adenine residues undergo methylation in host-specific modification. It is also reasonable to assume that the same specific sites are recognized in modification and in restriction. As a matter of fact, both restriction and modification are exerted on DNA whose specific sites are not as yet methylated. Restriction and modification would then be very similar reactions at least with respect to the recognition of a specificity site. It is thus not improbable that one single gene product is involved in such site recognition and I would like to call this product S and the corresponding genetic information *s*. One step mutants in this information *(s⁻)* are expected neither to restrict nor to modify. Bacterial mutants which perform neither of these functions are indeed found to occur at frequencies typical for one step mutations (WOOD, 1966). It is not known yet, if the active restriction and modification enzymes are composed of more than one different subunits, one of which might be the product S, or if they are formed by a single gene product, a part of which could assume the site recognition function. These questions should be resolved soon experimentally in complementation studies.

I would now like to make some predictions based on the presented model. Since the same DNA molecules may undergo in an independent way several different types of modification it is reasonable to admit that host specific base sequences occur at random in the DNA molecules, rather than to occupy special loci of the genomes. As far as the genome of phage $\lambda$ is concerned, we showed by analysis of genetic recombinants that specificity sites are distributed over the DNA molecule. A rough estimation on the average occurence of sites for one given type of host specificity was made elsewhere (ARBER, 1968) and gave about one site per each $10^4$ desoxyribonucleotides of a genome. Some 6 to 7 bases would then form the specificity site, which is a quite reasonable length for a segment of a DNA molecule that we postulate to be recognized specifically by an enzyme.

A genetic mutation occurring within a specificity site of a DNA molecule, for example by a base substitution, would presumably alter the base sequence so that it is no longer recognized as specificity site. Working with phage fd, we were indeed able to isolate mutants that are neither restricted nor modified in strain B (ARBER and KÜHNLEIN, 1967). Phage fd. 0 grows with a probability of $7 \times 10^{-4}$ on B. We enriched less restricted mutants occurring in stocks of fd in growing the phage for limited times alternatively on strain B and on strain 0. Each infection of B provided a relative enrichment for phages less restricted than wild type fd, while the growth on strain 0 was necessary in order to reobtain phages that did not carry B-specific modification. In this way we repeatedly found mutants that grow with an efficiency of about $3 \times 10^{-2}$ on B. Submitting these mutants again to the same enrichment procedure as before, we found another mutant type which is not restricted at all by B, although its restriction by the unrelated P 1-specificity is unchanged, hence still the same as for wild type fd. We think that the fd genomes that are unrestricted in B are double mutants, since they were isolated in two discrete steps, and that they result from consecutive loss of both of two B-specific sites. We hope to be able to locate these sites on the genetic map of fd and to find thus definite confirmation for the existence of two well defined B-sites on the fd DNA molecule, as we had first postulated on the basis of the methylation measurements.

One of the unrestricted two step mutants of fd was grown on B and its methylation measured as was described above. As we had expected the level of 6 MAP was found to be the same as that of fd. 0, definitely lower than that of fd. B. It seems then that this double mutant has not only the susceptibility to B-specific restriction, but that it also has lost the ability to accept the B-specific methylation, i.e. it does not undergo B-specific modification.

Finally, I would like to comment briefly on in vitro experiments that Dr. S. LINN is working on in our laboratory. He undertook to prepare bacterial extracts, to fractionate them and to search for restricting and modifying activities using the double stranded replicative form (RF) of fd as substrate. The fdRF can be assayed for its biological activity by incubation with host cell spheroplasts, which give rise to the production of a phage progeny if they are successfully infected with an fd genome. It was found that extracts of B strains contain a restricting activity that reduces the biological activity of fdRF. 0, isolated from cells of strain 0 infected with fd, by a factor of several hundred fold, while it leaves fdRF. B unaffected. Extracts of strains of type 0 do neither restrict fdRF.0 nor fdRF.B, but those of P1 lysogenic strains restrict both, phage fd being restricted by cells carrying P1 prophage. The restriction *in vitro* is seen only in presence of $Mg^{++}$, S-adenosylmethionine and ATP. The requirement for S-adenosylmethionine could perhaps indicate that methylation plays also some role in the restriction reactions. It is to hope that this work will soon permit to define precisely the conditions under which restriction as well as modification act *in vitro* and thus open new approaches to solve many still remaining questions posed by the host-specific restriction and modification.

### References

ARBER, W.: Host-controlled modification of bacteriophage. Ann. Rev. Microbiol. 19, 365–378 (1965).
— Host specificity of DNA produced by *Escherichia coli*. 9. Host-controlled modification of bacteriophage fd. J. molec. Biol. 20, 483–496 (1966).
— Host-controlled restriction and modification of bacteriophage. Symp. Soc. Gen. Microbiol. 18, 295–314 (1968).
—, u. U. KÜHNLEIN: Muationeller Verlust B-spezifischer Restriktion des Bakteriophagen fd. Path. Microbiol. 30, 946–952 (1967).
BERTANI, G., and J. J. WEIGLE: Host-controlled variation in bacterial viruses. J. Bact. 65, 113–121 (1953).
HOFFMANN-BERLING, H., D. A. MARVIN u. H. DÜRWALD: Ein fädiger DNS-Phage (fd) und ein sphärischer RNS-Phage (fr), wirtsspezifisch für männliche Stämme von *E. coli*. 1. Präparation und chemische Eigenschaften von fd und fr. Z. Naturforsch. 18b, 876–883 (1963).
WOOD, W. B.: Host specificity of DNA produced by *Escherichia coli*. 7. Bacterial mutations affecting the restriction and modification of DNA. J. molec. Biol. 16, 118–133 (1966).

### Discussion

*Hoffmann-Berling:* You assume that a complex structure accomplishes both the modificative and the restrictive function. Such a complex would act on newly synthesized DNA in a (predominantly) modifying, on incoming DNA in a predominantly restricting manner. How do you interpret the difference?

*Arber:* One can forward various hypotheses to answer this question, as for example that restriction is preferentially exerted with enzyme at the cell surface. On the other hand it is possible that the chance is quite small to cut DNA while it is being synthesized and that under such conditions modification is quite effective.

*Winkler:* Did you measure the number and size of DNA fragments after *in vitro* restriction of fd-DNA?

*Arber:* Yes, indeed we started to do such measurements in the electron microscope and we seem to obtain two fragments of unequal size.

*Hofschneider:* Having only two sites of restriction you can consequently expect only two hits in fd RF under restrictive conditions. In your *in vitro* experiments RF is cut into two pieces of 1/3 and 2/3 of original length. Since 2 single strand hits could hardly give this result, does this mean that you have double-strand breaks or have you studied single-stranded products?

*Arber:* We do not know if the observed cutting in the restriction is a one-or multistep-reaction.

*Beckwith:* Is the double mutant of fd which is nonrestricted by B restricted by K?

*Arber:* No, and wild type fd is not restricted either by K12, and we interpret this finding as indicating, that fd DNA has no K-specific site. Sites for K- or B-specific modication seem indeed to be different from each other.

*Beckwith:* What explanation is there for Hurwitz's findings that methylation is not responsible for modification of $\lambda$.

*Arber:* These experiments were done under conditions of superinfection with UV irradiated T3 and it is uncertain how this could influence the results of the experiments.

*Ristow:* How do you prevent DNA degradation by nucleases in your *in vitro* system and what ionic strength do you use in your assay?

*Arber:* We first used Dr. HOFFMANN-BERLINGS endonuclease I — less strains and then learned to find experimental conditions to minimize nonspecific DNA degradation. Unfortunately I cannot remember the precise optimal conditions used by Dr. LINN in his assays.

# Genetic Recombination Induced by Ultraviolet Light

By

Paul Howard-Flanders, Brian M. Wilkins, and W. Dean Rupp

Departments of Radiology and Molecular Biophysics
Yale University, New Haven, Connecticut, U.S.A.

## Introduction

When bacteria are exposed to ultraviolet (UV) light, the principal photoproducts to be formed in their DNA are pyrimidine dimers. These dimers are of the cyclobutane type and are formed between adjacent pyrimidine bases in the same single strand (SETLOW, 1966). It is also known that wild type cells survive exposure to UV light by means of a recovery mechanism in which the pyrimidine dimers are excised from the DNA forming single strand gaps. The twin helix is then reconstructed by DNA synthesis using the intact opposite strand as template. Certain strains that carry mutations in loci designated *uvr* are unable to excise pyrimidine dimers, and are abnormally sensitive to the lethal effects of UV light (SETLOW and CARRIER, 1964; BOYCE and HOWARD-FLANDERS, 1964).

UV light is also known to induce genetic recombination. JACOB and WOLLMAN (1955) found that, in crosses between mutants of bacteriophage, the proportion of progeny recombinant for closely spaced markers could be increased if the phage were exposed to UV light. The effect of UV-irradiation upon recombination between mating strains of *E. coli* K 12 has also been investigated (JACOB and WOLLMAN, 1958; JOSET and WOOD, 1966; and DOUDNEY and BRUCE, 1966), but the effects are complicated and have not led to a clear understanding of the molecular mechanisms involved.

One possible mechanism for UV-induced genetic recombination is as follows. The temporary gaps formed by dimer excision might promote pairing with homologous DNA and so initiate genetic exchanges; accordingly, excision would be a necessary preliminary step in UV-induced recombination. It was found, however, that UV was less effective in wild type than in excision-defective bacteria in causing exchanges between linked markers (HOWARD-FLANDERS and BOYCE, 1966). Thus, far from finding that excision is a necessary preliminary step in UV-induced recombination, these results showed that excision diminishes the effect.

*What UV-induced Structure is Responsible for Initiating Recombination?*

The above experiments led us to ask what UV-induced structure in the DNA might be responsible for initiating recombination. Since the results indicated that recombination might be due to unexcised UV photoproducts, we had to consider alternative explanations to the one discussed above. One possibility was that pyrimidine dimers might be effective because they distort the DNA in the region of the dimer. Alternatively, recombination might be induced only following replication, when a new structure might be formed. This led us to investigate what happens when DNA containing pyrimidine dimers is replicated. It has been suggested that dimers block DNA replication, or that the polymerase replicates slowly past a dimer, perhaps inserting bases at random in this region (Bollum and Setlow, 1963; Swenson and Setlow, 1966). It is also possible that the polymerase leaves gaps in the strand opposite the dimers. This question is open to experimental investigation. If there are gaps opposite pyrimidine dimers, the DNA synthesized in UV-irradiated cells should be of lower molecular weight than the template DNA. We therefore set out to investigate the molecular weight distribution of single strand DNA making use of a method in which bacterial protoplasts are lysed in the surface layer of an alkaline sucrose gradient so that shear degradation is reduced to a minimum (McGrath and Williams, 1966).

*Molecular Weight Distribution of Single-strand DNA Synthesized in UV-irradiated Cells*

Before the molecular weight distribution of newly synthesized DNA in UV-irradiated cells can be investigated, it must first be shown that the DNA containing photoproducts is of sufficiently high molecular weight to be a satisfactory template, and that it is stable during the incubation period necessary for the proposed experiments. The test consists of labeling excision-defective cells with tritiated thymidine and comparing the sedimentation patterns of denatured DNA, from control and irradiated cells, after a 40-min incubation in nonradioactive medium. Figure 1 shows the distributions of the radioactive DNA when sedimented in alkaline sucrose. Sedimentation is from right to left, and the positions of intact strands of T2 and $\lambda$ bacteriophage DNA centrifuged under identical conditions are indicated at the top of the figure.

It is evident from the similarity of the positions of the two peaks that the molecular weight distribution of the DNA remains approximately the same, with a peak at about $2 \times 10^8$ daltons in both the irradiated and control cells. Thus, the DNA in these irradiated cells is of sufficiently high molecular weight to serve as a satisfactory template for the proposed experiments on DNA replication.

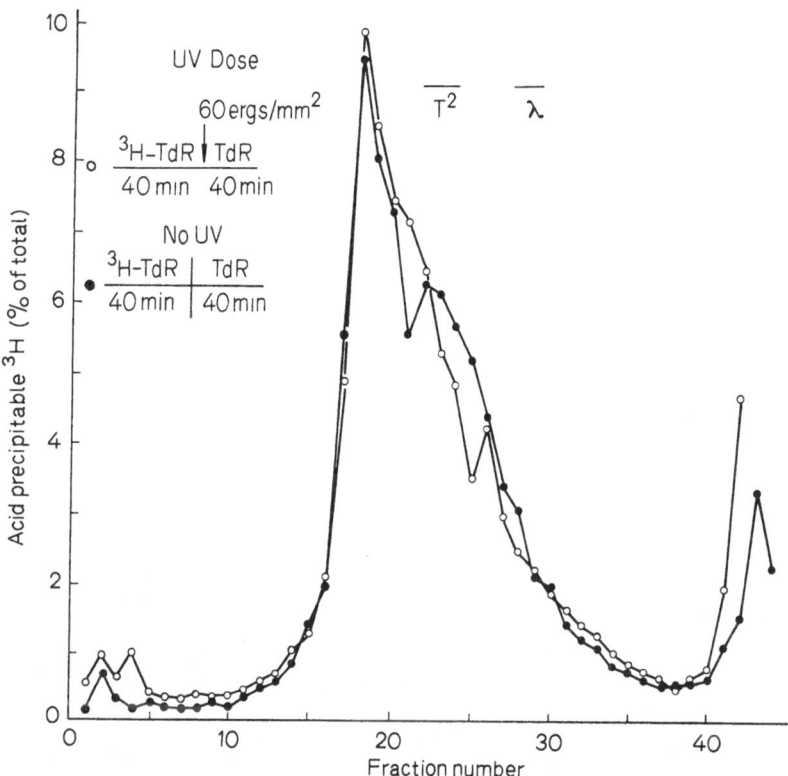

Fig. 1. Sedimentation pattern of DNA extracted from *E. coli* AB2500 *uvrA* bacteria after UV-irradiation. Log phase cells were grown in glucose-salts-casamino acids medium containing ³H-thymidine for 40 min. One sample was exposed to a UV dose of 60 ergs/mm², and then transferred to a nonradioactive medium and incubated for a further 40 min. A control sample was not irradiated before transfer to the cold medium. After incubation, the cells were converted to spheroplasts by treatment with lysozyme and EDTA at 0° and then layered on an alkaline 5—20% sucrose gradient (pH 12) and centrifuged in a Spinco SW50 rotor for 120 min at 30,000 rpm at 20°. The positions of intact strands from T2 and λ centrifuged under identical conditions are indicated. No difference between the sedimentation patterns for control and irradiated cells was detected in similar experiments omitting the 40 min incubation between exposure to UV and lysis

To investigate the molecular weight of the DNA synthesized *after* irradiation, the cells were exposed to UV light and then incubated for ten minutes with tritiated thymidine. The sedimentation patterns obtained in alkaline sucrose gradients are shown in Fig. 2. Compared with the

11*

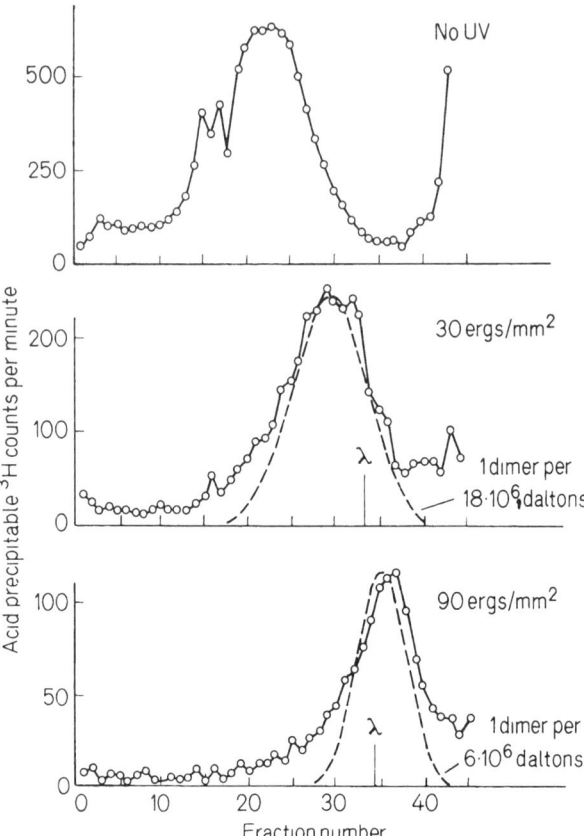

Fig. 2. The ³H-labeled acid precipitable DNA from UV-irradiated *E. coli* in each fraction is plotted against the number of the fraction collected from the sucrose gradients at pH 12. The excision-defective strain of *E. coli* K 12, AB 2500 *uvrA*, was grown in a glucose-salts-casamino acids medium containing 10 μg/ml thymidine. Exponentially growing cells were washed, exposed to UV-light and then incubated in the same medium supplemented with 2 μg/ml thymidine for 10 min. The cells were then transferred to the same medium containing 100 μC³H-thymidine/ml and incubated for 10 min. The cells were immediately converted to spheroplasts by treatment with lysozyme and EDTA at ice temperature, layered on top of an alkaline sucrose gradient and centrifuged for 2 h at 30,000 rpm at 25° C in a Spinco SW 50 rotor. Fractions were collected on filter discs which were then washed with trichloroacetic acid, followed by 95% ethanol and acetone. When dry, the discs were counted in a liquid scintillation counter. The broken lines show the theoretical sedimentation patterns calculated if the DNA had random single strand breaks occurring at a frequency equal to that of the pyrimidine dimers in the parental DNA, calculated from published yields for pyrimidine dimer formation. The mean spacings between pyrimidine dimers along the DNA are 6 and 18 million daltons following exposure to 90 and 30 ergs/mm² UV light at 2537 Å. The constants used for calculating the distribution are based on the sedimentation rate of DNA from bacteriophage λ under identical conditions. Single strand λ DNA was assumed to have a molecular weight of 16.8 million daltons, and to sediment at an average of 11.4 fractions from the meniscus in the gradient yielding 44 fractions

unirradiated control shown at the top, it can be seen that a UV dose of 30 ergs/mm² considerably reduced the sedimentation rate in alkaline sucrose, and that 90 ergs/mm² reduced it still further. Evidently, the molecular weight of the newly synthesized DNA is reduced by the presence of UV photoproducts in the template strand.

### Estimation of Molecular Weight

Because of the possibility that the newly synthesized DNA had been interrupted at each pyrimidine dimer in the template, it was of interest to compare the average molecular weight between randomly spaced pyrimidine dimers with the average molecular weight of the newly synthesized DNA as determined by sedimentation. To do this, the sedimentation rate of the bacterial DNA was first compared with that of ¹⁴C-labeled DNA from bacteriophage λ. Using the relationship between distance sedimented and molecular weight (ABELSON and THOMAS, 1966), and the published values for the yield of pyrimidine dimers, a theoretical distribution in the sedimentation gradient was then calculated for the DNA between successive randomly spaced pyrimidine dimers. The results of the calculations for each UV dose are shown as the broken lines in Fig. 2. The agreement obtained between the distribution calculated from the dimer yield and those obtained experimentally, indicates that the molecular weight of the newly synthesized single strands of DNA is approximately equal to the molecular weight between the UV-induced pyrimidine dimers. In these experiments, the total amount of DNA synthesized during the ten-minute labeling pulse was almost 20 times greater than the mean molecular weight obtained in these distributions, which suggests that about ten fragments must have been synthesized on each single strand of the damaged bacterial chromosome (RUPP and HOWARD-FLANDERS, 1968).

These results show there to be alkali labile bonds, or discontinuities of some kind, in the newly synthesized DNA at spacings approximately equal to the spacing between pyrimidine dimers. Such gaps could be formed if the DNA polymerase replicating the bacterial chromosome failed to insert bases opposite the pyrimidine dimers, but started again either adjacent to, or some distance beyond the dimer, where hydrogen bonding is again possible. It is already known that the E. coli DNA polymerase can replicate the single stranded circular DNA of bacteriophage M 13, and is able to start at random in the circle without a hydrogen-bonded 3'-hydroxyl terminus acting as primer (MITRA, REICHARD, INMAN, BERTSCH, and KORNBERG, 1967).

It is also possible that the DNA polymerase might insert special links, which are subject to action by a nuclease so that gaps are formed indirectly, or that the DNA is formed with alkali-labile bonds, which are broken only in the sucrose gradient. However, there is genetic evidence, to be discussed later, that the gaps are present in the DNA while still inside the cells. We therefore regard this third possibility as unlikely.

## Experiments with Mating Bacteria

To obtain further evidence of the existence of these daughter strand gaps, and their position relative to the pyrimidine dimers, we have turned to experiments with mating bacteria. The advantage of a conjugating system is that the bacterial DNA appears to be replicated during transmission from donor to recipient bacteria (Jacob, Brenner, and Cuzin, 1963; Ptashne, 1965; Gross and Caro, 1966; Bonhoeffer, 1966). Our experiments were designed to answer three questions.

(1) Are chromosomes containing UV photoproducts transmitted?

(2) If so, are there gaps in the newly synthesized strands?

(3) If gaps are formed, are they opposite the pyrimidine dimers?

It seemed a logical approach to this problem to use excision-defective Hfr or F' donor cells and to UV-irradiate these before mating. The advantage of this procedure is that a known number of pyrimidine dimers can be induced in the template strand of the chromosome to be transferred during conjugation. The rate of transfer of the Hfr chromosome from a UV-irradiated donor was investigated in a cross employing an excision-defective Hfr strain, which injects markers in the sequence *pro, leu, thr* and *arg*. These donor cells were mated with genetically marked F⁻ bacteria. After mixing the cells, mating was interrupted at various times, and the bacteria were plated on selective agar to detect the entry of various markers. There was remarkably little delay in the time of entry of the early marker *pro*, or the more distal marker *arg*. Exposure of the donor cells to 200 ergs/mm² delayed the transfer of the gene *arg*, which normally enters at about 35 mins, by less than three mins, and only reduced the total yield of recombinants to one third of control value. This UV dose reduces the colony-forming ability of the Hfr *uvr* strain by more than six logarithms, and would induce over 150 dimers in the single strand transmitted before the arginine marker. Thus, the delay in transmission may be only one or two seconds per transferred dimer.

## Are there Single Strand Gaps in the Transmitted Chromosome?

We next tried to find out if there are discontinuities opposite the dimers in the transmitted chromosome by employing the following rationale. Since DNA repair by excision is believed to depend upon the opposite strand for use as template for reconstructing the twin helix, it would be expected that a discontinuity opposite a pyrimidine dimer would effectively prevent such repair. If, however, discontinuities are

displaced from the pyrimidine dimers — perhaps distributed at random
in the chromosome — then repair by excision should frequently occur in
the recipient.

F factor from UV-irradiated donor

Fig. 3. Diagram of the postulated structure of the F' *lac* episome following its
transmission during conjugation from a UV-irradiated, excision-defective, donor
bacterium. The transmitted DNA contains dimers in one strand and gaps opposite
the dimers in the newly synthesized daughter strand

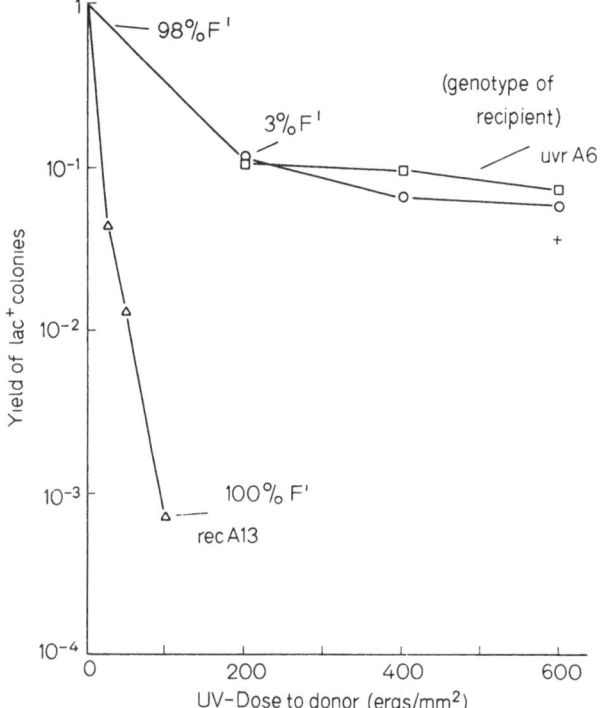

Fig. 4. The relative numbers of Lac+ colonies formed are plotted as a function
of the UV dose to the excision-defective F' *lac*+ donors before mating to F− *lac*
recipients of the genotype shown. The bacteria to be mated were grown to about
$2 \times 10^8$ cells per ml in yeast extract tryptone broth with 1% NaCl. The F' *lac*+ donor
GY854 *ura lac uvrB* was centrifuged, resuspended in buffer at pH 7 and exposed to
various doses of UV light of 2537 Å. These cells were resuspended in broth and
mated with equal volumes of the F− strains; AB1157+, AB1886 *uvrA 6* and AB2463
*recA 13*. After 30 min at 37°C, the cultures were again centrifuged, resuspended
in buffer and plated on media selective for Lac+ Ura+ colonies. They were incubated
for 48 h at 37°C. Certain of the colonies were picked and streaked to single colonies,
before testing them for their ability to act as F' *lac*+ donors. The number of Lac+
colonies is expressed as a fraction of the number formed in unirradiated controls.
The percentage of colonies found to be F' *lac*+ donors is indicated at three points

To investigate whether repair by excision occurs in the recipient, we crossed F' $lac^+$ donors with F$^-$ $lac$ recipient cells. The donor was excision-defective and was exposed to UV-irradiation prior to mating. The episomes transmitted in this cross may consist of one strand containing pyrimidine dimers and a second strand containing discontinuities opposite the dimers, as illustrated in Fig. 3. If this structure is correct, the same yield of Lac$^+$ colonies should be obtained in crosses with normal or excision-defective recipients. As seen in Fig. 4, this proved to be the case. At a dose of 600 ergs/mm², the yield of Lac$^+$ colonies was about 7 % of the value for the unirradiated control and was independent of the $uvr$ genes in the recipient which control dimer excision. The failure to detect any repair by excision in this system supports the hypothesis that there are discontinuities opposite the transmitted dimers.

In the crosses with the recombination-proficient recipients, only three percent of the Lac$^+$ colonies proved to be F' donors after 200 ergs/mm². This suggests that the transmitted $lac^+$ gene may have been integrated into the recipient chromosome and the F' character lost. This interpretation is supported both by the high UV-sensitivity of the yield of Lac$^+$ colonies obtained with a recombination-deficient recipient and by the finding that the Lac$^+$ colonies formed by $recA$ recipients harbor intact F' $lac^+$ episomes. As the $lac^+$ gene cannot be integrated into the bacterial chromosome in these recipients, it will be lost if it is not on an intact F' episome. To judge from the 37 % survival dose and the known dimer yield, one pyrimidine dimer in a single strand of molecular weight 50 million daltons prevents the survival of an intact transferred episome. A single dimer in the transmitted strand would inactivate a two-strand episome of about 100 million daltons, a result readily understood if there is a discontinuity opposite each transferred dimer, so that both strands are defective at the same point.

### Are Discontinuities Single Strand Gaps?

It has been suggested that hydrogen-bonded intermediate structures are necessary for the formation of recombinants (Watson and Crick, 1953; Levinthal, 1959). Such structures have been detected in bacteriophage T4 by Anraku and Tomizawa (1965), and in E. coli by Oppenheim and Riley (1966). Hydrogen-bonding between two homologous DNA molecules can presumably occur most readily in native DNA near the ends of single strands, or near single strand cuts. Thus, DNA containing gaps, with internal free ends stabilized by pyrimidine dimers in the opposite strand, might be expected to promote pairing and recombination. In view of these considerations, it seemed pertinent to investigate the types of recombinants formed when the DNA from one parent consisted of a single strand containing UV-induced photoproducts and a daughter strand containing discontinuities. Our experiments suggest that this type of structure would be formed during transfer between conjugating bacteria if the donor strain was exposed to UV light before mating.

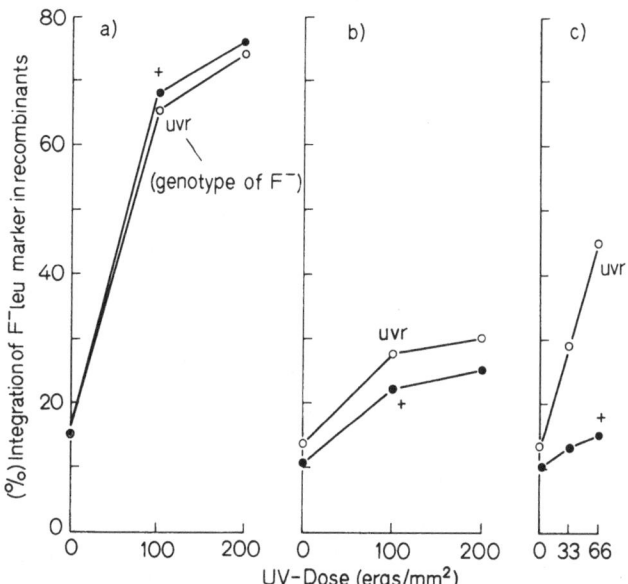

Fig. 5a. The frequency of double exchanges between *pro* and *thr* is plotted against the dose of UV light given to an excision-defective male before mating. AB3108 Hfr J2 *uvrB 5 thy tlr his* T6$^{S*}$ was UV-irradiated and mated at 37° with AB3105 F$^-$ *uvr$^+$ pro leu thr thy* T6$^R$ (•) and AB3106 F$^-$ *uvr B 5 pro leu thr thy* T6$^R$ (○). After 45 min, mating was stopped by the addition of phage T6. The bacteria were then incubated in broth + 30 μg/ml thymine at 37° for 3 h before LThy$^-$ Pro$^+$ Thr$^+$ His$^+$ T6$^R$ recombinants were selected. The percentage of these which carried the female *leu* marker was then determined by replica plating. The injection sequence for Hfr J2 is *pro, leu, thr, arg* . . .

Fig. 5b. The same experiment as that described for Fig. 5a, except AB3107 Hfr J2 *uvr$^+$ thy tlr his* T6$^S$ was the male parent

Fig. 5c. The frequence of double exchange between *pro* and *thr* is plotted against the dose of UV light given immediately after mating to merozygotes derived from either a *uvr$^+$* or *uvr* recipient strain. AB3108 Hfr J2 was mated with AB3105 F$^-$ *uvr$^+$* (•) and AB3106 F$^-$ *uvrB 5* (•) for 45 min. Mating was stopped by violent agitation in a vortex mixer and 1 ml samples of the mixture were diluted tenfold into cold buffer and UV-irradiated. The bacteria were harvested by centrifugation, treated with phage T6, and then incubated at 37° in broth + 30 μg/ml thymine for 3 h. LThy$^-$ Pro$^+$ Thr$^+$ His$^+$ T6$^R$ recombinants were selected and the percentage carrying the female *leu* marker was determined. *Genotypic (phenotypic) symbols denote: *pro* (Pro$^-$), *leu* (Leu$^-$), *thr* (Thr$^-$), *his* (His$^-$), *thy, thy tlr* (LThy$^-$); requirements for proline, leucine, threonine, histidine, high concentration of thymine, low thymine (2 μg/ml). *ara* (Ara$^-$); inability to utilize arabinose. (T6$^{S/R}$); sensitivity or resistance to phage T6. *uvr*; inability to excise pyrimidine dimers

---

* Genotypic and phenotypic symbols are described in the legend of Fig. 5.

The object of these experiments was to determine the frequency of double exchanges between two linked markers. An Hfr donor of origin J 2, which injects these markers early, was mated with genetically marked F⁻ recipients. In certain crosses, the Hfr strain was exposed to UV light before mating with the results shown in Fig. 5a. The frequency of double

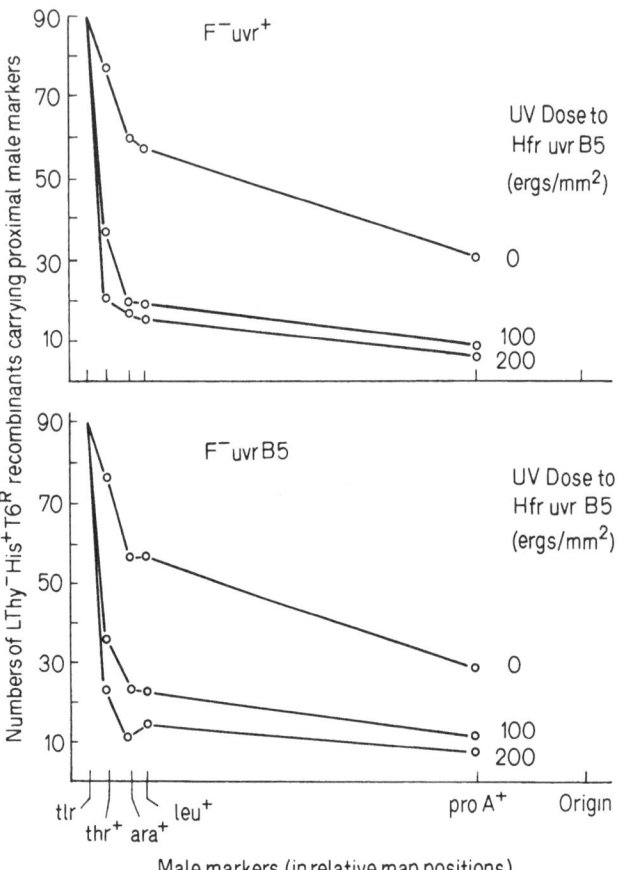

Male markers (in relative map positions)

Fig. 6. The numbers of unselected male markers incorporated in recombinants carrying the male distal marker *tlr\**, are plotted against the positions of the markers in the genetic map of *E. coli* K 12. These recombinants were derived from matings with untreated and UV-irradiated excision-defective Hfr bacteria. AB 3104 Hfr J 2 *uvrB 5 thy tlr his* T 6$^S$ was UV-irradiated with 100 and 200 ergs/mm² and mated with AB 3105 F⁻ *uvr⁺ pro leu ara thr thy* T 6$^R$ and AB 3106 F⁻ *uvrB5 pro leu ara thr thy* T 6$^R$. After 45 min, mating was stopped by the addition of phage T 6. The bacteria were then incubated at 37° for 3 h in broth +30 μg/ml thymine before LThy⁻ His⁺ T 6$^R$ recombinants were selected. 90 recombinants from each experimental class were then restreaked on master plates and tested by replica plating for the presence of each of the proximal male markers, *thr⁺*, *ara⁺*, *leu⁺* and *pro⁺*

exchanges between *pro* and *thr*, (as determined by the presence of the female *leu* marker in recombinants carrying the male *pro+* and *thr+* genes) is plotted against the UV dose. This frequency was low in crosses with unirradiated donors, but increased when the Hfr cells were exposed to UV light before mating. By comparing Fig. 5a and 5b, it may be seen that the increase in frequency of double exchanges was less marked when the irradiated donor was wild type rather than *uvr⁻*. This shows, as might be expected, that repair by excision occurred in the wild type donor.

The frequencies of double exchanges induced by the irradiation of donors is independent of the *uvr* genotype of the recipient, again indicating that the transmitted photoproducts could not be repaired by excision. However, as seen in Fig. 5c, photoproducts formed in the zygotes immediately after mating could also induce exchanges, but they differed in that they were subject to repair by excision. Thus, dimers formed in the zygote could be distinguished from transmitted dimers, as only the former were repaired by excision.

Additional data on the types of recombinants obtained in crosses with UV-irradiated donors are shown in Fig. 6. Recombinants were selected that carried a distal marker from the male, and were then analyzed for three clustered proximal markers. It is seen from this figure that the length of male chromosome integrated in the vicinity of the selected marker is sharply reduced by irradiation of the donor cells prior to mating. The yields of all classes of recombinants were again unaffected by the ability of the recipient to excise dimers. In unirradiated crosses, the most frequent class of recombinants carried male markers at all three loci, suggesting that long regions of male chromosome were frequently integrated. In the crosses with irradiated donors, however, this class was lost. The most frequent class contained female markers at most or all of these loci and there was a tendency for multiple short lengths of donor chromosome to be integrated. These lengths may have been shorter than the spacing between the markers employed, and their dimensions were not determined in these experiments (WILKINS and HOWARD-FLANDERS, in preparation).

### Conclusions and Summary

When DNA containing UV-induced pyrimidine dimers is replicated, the newly synthesized strands appear to be discontinuous. The spacing between these discontinuities is approximately equal to that between the pyrimidine dimers. These discontinuities were detected in experiments on DNA replication in *E. coli* K 12, both during vegetative growth and during DNA transfer between conjugating cells. When excision-defective Hfr or F⁺ donors are lethally UV-irradiated and crossed with F⁻ bacteria, the chromosomes or episomes are transferred almost normally. Recombinants are formed at 5 to 30% of the normal yield, and contain multiple short lengths of transferred DNA. The genetic activity of DNA transferred from a UV-irradiated donor is identical in normal and excision-defective recipients. Since an intact complementary strand is needed for repair by excision, this failure to detect repair in the recipient suggests that the complementary strand may be discontinuous opposite each pyrimidine dimer. Thus, recombination tends to occur following replication, and may be initiated by stabilized internal free ends so formed.

This work was supported by United States Public Health Service Grants CA 06519 and AMK 69397.

## References

Abelson, J., and C. A. Thomas: J. Molec. Biol. 18, 262 (1966).
Anraku, N., and J. Tomizawa: J. Molec. Biol. 11, 501 (1965).
Bollum, F. J., and R. B. Setlow: Biochim. Biophys. Acta (Amst.) 68, 599 (1963).
Bonhoeffer, F.: Z. Vererbungsl. 98, 141 (1966).
Boyce, R. P., and P. Howard-Flanders: Proc. Nat. Acad. Sci. (Wash.) 51, 293 (1964).
Doudney, C. O., and B. J. Bruce: Radiat. Res. 28, 597 (1966).
Gross, J. D., and L. G. Caro: J. Molec. Biol. 16, 269 (1966).
Howard-Flanders, P., and R. P. Boyce: Rad. Res. Suppl. 6, 156 (1966).
Jacob, F., S. Brenner, and F. Cuzin: Cold Spr. Harb. Symp. Quant. Biol. 28, 329 (1963).
—, and E. L. Wollman: Ann. Inst. Pasteur 88, 724 (1955).
— — Symp. Soc. Exp. Biol. 12, 75 (1958).
Joset, F., and T. H. Wood: Genetics 53, 343 (1966).
Levinthal, C.: In: The Viruses (Ed. F. M. Burnet, and W. M. Stanley), Vol. 2, p. 281. New York: Academic Press 1959.
McGrath, R., and R. W. Williams: Nature (Lond.) 212, 534 (1966).
Mitra, S., P. Reichard, R. B. Inman, L. L. Bertsch, and A. Kornberg: J. Molec. Biol. 24, 429 (1967).
Oppenheim, A. B., and M. Riley: J. Molec. Biol. 20, 331 (1966).
Ptashne, M.: J. molec. Biol. 11, 829 (1965).
Rupp, W. D., and P. Howard-Flanders: J. Molec. Biol. 31, 291 (1968).
Setlow, R. B.: Science 153, 379 (1966).
Setlow, R. B., and W. L. Carrier: Proc. Nat. Acad. Sci. (Wash.) 51, 226 (1964).
Swenson, P. A., and R. B. Setlow: J. Molec. Biol. 15, 201 (1966).
Watson, J. D., and F. H. C. Crick: Nature (Lond.) 171, 964 (1953).
Wilkins, B. M., and P. Howard-Flanders: In preparation.

## Discussion

*Starlinger:* What is the relation of the recovery dependent on recombination to the difference in sensitivily observed between strains B and K?

*Howard-Flanders:* I do not know why filament-forming strains of *E. coli* tend to be radiation-sensitive. As it occurs even in excision-defective strains, it does not depend upon excision.

*Trautner:* Do you have an idea about the size of the gaps produced? Can such gaps be healed by sealase?

*Howard-Flanders:* We have had no way to measure the numbers of bases missing in the gaps formed opposite dimers, but perhaps methods can be devised in the future. As they remain open for more than 10 min in bacteria, they are probably not joined by the polynucleotide ligase.

*Beckwith:* Couldn't destruction of DNA in the F⁻ have a lot to do with loss of linkage.

*Howard-Flanders:* These experiments were carried out in excision defective cells, and we have shown them to contain high molecular weight DNA after UV irradiation in tests with sedimentation in alkaline sucrose gradients.

*Schuster:* Photoreactivating enzyme (PRE) splits dimers in denatured DNA much more slowly than in native DNA and does not split dimers in small oligonucleotides. The question arises whether or not the PRE may act on thymine dimers in double-stranded DNA containing gaps (opposite the dimers).

*Howard-Flanders:* Dr. R. S. COLE tried this experiment, and obtained a substantial-photoreactivation of *F* episomes in *F*⁻ recipients carrying *rec A*. Moreover $\phi$X phage can be photoreactivated to a small extent (SETLOW, 1961).

*Arber:* With UV irradiated tranducing phage $\lambda$ dg one observes upon infection of *gal*⁻ recipient strains appearance of stable *gal*⁺ transductants rather than of heterogenotes carrying $\lambda$ dg as a prophage. Similarly with phages giving general transduction a shift from response by abortive transduction to stable integration occurs. In these system there is no obvious evidence for DNA replication previous to the recombination reactions.

*Howard-Flanders:* Our results indicate that stable internal free ends can be created at replication and that these can initiate genetic exchanges. But this should not be taken to restrict the ways in which recombination can be initiated — perhaps any free ends can serve this functions and the $\lambda$ dg may have free ends.

# Repair of DNA in a Mutant of E. coli Temperature-Sensitive in DNA Synthesis

By

J. Seehafer and H. Schuster

Max-Planck-Institut für Molekulare Genetik, Berlin/Germany

According to present hypotheses host cell reactivation (HCR) is regarded as a multi-step repair process of damaged DNA of cells or viruses [1]. The steps involved in the repair of double-stranded DNA are

1. enzymatic excision of the damaged bases,
2. widening of the resulting gap — i.e. release of nucleotides by the action of nucleases,
3. repair replication by which complementary nucleotides are inserted into the gap,
4. joining of the phosphodiester backbone after repair replication has been completed.

Cells lacking the capacity for this kind of excision repair (hcr⁻) still have a capacity to eliminate damages quite efficiently, provided that the gene(s) responsible for recombination is (are) present. From this it is concluded that recombinational as well as excision mechanisms contribute to recovery processes [1].

At the present time, our knowledge regarding the repair mechanisms on the molecular level is scanty. The difficulties involved in analyzing such mechanisms are obvious. For example, repair replication has to be distinguished from normal DNA replication. If repair of UV light induced damages is studied the latter may be suppressed by using high doses. However, a linear relationship between dose and the extent of DNA breakdown exists only within a certain dose range [1]. On the other hand excision and recombinational repair processes may operate at the same time, especially in those cases where the DNA is heavily damaged. Furthermore, most studies in the past were done with whole bacterial cells the DNA of which is difficult to isolate as an intact molecule. Therefore, we have looked for a system which might have some advantages compared with those already studied and which might satisfy at least two conditions:

1. Inherent capacity to differentiate between different repair processes and/or between different steps of one process.

2. The DNA to be repaired should be easier to handle and to characterize than the DNA of bacterial cells.

The system which we have chosen and which might fulfil the conditions, is the following:

1. a mutant of *E. coli* K 12 Hfr H (thy$^-$, str$^s$, T$^s_{DNA}$) the DNA synthesis of which is blocked at elevated temperatures ($\geq$ 41°C) [2],

2. phage lambda (clear plaque mutant $\lambda$ c) which does not give any progeny in this mutant at 41°C or higher temperatures whereas phage replication at 33°C is normal [3].

Preliminary experiments described in this paper reveal the continuance of the repair process, or at least part of it, under conditions where normal DNA synthesis is blocked.

## Host Cell Reactivation of Phage $\lambda$

Experiments to study the kinetics of repair of UV irradiated phage DNA are done in the following way. Cells of *E. coli* K 12 T$^s_{DNA}$ were grown in tryptone broth at 33°C, harvested in the stationary phase, and starved for 1 h in 0.01 M MgSO$_4$ at 33°C. UV irradiated $\lambda$ phages were adsorbed

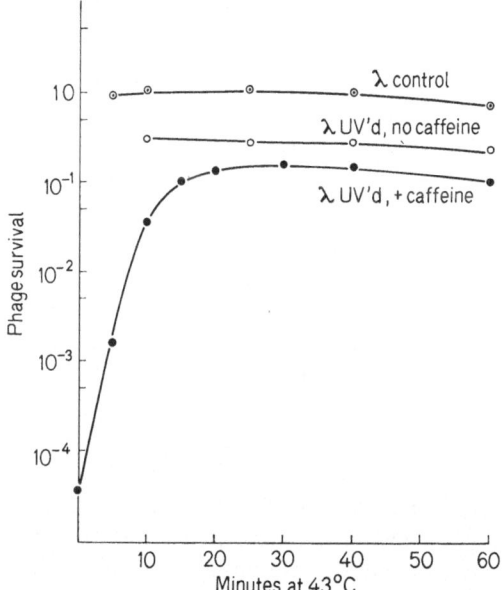

Fig. 1. Kinetics of repair of UV irradiated phage $\lambda$ c in *E. coli* K 12 T$^s_{DNA}$ at 43°C. Infective centers were incubated at a cell concentration of $5 \times 10^8$ per ml 0.01 M MgSO$_4$ at 43°C. (Exposure of the mutant cells to the elevated temperature at cell concentrations $< 10^7$ per ml results in a more or less rapid cell death and killing of infective centers [3].) For further explanation see text.

to the cells (multiplicity of infection = 0.01) and incubated in 0.01 M $MgSO_4$. Adsorption and incubation were done at 43°C. At different times aliquots were taken from the incubation mixture and mixed with caffeine (final concentration = 0.25 %) at the elevated temperature in order to block HCR [5].

The samples were plated at room temperature. Plating agar contained 0.25 % caffeine. As indicator strain E. coli K 12 AB 1886 uvr A-6 was used. Plates were evaluated for infective centers after overnight incubation at 37°C.

In one experiment the complexes of irradiated $\lambda$ and the $T^s_{DNA}$ mutant cells were incubated at 43°C for one hour, and the survival of infective centers followed. As can be seen from Fig. 1, the phage survival rises from $4 \times 10^{-5}$ (survival of irradiated $\lambda$ adsorbed to cells in the presence of caffeine) to approximately $10^{-1}$ within a period of 20—30 min (Fig. 1, "$\lambda$ UV'd + caffeine"; dose reduction factor [4] = 0.14). The survival of irradiated $\lambda$ remains fairly constant after incubation at 43°C at approximately 0.3 if infective centers are not treated with caffeine (Fig. 1, "$\lambda$ UV'd, no caffeine"). Similarly, the infective centers containing unirradiated $\lambda$ withstand the 43°C treatment (Fig. 1, "$\lambda$ control").

These results indicate that after a period of incubation under conditions where DNA synthesis is prevented caffeine does no longer effect host cell reactivation. Similar results were obtained with UV irradiated phage T 1 using the E. coli mutant B 94 (ad⁻, arg⁻) in which DNA and RNA synthesis can be blocked by adenine withdrawal [5].

## Dark Reactivation (DR) of E. coli K 12 $T^s_{DNA}$

The kinetics of recovery of colony forming ability of the irradiated $T^s_{DNA}$ mutant was followed in a similar manner to that described for survival of phage $\lambda$. Cells irradiated with a UV dose to yield 91 % survival in the presence of DR (Fig. 2, "UV'd") were incubated at 41°C for different periods of time. Since the unirradiated mutant cells are extremely sensitive to caffeine but much less affected by acriflavine (Fig. 2, "no UV, + acriflavine") DR was blocked by the addition of acriflavine [6] (final concentration = 5 μg/ml). Colony survival rises from $7 \times 10^{-3}$ to $1.5 \times 10^{-1}$ within 30 to 40 min at 41°C (Fig. 2, "UV'd, + acriflavine"; survival is measured relative to the number of non-irradiated cells which form colonies without acriflavine).

From these results we conclude that dark reactivation of lethal hits becomes insensitive to acriflavine during the time of exposure to 41°C, similar to the HCR of UV damages in phage DNA mentioned above. Obviously, the repair mechanism of the mutant active at 41°C does not discriminate between phage DNA and its own DNA. Similar results were obtained with the *E. coli* mutant B 94 (see above) in the absence of DNA synthesis although in this case the time for DR was much longer (4 to 5 h) [7].

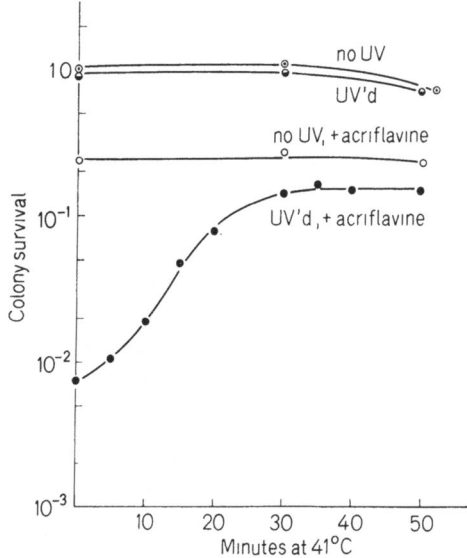

Fig. 2. Kinetics of recovery of colony forming ability of UV irradiated *E. coli* K 12 T$_{DNA}^s$ at 41°C. Cells ($2 \times 10^8$ per ml) are incubated in tryptone broth at 41°C. Samples are removed and plated on tryptone agar ("no UV", "UV'd") or after blocking DR on tryptone agar containing acriflavine ("no UV, + acriflavine"; "UV'd, + acriflavine"). For further explanation see text.

No conclusions can be drawn at the present time as to whether repair processes at 41—43°C are being completed or are only being initiated. The incorporation of thymine in the DNA at 41°C after UV irradiation of whole cells indicates that repair replication my take place in the absence of normal DNA synthesis [8].

## Host Cell Reactivation of UV Light Induced Phage Growth Delay

A delay of intracellular growth is caused by UV irradiation of phages. At a given UV dose this effect is considerably more expressed in hcr⁻- than in hcr⁺-cells. At a given phage survival level this delay occurs to the same extent in hcr⁺- and hcr⁻-cells [9].

These results may be interpreted in two ways.

1. UV irradiation causes, besides lethal lesions, non-lethal photo-products in the DNA, the latter being responsible for the growth delay. Both types of lesions can be eliminated efficiently by HCR [9].

2. The delay may be caused totally or partially by repair mechanisms other than HCR.

HCR of lethal lesions in phage $\lambda$ becomes insensitive to caffeine during infection of *E. coli* K 12 $T_{DNA}^s$ at 41 or 43°C. Assuming that this is also true for HCR of the hypothetical non-lethal lesions no growth delay is to be expected if one-step growth curves (at the non-restrictive temperature) are measured *after* preincubation at 41°C. If on the contrary growth delay is caused by another repair mechanism *dependent on DNA synthesis* pretreatment at 41°C should not influence the delay period at 33°C.

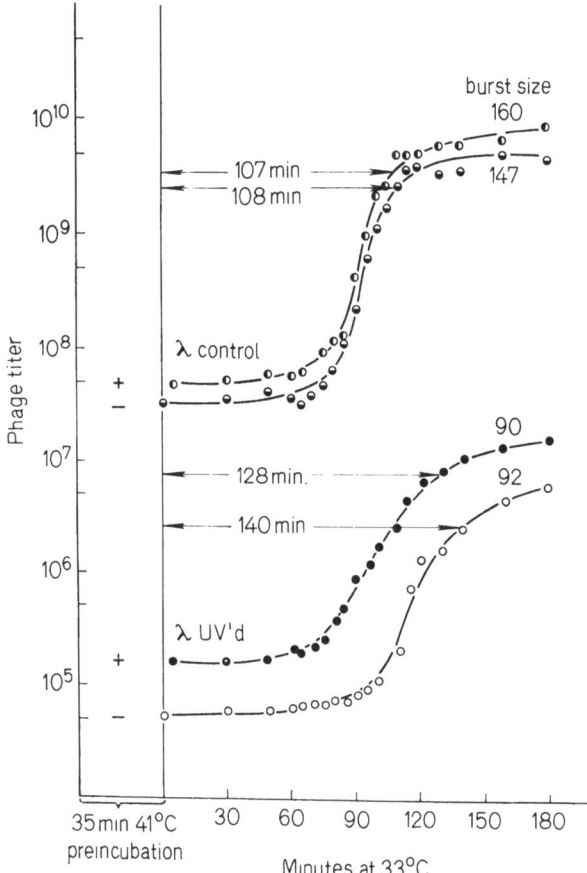

Fig. 3. One-step growth curves of UV irradiated and unirradiated phage $\lambda c$ in *E. coli* K 12 $T_{DNA}^s$. Phages are adsorbed to cells (moi = 0.1) at 33°C resp. 41°C. Infective centers are kept in tryptone broth at 33°C with and without preincubation for 35 min at 41°C. Cell concentration was kept at $5 \times 10^8$ per ml during preincubation at 41°C and diluted $10^{-1}$ to $10^{-4}$ after the temperature shift to 33°C. Eclipse periods are expressed as the time at 33°C necessary to yield phage titers equivalent to 50% of the final burst.

In Fig. 3 one-step growth curves of irradiated and unirradiated phage $\lambda$ in the $T^s_{DNA}$ mutant at 33°C with and without preincubation at 41°C are presented. Phages were irradiated to yield a survival of $1-3 \times 10^{-3}$. The eclipse period of unirradiated phages at 33°C with and without 35 min preincubation at 41°C is the same (Fig. 3, "$\lambda$ control"). With irradiated phages and without preincubation a growth delay of $140 - 108 = 32$ min is observed. On the other hand 41°C pretreatment shortens the growth delay period to only $140-128 = 12$ min (Fig. 3, "$\lambda$ UV'd").

These results indicate that growth delay is caused, at least partially, by HCR mechanisms. The residual growth delay remaining after the 41°C period ($128-107 = 21$ min) may be due to other repair processes dependent on DNA synthesis. To test this, experiments with $\lambda$ phages in hcr⁻ $T^s_{DNA}$ mutants under the conditions described are to be performed.

## References

1. HOWARD-FLANDERS, P., and R. P. BOYCE: Rad. Res. Suppl. 6, 156—184 (1966).
2. BONHOEFFER, F.: Z. Vererbungsl. 98, 141—149 (1966).
3. LANKA, E., and H. SCHUSTER: To be published.
4. JAGGER, J.: Bact. Rev. 22, 99—142 (1958).
5. SAUERBIER, W.: Biochem. biophys. Res. Commun. 14, 340—346 (1964).
6. FEINER, R. R., and R. HILL: Nature (Lond.) 200, 291—293 (1963).
7. METZGER, K.: Biochem. biophys. Res. Commun. 15, 101—109 (1964).
8. SCHALLER, H.: unpublished results (cited in reference [2]).
9. HARM, W.: Photochem. Photobiol. 4, 575—585 (1965).

## Discussion

*Howard-Flanders:* What is known of the actions of acriflavine and caffeine ?

*Schuster:* Not much is known with certainty of the actions of the two agents on the molecular level. Caffeine seems to inhibit the introduction of a first clip in the $\lambda$ phage DNA pre-exposed to ultraviolet light. [SHIMADA, K., and Y. TAKAGI: Biochem. biophys. Acta (Amst.) 145, 763—770 (1967).] On the other hand it is assumed that caffeine depresses the rate of repair incorporation of nucleotides into DNA without greatly affecting the release of oligonucleotides from irradiated DNA [GRIGG, G. W.: Mutation Res. 4, 553—557 (1967)]. The inhibitory effect of acriflavine on HCR may be attributed to the strong affinity of the dye to (unirradiated) DNA [LERMAN, L. S.: J. molec. Biol. 3, 18—30 (1961)].

# Prophage Attachment and Detachment: Comparison of Phages P2 and λ

By

G. BERTANI

Karolinska Institutet, Stockholm 60, Sweden

For several years our laboratory has been engaged in the study of various aspects of bacteriophage P2, and in particular its behavior in the establishment of lysogeny. At the same time, another temperate bacteriophage, λ, has been studied intensively in a number of laboratories, so much so that it has practically become paradigmatic in the study of lysogeny. We feel however that the continued detailed study of P2 is very desirable because this phage differs from λ in at least three properties, which we believe are of rather general biological interest: (a) P2 is not inducible, (b) P2 recombines with extremely low frequency, (c) P2 attaches itself at a variety of sites on the host-chromosome.

In what follows, after introducing the material and the problems, I review very briefly the more recent work on P2, and especially that which is currently in progress in our laboratory, comparing it with what is known for λ. For older work reference is made to an earlier review (BERTANI, 1958).

## Properties of Phage P2 and of its DNA

Bacteriophage P2 is one of the phages carried by the classical lysogenic strain, *Escherichia coli* Lisbonne, isolated some forty years ago.

Particles of phage P2 have a head (about 68 mμ across), probably icosahedral in shape, and a straight tail (about 140 mμ long) (ANDERSON, 1960). They contain protein and double stranded DNA, the latter representing about 40 % of dry weight, and being approximately equimolar (like the DNA of the host bacteria, *E. coli* or *Shigella*) in base composition.

P2 DNA is infectious (MANDEL, 1967) in the type of assay developed by KAISER and HOGNESS (1960) for λ DNA. MANDEL has followed the activities of the two DNAs after sucrose density gradient centrifugation. His data indicate for P2 DNA a molecular weight equal to approximately 70 % of that of λ DNA. They also suggest that P2 DNA like λ DNA can

form in vitro high molecular weight complexes, which regenerate monomers when exposed to high temperature. MANDEL'S observations are
consistent with the more precise results of current electron microscopical
work on P2 DNA by Drs. R. INMAN and A. KLEINSCHMIDT (personal
communication). "Melting maps" obtained by Dr. INMAN show furthermore that P2 DNA molecules form a homogeneous population in respect
to structure (i.e. they are not permuted in base sequence).

P2 DNA is unique in giving clearly three-step temperature denaturation curves, indicating a strong heterogeneity in the base ratio distribution along the molecule, a feature which might turn out to be a valuable
analytical tool.

## Behavior of phage $\lambda$[1]

In the establishment of lysogeny, phage $\lambda$ attaches normally to one
chromosite (place of prophage attachment on the chromosome of the host
bacterium), located between the genes for galactose utilization and biotine
synthesis. A large amount of information has accumulated indicating that
the insertion theory proposed by CAMPBELL (1962) is correct: the linear
$\lambda$ DNA, once in the cell, closes to form a ring and undergoes reciprocal
recombination with the host chromosome, thus becoming inserted line-
arly in the latter, but with a permuted phage gene sequence. The place
on the $\lambda$ chromosome where this recombination takes place, or episite, is
unique (SIGNER, 1964; ZICHICHI and KELLENBERGER, 1963).

Genetic recombination between vegetative $\lambda$ phages can take place
anywhere on the $\lambda$ chromosome. The recombination event that leads to
the insertion of the prophage, however, has been shown to be of a different
nature and to depend on the presence of a product ("attachment substance", "lysogenase") of a phage gene, usually repressed in the lysogenic
condition (CAMPBELL, 1965; SIGNER and BECKWITH, 1966; FISCHER-
FANTUZZI, 1967; CALEF, 1967). The concept of a special recombination
mechanism responsible for prophage attachment and detachment also
explains the old phenomenon of "curing" of lysogenic strains by superinfection with related, heteroimmune phages.

## P2 chromosites

Differently from $\lambda$, phage P2 can establish itself with ease at different
chromosites (BERTANI and SIX, 1958). In a host, Escherichia coli strain
K-12, which is also the commonly used host for $\lambda$, P2 can occupy with
equal probability one of two chromosites, called H (closely linked to the

---

[1] Given the nature of this article and the vast literature on phage $\lambda$, only leading
recent bibliographical references are given in this section. It should be noted that
some of the conclusions mentioned are based on the study of both $\lambda$ and another
phage, $\Phi 80$, that can recombine with it.

genes for histidine synthesis) and II (in the methionine and isoleucine genes region) (KELLY, 1963).

Strain K-12 is not however a particularly good host for P2, so that much work has been done with another strain of *E. coli*, called C. This strain is F⁻ and can recombine with K-12. It can also act as recipient for the F factor, thus becoming F⁺, and originate Hfr strains. It was thus possible to construct (in collaboration with B. KELLY, M. WIMAN, and I. SASAKI) a genetic map for strain C completely independent of that known for K-12, except for the use of the F factor, common to both strains. This map, based on 20−25 markers and a number of different Hfr derivatives (SASAKI and BERTANI, 1965), is very similar in its main lines to that of K-12 (see TAYLOR and THOMAN, 1964). There are however differences of detail, and one of these is demonstrated by the behavior of P2. In strain C phage P2 has a strong tendency to attach itself at a chromosite called I, also near the histidine genes, but not allelic with chromosite H in K-12. When doubly lysogenic strains are produced, one P2 prophage is as a rule attached in chromosite I, the second can be at one of a number of "second choice sites". One of these is a site which appears to be allelic with site II of K-12. Another, site III, has been localized near the tryptophan genes. Several others have been recognized, but not localized precisely on the host chromosome map (BERTANI and SIX, 1958; BERTANI, 1962; KELLY, 1963; SIX, 1961, 1963, 1966).

In still another host strain, a *Shigella*, the evidence available, although rather indirect, suggests that P2 behaves as in strain C. P2 can even lysogenize strains of *Serratia* (BERTANI, TORHEIM, and LAURENT, 1967), although here nothing is known concerning the map of the host, and therefore chromosite location. The *Serratia* case is however interesting, in that there is a marked difference in DNA base composition between P2 and *Serratia*. That true lysogeny is possible even here is however not surprising if the size of the chromosite (i.e. its length in terms of base pairs) is small. This point has been discussed in the paper referred to above.

Chromosites I and II are probably not identical. This is suggested by the extensive analysis by SIX (1963, 1966) of the behavior of the phage which originates from a prophage in site II. This phage can have a different site affinity from normal P2, *i.e.*, when used to lysogenize again strain C, it has a reduced preference for attachment in chromosite I. A reasonable interpretation, based on the CAMPBELL insertion model, and assuming similar, but not identical, chromosites, has been presented by SIX (1966).

One can ask however (a) whether CAMPBELL's model applies to P2, and, if so (b) whether the phage uses always the same episite in attaching to the various chromosites available to it. For this, however, a genetic map of the phage is necessary.

## Genetic Recombination in Phage P2

Although several plaque type and virulence mutants of P2 were already known fifteen years ago, the establishment of a genetic map has always been discouraged by the incredibly low frequencies of recombination obtained in mixed infection with this phage. Recently this problem has been taken up in a systematic way by G. LINDAHL. He has isolated a large number of temperature sensitive, conditionally lethal, mutants of P2 and, using selective techniques for the measurement of recombination frequencies, succeeded in establishing a map for this phage. The earlier known plaque type mutants have also been inserted in this map. The now existing map is apparently linear and consists of at least 14 complementation groups, each containing between one and seven independent mutations. The recombination frequency between the two farthest markers is about 0.15 % as compared with 15 % for phage $\lambda$. The reasons for the extremely low recombination ability of vegetatively multiplying P2 can only be surmised at this time.

LINDAHL'S map presents a most remarkable feature. It contains a relatively long region barren of markers. This situation could hardly be the result of a non-random distribution of temperature sensitive mutants. It could reflect however some features of the recombination mechanism of this phage. LINDAHL proposes (a) that the barren region includes the phage episite, and (b) that the hypothetical lysogenase works not only in recombining the phage with the host chromosome, in the course of lysogenization, but also in recombining phage with phage, more precisely episite with episite, in the course of vegetative multiplication. There would be, therefore, at least two recombination mechanisms operating in temperate bacteriophages: one, aspecific, working along the whole phage chromosome, the other, highly specific, working only at the episites. In P2, the first, aspecific, mechanism is — for reasons unknown — rather ineffective, and — as a consequence — the contribution of the second, region specific, mechanism is disproportionate and produces the "barren region" effect in the genetic map.

It has now been possible to perform transduction experiments (R. CALENDAR and G. LINDAHL) involving the cotransfer of P2 prophage markers with the histidine genes (which are closely linked to chromosites I and H). The results obtained to-date clearly indicate that the episite used by P2 in establishing lysogeny at chromosite I is indeed located in the "barren region" of the phage map. Work on other P2 chromosites is in progress.

## Induction and Prophage Detachment

P2 has long been the prototype of a non-inducible phage. Treatments such as UV irradiation, starvation for thymine, the addition of fluorodeoxyuridine or mitomycin, all known to induce phage $\lambda$, are completely ineffective for P2. P2-lysogenic strains are not detectably more sensitive to such treatments than non-lysogenic ones; neither do they show an increased phage production or, in the case of UV or thymine starvation, loss of immunity to superinfection.

Recently, Mrs. L. E. BERTANI has isolated and thoroughly investigated some temperature-dependent clear mutants of P2. These mutants produce clearer plaques at 42 C than at 30 C. Strains lysogenic for some of them, appear to be inducible by a 30° to 42°C temperature shift, but the induction is abortive. At 42°C these bacteria are inactivated and concomitantly lose immunity to superinfection, but, unless superinfected, they produce very little phage (about 1 phage per 100 bacteria). The amount of prophage type produced by these strains can be increased up to one or two phages per bacterium by superinfection with wild type P2, particularly if the superinfecting phage has received a small dose of UV (50 % of the phage still viable).

It can be shown by means of bacterial crosses that the failure of these lysogenic strains to produce phage at 42° C in the absence of superinfection, is contingent upon the failure of the prophage to detach from the bacterial chromosome. In these experiments, an Hfr strain, lysogenic for a temperature-dependent clear mutant at a given chromosite is incubated at 42°C for various lenghts of time and then mated with a recipient strain lysogenic for wild type P2 at the same chromosite. Recombinants are selected which have received from the Hfr parent a marker closely linked to the chromosite, and examined for the type of prophage carried. All recombinants are found to be lysogenic, and the frequency of transfer of the prophage type of the donor is more or less that expected from the linkage between the chromosite and the marker selected for. If an exactly analogous experiment is done with $\lambda$ a significant proportion of the recombinants are found to have neither the prophage of the donor, not that of the recipient: they are "cured". This suggests that incubation at 42°C while eliminating in both cases the immunity to superinfection of the bacterium, allows temperature dependent mutants of $\lambda$, but not of P2, to detach from the chromosome (which is then transfered without prophage to the recipient bacterium in the cross). Prophage detachment would thus be the critical step required for phage multiplication in the induction process. This conclusion is supported by the additional observation that cured recombinants in the type of experiment described above are found also in the P2 case, when the donor strain, during the high temperature treatment, is superinfected with (ultraviolet irradiated) wild type P2. Mrs. BERTANI proposes that the superinfecting phage supplies the function ("lysogenase") specifically required for prophage attachment and detachment, thereby suggesting that such a function, which is repressed in lysogenic bacteria, immediately develops in $\lambda$ following the disappearance of immunity, whereas in the P2 case it appears only following initiation of vegetative multiplication of the phage, as though it were a so-called "late" function. This is only a speculation for the time being, but it is consistent with the ability of phage P2 to establish multiple lysogeny in single infection (BERTANI, 1962), the phenomenon being presumably an indication that — on the average — a fair amount of vegetative multiplication usually takes place in P2 infection before prophage attachment occurs.

## Transduction and Eduction

It has been known for a long time that phage $\lambda$ can become associated with the bacterial markers to the right or to the left of its chromosite and thus transduce them. Phage $\Phi 80$ behaves similarly in respect to the tryptophan genes to which its chromosite is closely linked. One does not know whether this phenomenon can take place generally — even though at a low frequency — whenever an episome detaches from the host chromosome. Attempts to demonstrate transduction of the histidine genes by P2 have failed. However, very recently, KELLY and SUNSHINE (1967) have discovered in the P2 system a phenomenon that can be considered to be the inverse of specialized transduction, and for which the term eduction is being proposed. They find that strains of *E. coli* K-12 lysogenic for P2 in chromosite H, when being cured of their prophage, can undergo a permanent loss of the histidine genes (which are closely linked to chromosite H). They find furthermore that a similar loss takes place now and then as a result of the infection of sensitive K-12 bacteria, when these survive the infection without becoming lysogenic, as often happens with temperate bacteriophages. Clearly eduction represents a new powerful way by which a virus can modify the genetic structure of the host cell, and it should be very desirable to establish how generally it occurs.

## Acknowledgements

I wish to thank my colleagues L. ELIZABETH BERTANI, RICHARD CALENDAR and GUNNAR LINDAHL for permission to mention their unpublished results. Our work is supported by grants from the Swedish Medical and Natural Science Research Councils and the Swedish Cancer Society.

## References

ANDERSON, T. F.: On the fine structure of the temperate bacteriophages P1, P2 and P22. Proc. Eur. Reg. Conf. on Electron Microscopy, Delft 1960, Vol. II, p. 1008—1011.
BERTANI, G.: Lysogeny. Advanc. Virus Res. 5, 151—193 (1958).
— Multiple lysogeny from single infection. Virology 18, 131—139 (1962).

BERTANI G., and E. SIX: Inheritance of prophage P2 in bacterial crosses. Virology **6**, 357—381 (1958).
— B. TORHEIM, and T. LAURENT: Multiplication in *Serratia* of a bacteriophage, originating from *Escherichia coli:* lysogenization and host-controlled variation. Virology **32**, 619—632 (1967).
CALEF, E.: Mapping of integration and excision crossovers in superinfection double lysogens for phage lambda in Escherichia coli. Genetics **55**, 547—556 (1967).
CAMPBELL, A.: Episomes. Advanc. Genet. **11**, 101—145 (1962).
— The steric effect in lysogenization by bacteriophage lambda. II. Chromosomal attachment of the $b_2$ mutant. Virology **27**, 340—345 (1965).
FISCHER-FANTUZZI, L.: Integration of λ and λ $b_2$ genomes in nonimmune host bacteria carrying a λ cryptic prophage. Virology **32**, 18—32 (1967).
KAISER, A. D., and D. HOGNESS: The transformation of *Escherichia coli* with deoxyribonucleic acid isolated from bacteriophage λ *dg*. J. molec. Biol. **2**, 392—415 (1960).
KELLY, B.: Localization of P2 prophage in two strains of *Escherichia coli*. Virology **19**, 32—39 (1963).
KELLY, B. L., and M. G. SUNSHINE: Association of temperate phage P2 with the production of histidine negative segregants by *Escherichia coli*. Bioch. biophys. Res. Commun **28**, 237—243 (1967).
MANDEL, M.: Infectivity of phage P2 DNA in presence of helper phage. Molec. Gen. Genet. **99**, 88—96 (1967).
SASAKI, I., and G. BERTANI: Growth abnormalities in Hfr-derivatives of *Escherichia coli* strain C. J. gen. Microbiol. **40**, 365—376 (1965).
SIGNER, E. R.: Recombination between coliphages λ and Φ 80. Virology **22**, 650—651 (1964).
—, and J. R. BECKWITH: Transposition of the *lac* region of *Escherichia coli*. III. The mechanism of attachment of bacteriophage Φ 80 to the bacterial chromosome. J. molec. Biol. **22**, 33—51 (1966).
SIX, E.: Inheritance of prophage P2 in superinfection experiments. Virology **14**, 220—233 (1961).
— Affinity of *P2 rd l* for prophage sites on the chromosome of *Escherichia coli* strain C. Virology **19**, 375—387 (1963).
— Specificity of P2 for prophage site I on the chromosome of *Escherichia coli* strain C. Virology **29**, 106—125 (1966).
TAYLOR, A. L., and M. S. THOMAN: The genetic map of *Escherichia coli* K-12. Genetics **50**, 659—677 (1964).
ZICHICHI, M. L., and G. KELLENBERGER: Two distinct functions in the lysogenization process: the repression of phage multiplication and the incorporation of the prophage in the bacterial genome. Virology **19**, 450—460 (1963).

## Discussion

*Novich:* Could you not alternatively explain the failure to get induction of the temperature-sensitive P2 mutant except when super-infected by assuming that the repressor (immunity substance) is held very tightly by the operator (DNA) and is stabilized when bound. The super-infecting phage could accept the repressor, allowing the prophage to be expressed.

*Bertani:* Best evidence against that is the fact that the superinfecting phage needed to produce P2 prophage detachment can even be a heteroimmune phage, i.e. one which most likely does not bind the P2 immunity repressor.

# Mutants of Serratia Marcescens Defective or Superactive in the Release of a Nuclease*

By

U. WINKLER

Institut für Mikrobiologie. J. W. Goethe-Universität, Frankfurt/M.

During growth several microorganisms, belonging to different systematic groups, release nucleases into the medium. One of these is *Serratia marcescens* (JEFFRIES et al., 1957; ROTHBERG and SWARTZ, 1965). Bacteria of this species presumably excrete only a single nuclease which degrades both, DNA and RNA, as revealed by biochemical studies (EAVES and JEFFRIES, 1963). Since extracellular nucleases are expected to be troublesome in certain experiments, e.g. transfections, we were interested in mutants of *S. marcescens* showing no more extracellular nucleolytic activity. While isolating these socalled $nuc^-$ mutants we found accidentally some others $(nuc^{su})$ which were superactive in excreting nuclease. This paper describes some of the properties of these $nuc^-$ and $nuc^{su}$ mutants. Preliminary results were already presented at the annual DGHM-meeting at Kiel (WINKLER, 1967).

## Material and Methods

### Bacterial Strains and Bacteriophages

*S. marcescens* HY    prototrophic (LABRUM and BUNTING, 1953)

*S. marcescens* W 263 $his^-$ a 56, histidine requiring mutant of strain HY (KAPLAN, unpubl.)

*S. marcescens* W 200 $his^- phr^-$, mutant of strain W 263, more or less unable for photoreactivation of UV-lesions (WINKLER and HEIL, 1968). All newly isolated *nuc* mutants described in this paper were derived from this strain. They are named $nuc^-203$, $nuc^{su}207$, and $nuc^{su}275$. Correspondingly, W 200 is mostly called $nuc^+200$.

*S. marcescens* CVrc3 prototrophic (KAPLAN, unpubl.)

* Part of the work has been done at the Institut de Biologie Moleculaire, Université de Genève, Geneva, Switzerland.

S. *marcescens* CN    prototrophic (WINKLER, 1963)
E. *coli*        B       from the stock collection of Dr. A. BOLLE, Genève
E. *coli*        HfrC    from the stock collection of Prof. Dr. R. W. KAP-
                         LAN, Frankfurt/M.
*Kappa*                  temperate phage (ELLMAUER and KAPLAN, 1959),
                         growing on all *Serratia* strains mentioned above
                         except CVrc3.

## *Media*

Broth (NB):  8 g Bacto Nutrient Broth, 5 g NaCl, 1000 ml Aqua dest.
NB-agar    : 15 g Bacto Agar, 1000 ml Broth
NB-topagar: 5 g Bacto-Agar, 1000 ml Broth
M9         : 7 g $Na_2HPO_4 \cdot 2 H_2O$, 3 g $KH_2PO_4$, 1 g $NH_4Cl$, 0.5 g
             NaCl, 4 g Glucose, $10^{-3}$ M $MgSO_4$, $10^{-4}$ M $CaCl_2$, 1000 ml
             Aqua dest.
M9s        : 1000 ml M9, supplemented with 12.5 ml of a 20 % solution
             of Bacto Casamino Acids
P-buffer   : 7 g $Na_2HPO_4 \cdot 2 H_2O$ (SÖRENSEN), 3 g $KH_2PO_4$ (SÖREN-
             SEN), 4 g NaCl, 1000 ml Aqua dest.
DNA-agar   : 3 g DNA of low molecular weight (FLUKA AG, Buchs),
             1000 ml NB-agar. The DNA was dissolved in steril Aqua
             dest. before mixing it with autoclaved NB-agar. The final
             pH was adjusted to 7.2.

### *Assay of Deoxyribonuclease (DNase) and Ribonuclease (RNase)*

The procedure was as described by EAVES and JEFFRIES (1963). Several tubes were equally prepared by mixing 0.5 ml of 0.4 M Tris-HCl buffer of
pH 8.8 containing 0.04 M $MgCl_2$, 1 ml of substrate solution, and 0.5 ml of
supernatant of a bacterial culture. After different times of incubation at
30° C the enzyme reaction was stopped by adding 0.5 ml of 0.75 % uranylacetate in 25 % perchloric acid. The precipitate formed within few
minutes in an ice bath was removed by 15 min of low speed centrifugation. 0.1 ml of each of the supernatants were mixed with 3 ml of distilled
water and placed in a *Zeiss* spectrophotometer (type M4 Q III). The
optical density at 260 nm was due to fragments of nucleic acid which are
not precipitable by the uranylacetate reagent.

The substrate in the DNase-assay was a 0.2 % solution of highly
polymerized Salmon sperm DNA (Calif. Biochem. Corp.) in distilled
water. In the RNase-assay a similar solution of 1 % yeast RNA has been
used. For that purpose commercially available RNA of low molecular
weight was purified according to FRISCH-NIGGEMEYER and REDDI, 1957[1].

---

[1] Most of the purified RNA used in this work was a generous gift of Prof.
P. F. SPAHR, Genève. The rest was RNA purified by R. A. WINKLER.

The O.D.$_{260}$ of 1 % solutions of DNA and RNA in distilled water were 206 and 204, respectively.

## Assay of Protease

A 1 % stock solution of casein "reinst nach Hammarsten" (SIEGFRIED, Zofingen) was prepared in 0.1 M SÖRENSEN phosphate buffer of pH 7.6 according to KUNITZ (1946). Equal volumes (1.0 ml) of this solution and of supernatant of a centrifuged bacterial culture were mixed in several tubes at 37°C. After different times of incubation hydrolysis was stopped by adding 3.0 ml of cold 10 % TCA (CASTAÑEDA-AGULLO, 1956). The precipitate formed within 30 min at room temperature was removed by 20 min of low speed centrifugation. The supernatant, however, was undiluted placed into a Zeiss spectrophotometer. The optical density at 280 nm gave an estimate of the amount of TCA-soluble material enzymatically released from casein.

## General Biological Techniques

Plating of bacteria or phages was done by the softagar technique (see ADAMS, 1959). Dilutions were all made in P-buffer. Infected agar plates or culture tubes were incubated at 30°C if not otherwise stated. For aeration of broth- or M9s-cultures during incubation the tubes were put into a rolling apparatus. Phage stocks were produced by the softagar technique and purified by low speed centrifugations.

## Isolation of Nuc Mutants

Bacteria of the late log-phase (about 10⁹/ml) were washed and treated with N-methyl-N'-nitro-N-nitrosoguanidine (Light and Co., Colnbrook; abbrev. NG). The conditions were 175 mg/ml NG in P-buffer of pH 6.0 for 15 min at 30°C. The treatment was stopped by dilution and the bacteria were plated on NB-agar. Next day petri dishes with about 10⁴ colonies were seperately washed off and the suspensions of cells were spread on DNA-agar to give 150—200 colonies per plate on incubation. These colonies were finally replicated on NB-agar (LEDERBERG and LEDERBERG, 1952) in order to reserve them because the DNA-agar was now treated with vapor of 37 % HCl for 3 min. All undegraded DNA was precipitated. Therefore, nuc⁻ colonies were located in areas where the agar was white and turbid. Colonies of the wildtype and of the nuc$^{su}$ mutants, however, had a clear halo of small or large diameter, respectively. The use of nutrient agar containing DNA was first described by JEFFRIES et al. (1957), but these authors poured HCl on the plates.

## Test for Spontaneous Mutability

Bacteria: Single colonies of the strains to be tested (which were all his⁻ sm$^s$) were incubated in NB for 16 h and then plated on NB-agar (for the total number of viable cells), on NB-agar containing 5 μg/ml streptomycinsulfate (for the $sm^r$ mutants) and on M9-agar (for the $his^+$ mutants). At the beginning the average number of cells was $2.10^7$ on the strept.-plates and $7.10^8$ on the M9-plates. $Sm^r$ and $his^+$ colonies were counted after 2 and 3 days of incubation, respectively.

Phage *Kappa*: Wildtype plaques developed on strain HY were used separately for producing phage stocks on each of different host strains. These stocks were then plated on strain HY. Plaques with a narrow violet halo (*e* mutants) or a clear center (*c* mutants) were scored after overnight incubation on plates with up to 350 or 3500 plaques, respectively.

### Ultraviolet-Irradiation

Phages diluted at least $10^{-2}$ in P-buffer were irradiated in a petri dish with a germicidal lamp (Osram HNS-12) emitting mainly radiation of wavelength 253.7 nm. The intensity at the working distance was about 10 ergs $mm^{-2} sec^{-1}$ as measured according to LATARJET et al. (1953). Irradiated phages were plated and incubated under yellow light in order to prevent photoreactivation.

### Test for Host-Controlled Modification and Restriction

Phages grown separately on *nuc* mutants were plated parallel on strain $nuc^+ 200$ and the one used for making the phage stock. As long as the efficiency of plating (EOP) was equal under both plating conditions strains did not differ in their ability to modify phage *Kappa*. In order to compare the restriction ability of *nuc* mutants with that of their parent strain $nuc^+ 200$ phage *Kappa* was used which had been grown on strain CN before (STEIGER, 1964).

## Results

### Isolation of Nuc Mutants

Bacteria of strain $nuc^+ 200$ were treated with NG as described under methods. In two independent experiments only few cells were inactivated but the mutagenic effect of NG was very strong as indicated by the appearence of many colonies ($= pig^-$ mutants) showing no or a different pigmentation than the wildtype (Table 1). Since most of the cells were in the growing phase when they were exposed to the mutagen many of these $pig^-$ mutant colonies were not pure white or pink clones but sectored or sprinkled.

For identifying *nuc* mutants the mutagen-treated bacteria were not plated directly on DNA-agar but first incubated on NB-agar (see methods). This intermediate step was necessary for allowing phenotypic expression of the new genotypes and segregation of multinucleated cells.

On DNA-agar treated with HCl 18 h old colonies of the parent strain $nuc^+200$ were surrounded by clear halos of about 2 mm in width. Corresponding colonies of $nuc^{su}$ strains showed halos with approximately 7 mm in width and those of $nuc^-$ bacteria did not have any halo except the

Table 1. *Lethal and mutagenic effect of NG on S. marcescens $nuc^+200$. Colony pigmentation ($pig^-$) mutants were scored on NB-agar the day after the cells were treated with NG. One day later nuclease (nuc) mutants were identified on DNA-agar (for details see* METHODS*). In both experiments the frequency of spontaneous $pig^-$ mutants was less than 0.1%*

| Time of | $N/N_0$ | Fraction of mutants | | |
|---------|---------|--------|--------|--------|
| experiment | % | $pig^-$ | $nuc^-$ | $nuc^{su}$ |
| Aug. 66 | 85 | 76/637 | 5/6000 | 2/6000 |
| May 67 | 100 | 58/824 | 7/6800 | 14/6800 |
|  | 93% | 9.16% | 0.094% | 0.125% |

leaky mutants. Among 12,800 colonies tested on DNA-agar a total of 12 $nuc^-$ and 16 $nuc^{su}$ mutants was found (Table 1). At least 9 strains of each of the two mutant groups arose by nonidentical mutational events as known from the experimental procedure. Most of the $nuc$ mutants produced red pigment and were monoauxotrophic for histidine like the parent strain W200. Other strains were disregarded. The mutants mainly described in this paper are the strains $nuc^-203$ and $nuc^{su}207$.

In addition to W200 several other strains of *S. marcescens* were tested on DNA-agar, e. g. CVrc3, CN, and a subculture of W200 lysogenic for phage *Kappa*, but none of these differed in its ability to release nuclease.

Usually bacteria of *E. coli* do not excrete nuclease into growth medium. However, since this species belongs to the *Enterobacteriaceae* as *S. marcescens* does we tried to induce mutants of *E. coli* HfrC which release nuclease. Bacteria were treated with NG as already described. About 50% of the cells survived. None of 11,600 colonies tested showed any nucleolytic activity on DNA-agar.

### Extracellular DNase, RNase and Protease Activity

Broth agar containing DNA was sufficient to isolate $nuc$ mutants of *S. marcescens* but for characterizing them optical tests were used. Resting cultures of $nuc^-203$, $nuc^{su}207$, and parent strain $nuc^+200$ were centrifuged and supernatants were allowed to react with DNA, RNA, or casein as substrate until the reaction was stopped by some reagent precipitating all substrate molecules which were not or only slightly degraded until then. After removing the precipitate by centrifugation the optical density of the supernatant was measured at 260 or 280 nm, depending on the test (see methods). As shown in Fig. 1 and 2 both *nuc* mutants differ from the parent strain in a simultaneous change in the ability to attack DNA as

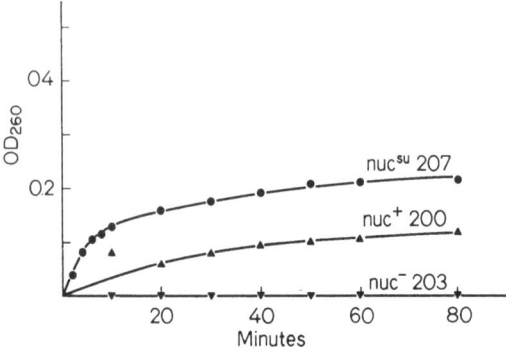

Fig. 1. Extracellular DNase activity of different strains of *S. marcescens*. The formation of products of DNA breakdown which are soluble in uranylacetate-perchloric acid is shown as a function of time. Bacterial cultures used in this test were grown in M9s for 36 h before removing the cells by centrifugation. The supernatants were assayed without further dilution (see METHODS). All samples were measured against a blank of distilled water and corrected for the absorption at $t = 0$

well as RNA. Regarding this property all other 18 *nuc* mutants behaved identical, therefore excluding the possibility that the two phenotypic differences between parent strain and *nuc* mutants arose by more than a single mutational event. Since casein was hydrolysed equally well by cell-free cultural medium of $nuc^- 203$, $nuc^{su} 207$, and $nuc^+ 200$ (Fig. 3) it can be concluded that the *nuc* mutants did not originate by mutations disturbing unspecifically the release of all extracellular enzymes. Consequently, *S. marcescens* $nuc^+ 200$ seems to release only a *single* type of nuclease degrading nucleic acids independently of their shugar residue. This agrees with results of EAVES and JEFFRIES (1953) who studied partly purified extracellular nuclease of another strain of *S. marcescens*.

In M9s solution the extracellular nucleolytic activity of $nuc^{su} 207$ was about 7 times as high as that of $nuc^+ 200$, calculated from changes in $O.D._{260}$ at the beginning of the enzyme reaction (Fig. 1 and 2). In some other experiments this difference was slightly higher.

The amount of fragments[1] enzymatically produced from DNA or RNA during 80 min of incubation with cell-free cultural medium of $nuc^{su} 207$

---

[1] "Fragments" are those oligonucleotides or even smaller parts of nucleic acid which were not precipitated by the uranylacetate-reagent used for the test. The amount of fragments was measured spectrophotometrically at 260 nm.

was about 40—50 % of the total nucleic acid originally present in each test tube. However, this does not allow conclusions on the degree of degradation of the nucleic acids, because the optical test which we used did not distinguish between small breakage products of different size. Furthermore, the average molecular weight of the substrate molecules was unkown.

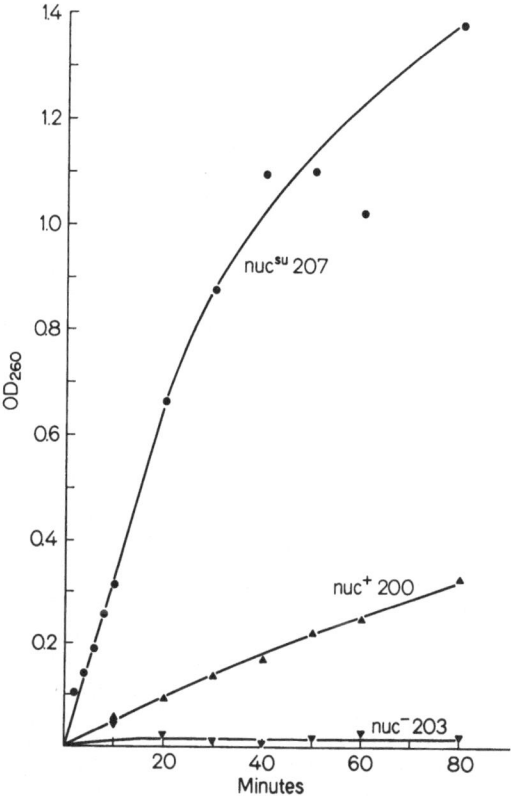

Fig. 2. Extracellular RNase activity of different strains of *S. marcescens*. Bacterial cultures used in this test were the same as in the DNase-assay of Fig. 1

No RNase inhibiting factor which might be present in cultures of $nuc^+200$ but not in those of $nuc^{su}207$ was found in experiments where the extracellular nucleolytic activity of $nuc^{su}207$ was measured in the presence of cultural medium of $nuc^+200$ (1:1 mixture).

In general the DNA used for DNase tests was dissolved in distilled water and was therefore denatured. However, DNA dissolved in $10^{-3}$ M Tris-HCl buffer (which preserves its native configuration) was an equally good substrate for the extracellular DNase of $nuc^+200$.

During growth of *S. marcescens* in M9s solution bacteria release material with a broad absorption maximum around 260 nm. Since bacteria of $nuc^- 203$, $nuc^{su} 207$, and $nuc^+ 200$ produced approximately equal

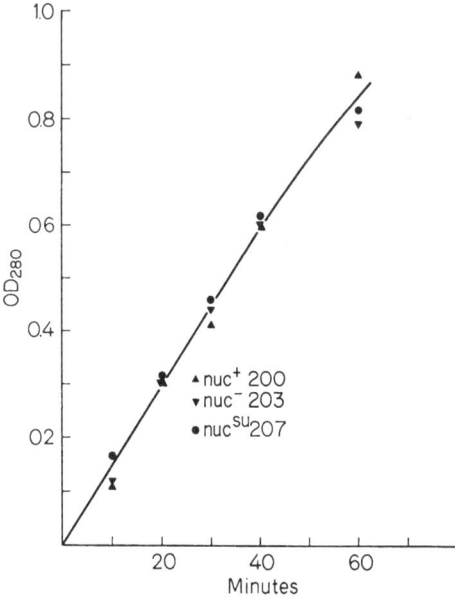

Fig. 3. Extracellular proteolytic activity of different strains of *S. marcescens*. The formation of TCA-soluble material from casein is shown as a function of time. Bacterial M9s cultures used in this test were 30 h old when centrifuged. The supernatants were diluted 1 : 2 in P-buffer of pH 7.6 and then assayed as described under METHODS. The blank consisted of $t = 0$ samples

amounts of the "260-material" this process seems to be unrelated to the release of nuclease. In agreement with this more than 90 % of the "260-material" was of low molecular weight because it was neither precipitable by 10 % TCA nor by the uranylacetate reagent.

*Extra- and Intracellular RNase Activity and Growing State of the Bacteria*

Nuclease which is present in the cell-free solution of an old *Serratia* culture might have been released only during the logarithmic phase of the bacteria or is simply due to autolysis of cells during stationary phase. The following experiments were performed in order to distinguish between these two possibilities. Bacteria of strain $nuc^{su} 207$ and $nuc^+ 200$, respectively, were grown in M9s solution and samples were taken at different times mainly for assaying the number of colony formers per ml and the extracellular RNase activity.

As shown in Fig. 4 99 % of the extracellular RNase activity which was present in a 36 h old culture of $nuc^{su}207$ was liberated during the late logarithmic phase, where the titer of viable cells increased from $3.10^8/ml$

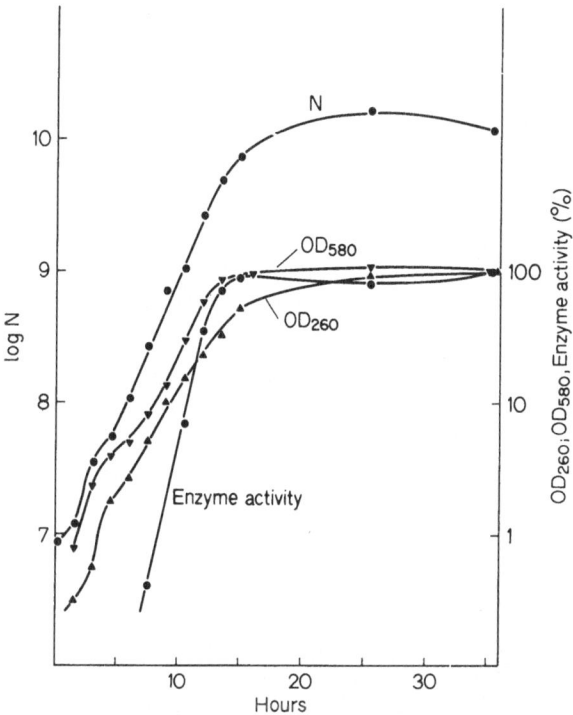

Fig. 4. Extracellular RNase activity in a M9s culture of $nuc^{su}$ 207 as a function of the growing state of the bacteria. $N$ gives the number of viable cells per ml. Each point of the RNase curve represents the slope of a different enzyme assay curve. The slope at 35.5 h, which is equal to 100% enzyme activity, corresponds to an increase in O.D.$_{260}$ from 0 to 0.1 within 32 sec.
OD$_{580}$ gives the optical density of culture samples at 580 nm *before* centrifugation. OD$_{260}$ gives the optical density of the supernatant of centrifuged culture samples. 100% of OD$_{260}$ and OD$_{580}$ correspond to absolute values of 5.65 and 3.57, respectively

to $7.10^9/ml$. Later on, during 20 h of stationary phase the titer of viable bacteria varied only by a factor of two and the cell-free RNase activity kept constant, i.e. most of the extracellular nuclease was probably not due to autolysis. Fig. 4 shows furthermore that bacteria of strain $nuc^{su}207$ excrete "260-material" at a slower rate than RNase. This agrees well with observations reported and discussed in the chapter before. Bacteria of strain $nuc^+200$ behaved similar to those of $nuc^{su}207$ in so far as they released most of their RNase shortly before reaching the resting phase (Table 2).

13*

The following experiments were performed to see how much intracellular RNase is present in a bacterial culture at the time when samples are taken for assaying the extracellular enzyme level[1]. For this purpose cultures of different age were always divided in two parts: One was immediately centrifuged in order to measure only the extracellular RNase activity in the supernatant. The other part, however, was vigorously mixed with toluene before sedimentation of the cells because this treatment is generally believed to liberate intracellular enzymes.

Table 2. *Total RNase activity and fraction of intracellular RNase in different bacterial cultures as a function of the titer of viable cells. Bacterial cultures grown in M 9 s were treated with toluene (final concentration 2—3%) at 30°C for 15 min. The total RNase activity is always expressed in per cent (rel. units) of the highest activity measured in each series of experiments. The numbers in brackets relate the $nuc^+$ 200 to the $nuc^{su}$ 207 experiment. All enzyme activities are calculated from the slopes of different enzyme assay curves*

| Strain No. | Incubation time h | Titer of viable cells per ml $\times\ 10^9$ | RNase | |
|---|---|---|---|---|
| | | | total activity (rel. units) | % released by toluene |
| $nuc^{su}$ 207 | 14.3 | 1.50 | 13.9 | 47.4 |
| | 17 | 3.01 | 49.7 | 49.4 |
| | 21.3 | 8.40 | 100 | 40.0 |
| $nuc^+$ 200 | 18 | 3.66 | 23.3   (3.3) | 52.4 |
| | 20.3 | 7.28 | 84   (11.8) | 0 |
| | 24.3 | 16.9 | 100   (14.1) | 0 |
| $nuc^-$ 203 | 16 | 1.82 | 0 | 0 |
| | 19.3 | 5.59 | 0 | 0 |
| | 22.8 | 13.20 | 0 | 0 |

Toluene treatment of $nuc^{su}$ 207 bacteria caused release of RNase activity almost as much as already was present extracellularly (Table 2). The constant ratio of intracellular to extracellular RNase activity in two samples taken at different times of the experiment indicates that during the interval formation *and* release of RNase had a similar kinetics. With increasing age of the culture the fraction of intracellular RNase among the total decreased slightly. This was confirmed by other experiments. The effect of toluene on $nuc^+$ 200 seemed to be much more dependent on the age of the culture than in the case of $nuc^{su}$ 207 (Table 2). Either "old" bacteria did really not contain intracellular RNase in amounts detectable by our assay or these cells were resistant to the attack of toluene.

---

[1] "Intracellular RNase" means here RNA degrading enzyme which is not extracellular. Part of it might not be truly intracellular but might only be bound to the cell wall.

In cultures of strain *nuc⁻* 203 no RNase could be liberated by treating the bacteria with toluene (Table 2). This suggests that the intracellular RNase detected in corresponding experiments with *nucˢᵘ* 207 and *nuc⁺* 200 is probably of the extracellular type.

### Some Biological Properties of nuc Mutants

Mutants *nuc⁻* 203 and *nucˢᵘ* 207 were compared with the parent strain *nuc⁺* 200. Only some of those biological properties of bacteria were studied which might be influenced by mutational changes in the formation of a nuclease.

*Growth.* During the logarithmic phase bacteria of all three strains had the same generation time of 76 to 78 min in M9s solution. Usually all three cultures reached a final titer of $1.2 \cdot 10^{10}$ viable cells per ml.

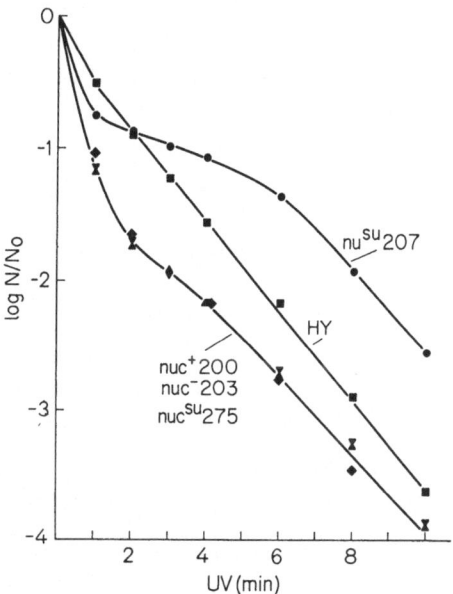

Fig. 5. Survival of extracellularly UV-irradiated phage *Kappa* as a function of dose (min) and the bacterial strain used as plating indicator

*Spontaneous mutability.* The strains studied were all *his⁻ smˢ*. The frequency of *his⁺* revertants in NB-cultures of the three strains varied only between 4.1 and $5.9 \cdot 10^{-9}$. The corresponding values for *smʳ* mutants resistant to 5 μg/ml of streptomycinsulfate in NB-agar were 1.1 and $1.7 \cdot 10^{-6}$. Stocks of phage *Kappa* obtained from growth on the three strains did not significantly differ in the fraction of *e* mutants (1.8 to $2.7 \cdot 10^{-3}$) and *c* mutants (1.6 to $5.0 \cdot 10^{-5}$; P = 0.60).

198 U. WINKLER:

*Host cell reactivation (HCR). S. marcescens* HY is able to reactivate in the dark more than 90 % of lethal damage induced in phage *Kappa* by extracellular UV-treatment (WINKLER, 1964). Strain $nuc^+200$ obtained from HY is slightly reduced in its ability to perform HCR (WINKLER and HEIL, 1968). Strain $nuc^-203$ and three different $nuc^{su}$ mutants (e.g. No. 275) showed the same HCR as strain $nuc^+200$ (Fig. 5). The abnormal behaviour of $nuc^{su}207$ as host of UV-irradiated phage might be due to a secondary mutation.

*Host controlled modification and restriction.* Strain $nuc^{su}207$ and 4 different $nuc^-$ mutants (e.g. No. 203) restricted phage *Kappa* grown on CN as HY does (WINKLER, 1963). The efficiency of plating varied only between 1.2 and $3.6 \cdot 10^{-3}$. No differences were found in the modifying ability of HY and the 5 *nuc* mutants mentioned above.

*Formation of bacteriocin. S. marcescens* HY produces a bacteriocin which inactivates *S. marcescens* CVrc3 as well as *E. coli* B and HfrC (WINKLER, 1967). All 16 $nuc^{su}$ mutants produced (or released) significantly more bacteriocin than the parent strain $nuc^+200$ did. Mutations from $nuc^+$ to $nuc^-$, however, were not correlated with a change in bacteriocin formation. Details on this subject will be reported elsewhere.

## Discussion

### *Defective Mutants (nuc⁻)*

After treatment of *S. marcescens* W200 with nitroso-guanidine several mutants were found more or less unable to release nuclease into their growth medium (Table 1 and 2). It is unknown so far whether the genetic information for the extracellular nuclease of *S. marcescens* W200 is carried on the bacterial chromosome or on a plasmid. Consequently, one might either assume that the $nuc^-$ mutants arose by mutation, for instance in a structural gene, or that they are due to accidental loss of a hypothetical plasmid. The idea that synthesis of extracellular enzymes might be governed by plasmids comes from NOVICK (1963), RICHMOND (1965), and some others who found in *Staph. aureus* that genes responsible for exo-penicillinase behave like extrachromosomal particles.

Like *S. marcescens*, several other microorganisms excrete nuclease into their growth medium too, for instance *Staph. aureus* (CUNNINGHAM et al., 1956), *Pseud. aeruginosa* (CATLIN, 1956), *Bac. subtilis* (NISHIMURA and NOMURA, 1958), and *Neurospora crassa* (TAKAI et al., 1966), but artifically induced $nuc^-$ mutants have never been reported so far. From *E. coli*, however, which usually produces only intracellular nucleases several mutants deficient in RNase I (GESTELAND, 1966) or endonuclease I (DÜRWALD and HOFFMANN-BERLING, 1968) were recently isolated by special techniques.

On the basis of biochemical studies by LESCHINSKAJA and coworkers (1966) it is uncertain so far whether cells of *S. marcescens* liberate at least two nucleases with different substrate specificity or only a single type of enzyme hydrolysing DNA as well as RNA. Our results fit best the latter hypothesis because 9 $nuc^-$ mutants isolated independently for DNase deficiency showed all a simultaneous loss of extracellular RNase activity (Fig. 1 and 2). The first hypothesis, however, was not excluded.

Treatment of $nuc^-$ bacteria of *S. marcescens* with toluene liberated no RNase in appreciable amounts (Table 2). Since it is unlikely that the cells were really free of any RNase this might be explained alternatively as follows:

a) The toluene-treatment was insufficient to disrupt the permeability barriers of the cells completely so that truly intracellular located RNase could not be liberated.

b) The action of toluene on the bacteria enabled intracellular enzymes to escape but our enzyme-assay (for instance the use of pH 8.8) was not suitable to detect typical intracellular RNase.

### Superactive Mutanis ($nuc^{su}$)

Several mutants were isolated from *S. marcescens* W 200 which show much higher extracellular *and* intracellular nucleolytic activity than the wildtype. This difference might be explained in at least 3 ways:

a) Regardless of one or several types of nucleases excreted by the wildtype of *S. marcescens*, cells of $nuc^{su}$ mutants release an additional type of nuclease.

b) Cells of the wildtype and $nuc^{su}$ mutants produce approximately equal amounts of a single type of extracellular nuclease, but the mutant enzyme has an increased activity per catalytic site; i.e. the change from $nuc^+$ to $nuc^{su}$ is due to a mutation in a *structural* gene.

c) Synthesis of extracellular nuclease occurs in bacteria of the wildtype at a slower rate or is more repressible than in $nuc^{su}$ cells. Thus $nuc^{su}$ mutants might have arisen by mutations in DNA-segments with *control* function (promoter; operator; i-gene etc.).

It is not possible yet to decide which of these hypothesis fits the facts; this requires purification and fractionation of the extracellular nuclease. At present the third hypothesis is most favoured because it was proved right recently by RADDING (1964, 1966) in the case of *E. coli* lysogenic for a $nuc^{su}$ mutant (No T 11) of phage lambda.

The limited correlation of bacteriocin formation and extracellular nuclease activity will be discussed elsewhere.

## Acknowledgements

I am greatly indebted to Professor E. KELLENBERGER and Professor W. ARBER for the opportunity to perform part of this work in their laboratories of the Université de Genève. I wish to thank Professor P. F. SPAHR and Professor R. KRETZINGER, Genève, for helpful advice and many discussions. Some of the biological experiments were done with the excellent technical assistance of Miss A. BECKER, Frankfurt.

## Summary

Cultural medium of *S. marcescens* W 200 contains at least DNase, RNase, and Protease-activity. From this strain mutants were isolated which were either defective ($nuc^-$) or superactive ($nuc^{su}$) in forming extracellular DNase. Since all these mutants showed simultaneous changes in their ability to produce extracellular RNase it was concluded that *S. marcescens* W 200 releases usually only a single type of nuclease degrading both, DNA as well as RNA. Casein was hydrolysed equally well by cultural medium of *nuc* mutants and strain W 200 thus indicating that the mutants did not arise by mutations disturbing unspecifically the release of all extracellular enzymes. Extracellular nuclease of strain W 200 and $nuc^{su}$

mutants (as measured by its Rnase-activity) was synthesized and liberated in the late logarithmic phase of cultural growth. During the resting phase there were no changes in the extracellular level of nuclease. Treatment of a $nuc^-$ mutant with toluene which is expected to open the cells did not liberate any RNase. Superactive mutants, however, treated similarly released almost as much RNase as already was present extracellularly. Mutations from $nuc^+$ to $nuc^{su}$ were always accompanied by a strong increase in the bacteriocin production of the cells. Other biological properties, as growth rate, spontaneous mutability, dark reactivation and restriction were unchanged in $nuc^{su}$ and $nuc^-$ mutants. Several hypothesis explaining how $nuc$ mutants might differ from the wildtype were discussed.

## References

ADAMS, M. H.: Bacteriophages, pp. 592. New York: Interscience Publ. Inc. 1959.
CASTAÑEDA-AGULLO, M.: J. gen. Physiol. 39, 369 (1956).
CATLIN, B. W.: Science 124, 441 (1956).
CUNNINGHAM, L., B. W. CATLIN, and M. J. PRIVAT DE GARILHE: Amer. Chem. Soc. 78, 4642 (1956).
DÜRWALD, H., and H. HOFFMANN-BERLING: J. mol. Biol. 34, 331 (1968).
EAVES, G. N., and C. D. JEFFRIES: J. Bact. 85, 273 (1963).
ELLMAUER, H., and R. W. KAPLAN: Naturwissenschaften 46, 150 (1959).
FRISCH-NIGGEMEYER, W., and K. K. REDDI: Biochem. biophys. Acta (Amst.) 26, 40 (1957).
GESTELAND, R. F.: J. mol. Biol. 16, 67 (1966).
JEFFRIES, C. D., D. F. HOLTMAN, and D. G. GUSE: J. Bact. 73, 590 (1957).
KUNITZ, M.: J. gen. Physiol. 30, 291 (1946).
LABRUM, E. L., and M. I. BUNTING: J. Bact. 65, 394 (1953).
LATARJET, R. P., P. MORENNE, and R. BERGER: Ann. Inst. Pasteur 85, 174 (1953).
LEDERBERG, J., and E. LEDERBERG: J. Bact. 63, 399 (1952).
LESCHINSKAJA, I. B., V. I. TANJASHIN, B. M. KURINENKO, and N. KALACHEVA: Mikrobiologia 35, 1094 (1966).
NISHIMURA, S., and M. NOMURA: Biochem. biophys. Acta (Amst.) 30, 430 (1958).
NOVICK, R. P.: J. gen. Microbiol. 33, 121 (1963).
RADDING, C. H.: Proc. nat. Acad. Sci. (Wash.) 52, 965 (1964).
— J. molec. Biol. 18, 251 (1967).
RICHMOND, M. H.: Brit. med. Bact. 21, 360 (1965).
ROTHBERG, N. W., and M. N. SWARTZ: J. Bact. 90, 294 (1965).
STEIGER, H.: Z. allg. Mikrobiol. 4, 367 (1964).
TAKAI, N., T. UCHIDA, and F. EGAMI: Biochem. biophys. Acta (Amst.) 128, 218 (1966).
WINKLER, U.: Z. Naturforsch. 18 b, 118 (1963).
— Virology 24, 518 (1964).
— Zbl. Bact. I. Orig., 205, 62 (1967).
—, and K. HEIL; Photochem. Photobiol., in press (1968).

## Discussion

*Starlinger:* How well established is the identity of the RNase and the DNase? Could it be that you deal with regulation mutants of an operon? Have the two activities been carried jointly through a fractionation procedure?

*Winkler:* To your first question: The existence of at least 9 independently isolated $nuc^-$ mutants which all simultaneously lost the ability to degrade RNA *and*

DNA is fairly good indirect evidence that wild type bacteria liberate only a single type of extracellular nuclease. To your second question: The complete lack of *nuc* mutants showing only a single defect (either DNase or RNase) is not compatible with your assumption that I am dealing with regulation mutants exept you make the additional assumption that the structural genes for the extracellular DNase and RNase are more or less completely resistant to the mutagenic effect of NG where as the regulation gene consists of more than a single NG-hot spot. To your third question: We did not fractionate yet the extracellular nucleolytic activity.

*Matthaei:* Do you know the mode of action and the degradation products of the nuclease excreted ? What are its ionic requirements ?

*Winkler:* We do not know yet exactly the mode of action, but it is presumably an endonuclease. We worked routinely at a $Mg^{++}$ concentration of $10^{-2}$ molar.

# In vivo Studies of Inducer-Repressor Interactions

By

AARON NOVICK

Institute of Molecular Biology, University of Oregon, Eugene, Oregon

The nature of the repressors which control gene expression and the mechanism of their action promise to be clarified very soon largely because of the recent successful development of *in vitro* assays for the *lac* repressor (GILBERT and MÜLLER-HILL, 1966) and for the repressor controlling the temperate phage $\lambda$ (PTASHNE, 1967). These developments have followed earlier experiments with whole cells which originally led to the postulation of repressors and which continued to provide further support for their existence. In the present report, I would like to describe some studies in whole cells of the interaction of galactoside inducers with the *lac* repressor.

In these studies it is generally assumed that under similar conditions of growth changes in rate of $\beta$-galactosidase formation directly reflect changes in level of repressor. That the rate of formation of $\beta$-galactosidase in uninduced bacteria — the so-called basal rate — is determined by the level of repressor present has now become quite clear. Early work showed that the basal level could not be accounted for by the presence of de-repressed (constitutive) mutants since the number of such mutants is typically two few. Further, in kinetic studies of the de-repression of certain mutants which become de-repressed following transfer to growth at high temperatures, it was seen that the rising rate of enzyme synthesis extrapolates back to the basal level at the time of transfer, consistent with the interpretation that the rate of enzyme synthesis is determined by the level of repressor at zero time as well as at later times (SADLER and NOVICK, 1965).

Although there were reports that the basal rate was the consequence of an escape from repressor control at some point during gene duplication (HANAWALT and WAX, 1964) it is now evident that no such leakage occurs. The apparent dependence of basal rate on DNA synthesis is not true of all bacterial strains (OVERATH and STANGE, 1966). In a further study, the formation of $\beta$-galactosidase was determined in a mutant in which the synthesis of DNA could be stopped abruptly by transfer from growth at 25° to 39° (FANGMAN, GROSS, and NOVICK, 1967). Despite the arrest in synthesis of DNA, protein and RNA continue to be made at a high rate for up to a twenty fold increase. During this time, $\beta$-galactosidase is synthesized at a constant rate (relative to total mass) while DNA synthesis (incorporation of $^{14}$C-thymidine) ceases. Similar results were obtained in studies of the formation of ornithine transcarbamylase.

It could be argued that those cells which were hypothetically de-repressed at the instant synthesis of DNA was stopped, were locked in the de-repressed state. The synthesis observed after arrest of synthesis of DNA would be thus entirely due to these cells, which in the case of $\beta$-galactosidase would represent 1/2000 of the population (since the basal rate is 1/2000 of the maximum rate). To test this possibility the distribution of $\beta$-galactosidase among individual bacteria was studied, using the technique developed by Professor BORIS ROTMAN (ROTMAN, 1961), and it was found that the level of $\beta$-galactosidase rises uniformly among all cells during the block in synthesis of DNA.

These results make it clear that the basal rate has nothing to do with DNA synthesis and is in all likelihood determined by the level of repressor present. From the fact that the basal rate remains the same following the stop in synthesis of DNA, one can conclude in addition that there is no appreciable transcription associated with replication, *i.e.*, the repression mechanism allows no detectable leak during replication.

That the *lac* repressor must be a protein has been generally accepted for same time (SADLER and NOVICK, 1965), and recently, of course, this conclusion has been verified directly by the work of GILBERT amd MÜLLER-HILL (GILBERT and MÜLLER-HILL, 1966). That repressors are polymeric proteins is likely on theoretical grounds and has been argued to explain the gene dosage effects observed in mutants in which the synthesis of repressor is temperature-sensitive (SADLER and NOVICK, 1965). Study of the stabilization of repressor formation by inducers in these mutants, called $i^{TSS}$, has given further reason to believe that the *lac* repressor is polymeric. In addition these studies have given added corroboration that the substance identified by GILBERT and MÜLLER-HILL is the $i$ gene product, and as argued earlier, is the agent responsible for gene repression.

Stabilization of formation of repressor by inducers in $i^{TSS}$ strains can be dramatically demonstrated in the following way. When a culture of an $i^{TSS}$ strain is incubated at 41°, $\beta$-galactosidase synthesis occurs at maximum rate. If in addition $10^{-3}$M IPTG is present, synthesis of $\beta$-galactosidase again continues at maximum rate. But, if the IPTG is removed and the culture incubated further at 41°, there is an abrupt repression of synthesis of $\beta$-galactosidase. The normal rate of synthesis is not resumed until after several doublings of growth at 41°. In fact, the behavior of the culture after removal of IPTG is just like that of one which was grown at 30° and then transferred to 41°. Apparently IPTG restores the rate of formation of repressor in this $i^{TSS}$ strain to the full value found at low temperatures. But the repressor present is not manifest, of course, until the IPTG is removed.

Another demonstration of stabilization of formation of repressor in $i^{TSS}$ cultures can be obtained by growth of the culture with varying concentration of inducer at intermediate temperatures. Here it is possible to find conditions where there is an apparent repression of synthesis of $\beta$-galactosidase because the increased formation of repressor in the presence of inducer outweighs the inducing effect. For example, in Fig. 1

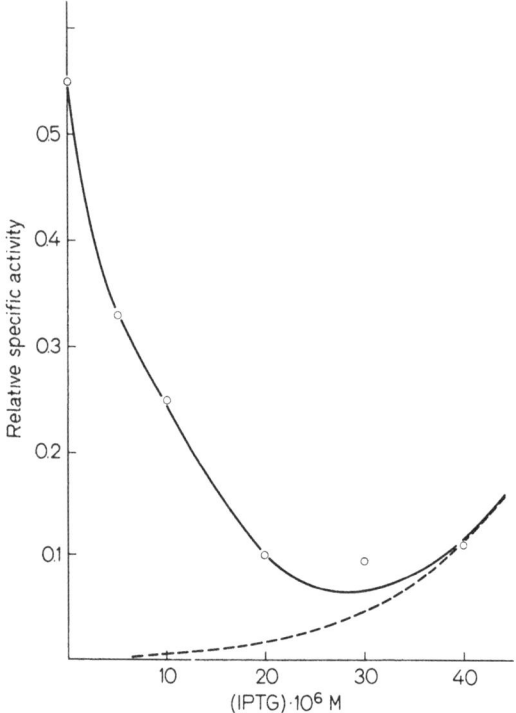

Fig. 1. The relative specific activities of a series of cultures of an $i^{Tss}$ strain (E 321) grown at 38.5° with variuos concentrations of IPTG, indicated on the abcissa, is given. A fully derepressed strain has a specific activity of 1.0. The dotted line represents the specific activities observed with wild type strains

is given the $\beta$-galactosidase level observed in a series of cultures grown at 38° at various concentrations of IPTG. It can be seen that at lower concentrations the rate of formation of $\beta$-galactosidase decreases with increase in concentration of IPTG. But as the concentration of IPTG approaches that which induces wild type *E. coli* there is the expected rise in rate of synthesis with increase in concentration of IPTG.

This stabilization of formation of repressor by IPTG in the $i^{TSS}$ strains provides an explanation for the otherwise curious phenomenon of the synthesis of a protein being temperature sensitive. For example, one can assume that the rate of synthesis of repressor falls toward zero at higher temperatures because an intermediate in the synthesis of repressor is thermolabile and that this intermediate can combine with IPTG and be stabilized by it.

From studies of the dependence of stabilization of formation of repressor in $i^{TSS}$ strains on concentration of IPTG, estimates of the binding constant of repressor for IPTG and other galactosides can be obtained. Such estimates can be based on the assumption that binding to IPTG stabilizes the temperature-sensitive precursor. This leads to an increase in rate of formation of repressor, and therefore, of level of repressor, which apparently more than offsets the reduction of effective repressor by the inducing action of these levels of IPTG. For example, consider the following scheme:

$$\text{Inactivated precursor}$$

$$\xrightarrow{\quad S \quad} \text{Procursor } (m) \xrightarrow{\qquad\qquad} \text{Repressor } (R)$$

with $\beta \uparrow$ above Procursor, and

$$m + I \overset{K}{\rightleftharpoons} mI$$

where $m = $ level of precursor, $R = $ level of repressor, the rate of thermal inactivation of repressor $= \beta m$, $I = $ concentration of IPTG, $mI = $ a complex of $I$ and $m$ which is thermostable, and $K = (m)(I)/(mI)$, the binding constant.

It is further assumed that the precursor-IPTG complex, $mI$, can equally well serve as a precursor of repressor. At low levels of repressor, most of the precursor is inactivated and its steady state concentration is given by:

$$m = \frac{S}{\beta},$$

where $S = $ rate at which the bacteria make the precursor and assuming $\beta \gg \alpha$, the bacterial growth rate constant. The rate of formation of repressor is then a function of the concentration of $m + mI$ which is given by $m(1 + I/K)$. To be general one could assume that the rate of formation of repressor is proportional to: $m^n [(1 + I/K)]^n$.

At lower levels of $I$ this can be approximated by:

$$m^n \left(1 + n \frac{I}{K}\right).$$

If one neglects the inducing effect at the lower concentration of inducer, the relative level of repressor present is given by the reciprocal of the rate of $\beta$-galactosidase formation (SADLER and NOVICK, 1965). Therefore, when

$I = K/n$, the rate of $\beta$-galactosidase formation is one-half the rate when
$I = 0$. Thus, the binding constant can be obtained from:

$$K = I^*n$$

where $I^*$ = the inducer concentration which reduces the rate of $\beta$-galac-
tosidase formation to one-half.

These expectations can only be realized at intermediate temperatures
like 38°, at lower temperatures the repressor level is already near maximum
and at higher temperatures the inducing effect predominates before the
level of repressor becomes appreciable.

Studies have been made of the stabilizing effect of a number of galactosides on formation of repressor in $i^{TSS}$ and in $i^{S, TSS}$ strains and the concentrations which reduced the rate of $\beta$-galactosidase production to one-
half are presented in Table 1. It can be seen that as expected from earlier
observation that $i^S$ strains are not inducible (WILLSON et al., 1964) much
higher concentrations of IPTG are needed for stabilization of formation
of repressor in $i^S$ strains.

Table 1. *Dissociation constants for various galactosides*

|  | Stabilization Expts. | | GILBERT and MÜLLER-HILL | |
|  | $i^{+, TSS}$ | $i^{S, TSS}$ | $i^+$ | $i^i$ |
|---|---|---|---|---|
| IPTG | $7 \times 10^{-6}$ M | $3 \times 10^{-3}$ M | $1.3 \times 10^{-6}$ M | $6 \times 10^{-7}$ M |
| TMG |  | $15 \times 10^{-3}$ |  | $2 \times 10^{-6}$ |
| ONPF | $2.5 \times 10^{-4}$ |  |  | $2 \times 10^{-5}$ |
| TPG | $6 \times 10^{-3}$ |  |  |  |

In Table 1 are also given the values of the constants obtained by
GILBERT and MÜLLER-HILL from direct measurement of the binding of
these inducers by the $i$ gene product protein. It can be seen that the
relative values (*e.g.*, IPTG compared to ONPF) are in excellent agreemen
with the values found here. The relative values do not depend, of course,
on the value of $n$, since the kinetics of formation of repressor should be
independent of the inducer used. The disparity in the absolute value for
IPTG of about five could be attributed to $n$ having a value of this order of
magnitude. In fact, studies of the effect of varying the precursor level by
varying the gene dosage indicated that the rate of formation of repressor
varied with the fourth power, of precursor concentration.

These observations provide further support for the belief that the
protein identified by GILBERT and MÜLLER-HILL is the protein responsible for the repression phenomenon observed in whole cell studies. They
likewise appear to be consistant with the earlier conclusion from whole

cell studies that the *lac* repressor is a tetrameric protein formed of monomers having a binding site for inducers.

The present interpretations also account for the fact that conditions can be found, *e.g.*, the $i^{TSS}$ strain at 38°, where certain concentrations of inducer actually reduce the rate of formation of $\beta$-galactosidase. Evidently the extent of reduction of effectiveness of repressor varies with the second power of concentration of inducer (BOEZI and COWIE, 1961), probably because two molecules of inducer must be bound to render a repressor molecule inactive. On the other hand, only one molecule of inducer need be bound to stabilize the formation of repressor; and the relative increase in repressor level can be substantial when most of it would otherwise be inactivated, as at 38°.

The assumption that the monomer of a polymeric protein is more unstable than the polymer is not without precedent. This has already been observed for $\beta$-galactosidase itself (D. PERRIN, personal communication). Precedents for the assumption that the rate of polymerization is proportional to the fourth power of the concentration of monomer are less available since so little is yet known of the mechanism of aggregation of polymeric proteins.

Another conclusion which is suggested by the present experiments is that IPTG does not produce a depolymerization of repressor, *i.e.*, the allosteric transition in repressor does not involve depolymerization. The kinetics of formation of repressor observed here indicate that there is no monomeric entity between the thermosensitive precursor and the oligomer. Therefore, if depolymerization were produced by IPTG, it should also sensitize $i^{TSS}$ repressor toward thermal inactivation; and this was not found in experiments designed to test this possibility.

Thus, the present results support the simple picture which has had general acceptance for some time and which soon promises to be tested directly *in vitro*. In this view the repressor is an allosteric protein which in the absence of inducer assumes a form which combines with operator and which in its presence assumes a form where this affinity is lost, leaving the operator free to perform its function. This expectation has now been verified by direct observation [GILBERT and MÜLLER-HILL, Proc. nat. Acad. Sci. (Wash.) 58, 2415 (1967)].

This research was supported by research grants from the National Science Foundation (U.S.) (N.S.F. 1068-GB5679) and from the U.S. Public Health Service (P.H.S. 868-A1-02808).

## References

BOEZI, J. A., and D. B. COWIE: Biophys. J. 1, 639 (1961).
FANGMAN, W., C. GROSS, and A. NOVICK: J. molec. Biol. 29, 317 (1967).
GILBERT, W., and B. MÜLLER-HILL: Proc. nat. Acad. Sci. (Wash.) 56, 1891 (1966).

HANAWALT, P., and R. WAX: Science **145**, 1061 (1964).
OVERATH, P., and H. STANGE: Z. Vererbungsl. **98**, 71 (1966).
PTASHNE, M.: Proc. nat. Acad. Sci. (Wash.) **57**, 306 (1967).
ROTMAN, B.: Proc. nat. Acad. Sci. (Wash.) **47**, 1981 (1961).
SADLER, J., and A. NOVICK: J. molec. Biol. **12**, 305 (1965).

## Discussion

*Starlinger:* How can segregation occurs considerably in 2 generations, if there are 10 repressor molecules/gene copy.

*Novick:* I am not sure how accurate this number is. Perhaps it is actually lower. Also Carol Gross in our laboratory has found that the TSS repressor is slightly temperature-sensitive and could fall to $1/_4$ to $1/_2$ in an hour or so.

*Starlinger:* Why does the PJM paradox not work in alkaline phosphatase?

*Novick:* The absence of a lag in onset of repression for alkaline phosphatase could be explauned of course if there were three times or more as much repressor there as for *lac*. But perhaps in view of the fact that the phosphatase system also has positive control it may be premature to try explain this.

# Strong-Polar Mutations in the Galactose Operon of E. coli

By

PETER STARLINGER, ELKE JORDAN, and HEINZ SAEDLER

Institut für Genetik der Universität zu Köln

In bacteria, genes are often organized in operons. If a mutation occurs in one of the genes of an operon, in some cases the activities of the other genes are influenced. The influence on the activity of the genes distal to the mutation with respect to the operator is most pronounced. The mutations of this type are therefore called polar. They have been most thoroughly described in the lactose and tryptophan operons in E. coli and in the histidine operon of S. typhimurium (NEWTON, BECKWITH, ZIPSER, and BRENNER, 1965; MARTIN et al., 1966; YANOFSKY and ITO, 1966).

Many polar mutations have been shown to be of the nonsense type. Since all ribosomes are believed to lose the growing peptide chain at the site of the nonsense mutation, models had to be designed which were able to explain, why the effect of this mutation on the expression of the distal genes was not of the all or none type. The most commonly accepted model assumes that the ribosomes continue to travel along a messenger molecule after the loss of the peptide chain. If they arrive at the next gene border, they resume peptide synthesis. Some ribosomes, however, do not arrive at this gene border, since empty ribosomes have a certain probability of falling off the messenger. This probability is thought to be a function of the distance which the ribosomes travel in the empty state.

This model is able to explain most features of polarity, especially the dependence of the degree of polarity on map position. Mutations of the amber type, one of the three nonsense classes, have been shown to occur in the galactose operon also. They were studied in the transferase gene, the second gene of the operon, by JORDAN and SAEDLER (1967) with respect to polarity and suppression of polarity for the function of the kinase gene, the last gene of the operon. The transferase ambers behave like all known nonsense mutations in other operons, thus showing that the galactose operon is not different with respect to genetics or physiology from other operons.

In addition to the nonsense mutations, frameshift mutations also have been shown to be polar. They resemble the nonsense mutations very closely in their polarity properties. This coincidence does not seem to be fortuitous, since it is now believed that the polarity of the frameshifts is caused by nonsense codons, which the ribosomes encounter in the shifted reading frame (Martin, 1967).

In the galactose operon, we have studied another class of polar mutations, which differs drastically from the polar mutations described so far. They are much more polar than nonsense mutations. While the latter produce the enzymes of the distal genes in amounts of several to many percent of the wildtype level, the strong-polar mutants, as we call our mutants, produce the "distal enzymes" only in amounts of about $10^{-3}$ of the induced wildtype level.

Strong-polar mutants are either of the $0°$ type with greatly depressed levels of all three enzymes of galactose metabolism, or they occur in the transferase gene. The latter have no transferase, strongly reduced kinase, but the first enzyme, the epimerase is unaffected. (Saedler and Starlinger, 1967; Jordan, Saedler, and Starlinger, 1967.)

Similar mutants have been isolated by Shapiro in London (Shapiro, 1967).

I want to confine my talk now to the strong-polar mutations in the transferase gene, since these can be unambiguously mapped by deletion mapping (Pfeifer and Oellermann, 1967), while we do not have a mapping procedure yet for an unambiguous mapping of the $0°$'s. With respect to mutability and physiology, however, the $0°$ class resembles closely the $t^{sp}$ class, as we call the strong-polar mutations in the transferase gene.

The $t^{sp}$ map throughout the transferase gene. We have found mutations of this class in 6 out of 12 deletion groups without extensive searches. The $t^{sp}$ revert spontaneously in most instances, and do thus not belong to the class of large deletions.

They cannot be induced to revert by mutagens believed to cause base substitutions, including 2-aminopurine, ethyl methane sulfonate and N-methyl-N'-nitro-N-nitrosoguanidine. From this, one is tempted to conclude that they have not arisen as a consequence of base substitution. This makes it unlikely that they belong to the known nonsense classes. This latter conclusion is supported by the lack of suppressibility by known nonsense suppressors and by the failure to find suppressor mutations unlinked to the galactose region among reversions to the Gal$^+$ phenotype.

It remained to be tested, whether the $t^{sp}$ are frameshift mutations. We compared them with the only well characterized frameshift mutations in bacteria, which were characterized by Martin (1967) in the C gene in the histidine operon. Our mutations differ from Martin's (1) in their higher rate of spontaneous reversion, (2) in their having transferase in wildtype amounts in all reversions tested, (3) in their lack of response to the mutagen

ICR 191. The last finding by itself is not very strong. It is believed that the reversion of a frameshift mutation often occurs at a second site, leaving a short altered amino acid sequence in the protein. If the transferase were very sensitive to even small structural changes, second site reversion might well lead to a non-functional protein in all cases investigated. − To exclude this possibility, we introduced the mutations on F gal particles into a complementing background, where the transferase was supplied by the host and only the reversal of polarity for kinase production was tested. Even under these conditions, no stimulation by ICR 191 was found, thus strengthening our argument that the $t^{sp}$ do not belong to the frameshift class.

A last possibility considered was that the $t^{sp}$ arise as a consequence of a chromosomal rearrangement, like inversions or insertions of foreign genetic material. The latter has been demonstrated in several instances in bacteria to lead to an inactivation of the corresponding gene. We do not have a way to exclude this with certainty at present. If one considers it likely, however, that processes which lead to chromosomal rearrangements have something to do with recombination, one expects changes in the rate of backmutation, if the mutation is measured in a recombinationless and in a normal background. This has been done, and no difference in the rate of backmutation was found.

Thus we do not know which changes in the DNA lead to the $t^{sp}$ phenotype. We think it most likely that conclusions from the mutagen-induced reversion as to the nature of the mutation are presently somewhat overestimated. Possibly such mutagens act differently on different parts of the DNA, depending on the DNA in the region in question. A relevant observation was recently published by Fränkel-Conrat on RNA, where the action of nitrosoguanidine was found to be very conformation specific.

Another difference between strong-polar mutations and nonsense mutations is the response to inducer. Polar nonsense mutants are normally inducible. Strong-polar mutants, however, make the small residual amount of the "distal enzymes" constitutively. We think that this observation can be best explained by the assumption that the residual synthesis is due to some rare event overcoming the effect of the mutation, and that this event becomes rate-limiting for the whole process. This statement seems to exclude some possibilities. If m-RNA production is controlled by the repression system, and mutation acts at the level of translation, then one would expect that production of more messenger molecules on induction would lead to the production of more enzyme, even if the probability for the single messenger to be read successfully would remain low.

The non-inducibility of the $t^{sp}$ mutations can then only be explained by inserting a further assumption, namely that a stop in the synthesis of

14*

protein causes a corresponding decrease in the rate of RNA synthesis. This has been proposed a long time ago, since in 0° mutants no messenger is found (AMES and HARTMAN, 1963; STENT, 1964). But it was always possible to argue that the failure to find the messenger was due to quick degradation of unread molecules.

Another, and possibly simpler, assumption to explain the rate-limitation of the $t^{sp}$ mutations for the induction process is that these mutations act at the level of transcription. If the mutation would cause the polymerase to stop, and if this stop could only rarely be overcome, the whole process might easily become dependent on the probability of this event alone.

In order to test this last assumption, experiments are clearly required to investigate the production of mRNA in the $t^{sp}$, preferably in vitro in order to circumvent the difficulty of all relevant in vivo experiments that missing messenger might as well have been degraded as not produced at all. We are presently trying to do such an experiment.

## References

AMES, B., and P. HARTMAN: Cold Spr. Harb. Symp. quant. Biol. **28**, 349 (1963).
JORDAN, E., and H. SAEDLER: Molec. Gen. Genet. **100**: 283 (1967).
— —, and P. STARLINGER: Molec. Gen. Genet. **100**: 296 (1967).
MARTIN, R. G.: J. molec. Biol. **26**, 311 (1967).
— H. J. WHITFIELD, W. B. BERKOWITZ, and M. J. VOLL: Cold Spr. Harb. Symp. quant. Biol. **31**, 215 (1966).
NEWTON, W. A., J. R. BECKWITH, D. ZIPSER, and S. BRENNER: J. molec. Biol. **14**, 290 (1965).
PFEIFER, D., and R. OELLERMANN: Molec. Gen. Genet. **99**, 248 (1967).
SAEDLER, H., and P. STARLINGER: Molec. Gen. Genet. **100**: 178, 190 (1967).
SHAPIRO, J.: Ph. D. Thesis, Cambridge University 1967.
STENT, G.: Science **144**, 816 (1964).
YANOFSKY, C., and I. ITO: J. molec. Biol. **21**, 313 (1966).

## Discussion

*Bertani:* Could your "strong polar" mutants be one to a transposition of the K cistron to another region of the chromosome, away from the *gal* operator control?

*Starlinger:* We do not believe that the K is transposed away, because some of our mutations still can be induced by at most a factor of 2 with D-fucose. This usual effect would make it unlikely that they have been removed from the *gal* operator completely. — Besides it would be difficult to explain that the mutation revert readily to wild type levels of transferase. There exists the possibility of an insertion of DNA into the transferase gene. This should inactivate the transferase, but it would have to be explained, why kinase is produced at so low a rate.

# The Role of the Promoter Site in Gene Expression

By

R. Rita Arditti, J. H. Miller*, J. Scaife, and J. R. Beckwith

Department of Bacteriology, Harvard Medical School, Boston, Massachusetts

## The Role of the Promoter Site in Gene Expression

There are three structural genes in the *lac* (lactose) operon (Jacob and Monod, 1961). The *z*-gene codes for the enzyme β-galactosidase, the *y*-gene determines a component in the permeation system for galactosides (permease) and the *a*-gene codes for thiogalactoside transacetylase. The order of these genes is firmly established to be: *z-y-a* (Fig. 1). In addition we know that the *i*-gene, the structural gene for the repressor which regulates induction of the *lac* operon is located to the "left" of the *z*-gene (Fig. 1). Between these two determinants lies the operator (*O*), the site at which the repressor acts (Jacob and Monod, 1965; see Beckwith, 1967).

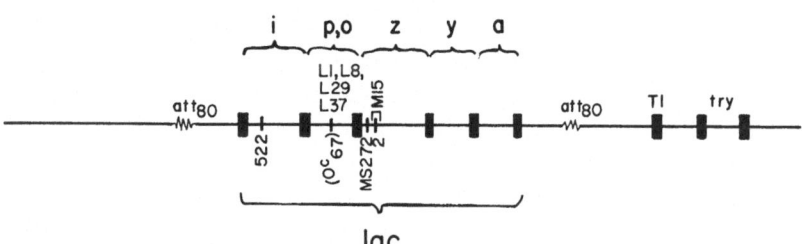

lac

Fig. 1. The *lac* operon transposed to the φ80 attachment site. The transducing φ80 *dlac* is shown integrated in the chromosome, between the two φ80 attachment sites (wavy lines). The mutations discussed are indicated in the *lac* operon. Many mutations of the locus (T1) determining sensitivity to the phage *T*1 are deletions also removing components of the tryptophan operon (*try*) and the *lac* operon. MS272 was isolated by Dr. M. Malamy (Malamy, 1966).

Recent studies *in vitro* show that the repressor binds directly to the DNA of the operator (Gilbert and Muller-Hill, 1967). It may therefore be supposed that inducers of the operon unstabilize the association between the repressor and the operator DNA. As a result, messenger RNA begins to be synthesized on the template of the structural genes, implying

---

* Department of Biochemistry and Molecular Biology, Harvard University, Cambridge, Massachusetts.

that the role of the repressor is to prevent the RNA polymerase from binding to the DNA or from initiating messenger synthesis. It follows that the site in the *lac* operon to which the RNA polymerase binds should be located so as to place the binding or transcribing activity of the enzyme under repressor control. This conclusion raises two important questions.

To what point in the *lac* operon does the RNA polymerase bind to initiate synthesis?

Which process is blocked by the repressor?

The binding site for RNA polymerase appears to be outside the operator since operator mutations do not affect the maximal level of gene expression (Jacob, Ullman, and Monod, 1964). There must therefore be a separate site in the operon, to which RNA polymerase binds. The term *promoter* introduced by Jacob et al. (1964) would describe such an element since it is defined as a site essential for the expression of an operon.

The same term would also cover the properties of another region which may also exist in an operon whose function would be to determine the binding site for ribosomes on the *lac* messenger. However we propose that the use of the term *promoter should be restricted to the binding site for RNA polynerase*, thereby avoiding confusion between two sites having completely different roles in *lac* operon expression.

We may thus restate our questions in genetic terms as follows:

1. Can promoter sites be identified?
2. What is the structure of these sites?
3. Can they be distinguished from ribosomal binding sites?
4. Where are they located with respect to the operator? This question is of central importance to our understanding of the regulation of the *lac* operon. As we shall see later it has a direct bearing on how we envisage the physical mechanisms by which they are realized.

In looking for answers to these questions we have sought mutants in which the promoter site has been altered. At least two types of mutation are expected; mutants should arise in which the promoter has been completely removed and have thus lost the ability to express their *lac* genes. In addition there should be mutants in which the site is still present but in an altered form. By the same reasoning we may also expect mutants with an altered binding site for ribosomes. Both types would have a maximum level of expression of the *lac* operon which is lower than normal (Scaife and Beckwith, 1966).

Mutants have been isolated in which the maximal expression of the *lac* operon has been coordinately reduced (Scaife and Beckwith, 1966). Their properties are completely consistent with those expected for a strain with an altered promoter, although at the present moment we cannot exclude the possibility that they have an altered ribosomal binding site.

In this paper we shall present evidence on the genetic location of the lesion in the mutants. We shall also present the results of reversion studies which have a bearing on the genetic nature of their mutation.

## Results

### The Genetic Location of the Mutations L8 and L37

We have previously described the properties of strains carrying the mutations L8 or L37 (SCAIFE and BECKWITH, 1966). They have all the properties expected for promoter mutants. Their $i$-gene and operator are intact but their $lac$ genes only function at about 6 % of the normal induced rate. We now know that the lesion in these mutants is located between the $i$-gene and the earliest mutable site known in the $z$-gene.

We have oriented the promoter mutations with respect to the $z$-gene by using deletion mutants whose lesion extends from outside the $lac$ region and ends inside the $lac$ operon itself. The method provides an unequivocal test by which we can locate mutations in the $lac$ operon.

The deletion mutants have been isolated from a strain (X5144) carrying the defective transducing phage $\phi$80d $lac$ (BECKWITH and SIGNER, 1966). The $lac$ region has been deleted from its normal chromosomal site in this strain. Figure 1 shows the structure of the chromosomal region carrying $\phi$80d-$lac$. It will be seen that since the phage has integrated at the normal site for $\phi$80 attachment the $lac$ genes are now close to the gene determining sensitivity to the phage T1 and the nearby tryptophan ($try$) operon. T1-resistant derivatices were isolated from this strain. T1 resistant mutants are known to include a large proportion of extensive deletions (YANOFSKY and LENNOX, 1959). In a strain lysogenic for $\phi$80 the deletion can cut into the prophage (FRANKLIN, DOVE, and YANOFSKY, 1965). The same effect is observed in a $\phi$80d-$lac$ lysogen (X5144) and in fact a T1-resistant derivative which we use in these studies (X8554) has a deletion extending from the $try$ operon through the phage attachment site into the middle of the transducing phage (Fig. 1). The deletion in the mutant X8554 has resulted in the loss of the $a$, $y$ and $z$ genes, as shown by the inability to yield $lac^+$ recombinants with a series of $lac^-$ point mutants. In this manner it has been shown that the deletion covers all the tested sites in the $z$-gene including the site of its earliest known mutation, MS272. By contrast wild-type recombinants are obtained in crosses between this deletion and L8 and L37, leading to the conclusion that the site of these mutations lies outside the earliest mutant site in the $z$ gene. The above finding confirms and extends our previous conclusion that L8 and L37 map to the left of the $z^-$ mutation, $2$ (see Fig. 1) and provides a solid basis for assuming that they are in fact located in a region outside the structural gene for $\beta$-galactosidase. This conclusion is also suggested by the properties of revertant strains isolated from L 8 and L 37 which are described in the following section.

## The Position of the Promoter with Respect to the Operator

There is a class of regulatory mutants of the *lac* operon ($i^- o^c$) which in the absence of inducer synthesize the *lac* enzymes at the maximal rate (Jacob, Ullman, and Monod, 1964). These mutants are designated $i^- o^c$ because dominance tests indicate that both the operator and the *i* gene are inactivated. Their properties therefore suggest that they carry a deletion extending from the operator into (or perhaps beyond) the *i* gene. However recent recombination studies by Dr. Julian Davies have shown that this is not the case. He has found that different $i^- o^c$ mutants can recombine to give normal, fully inducible recombinants. This means that the lesions in these strains cannot overlap. In fact these studies have shown that the lesions in these strains behave like point mutations falling in two clusters at least one of which is actually inside the *i* gene.

In a previous study we reported that both L 8 and L 37 recombine with two $i^- o^c$ mutations (67 and 522, see map). On the basis of what we now know about the $i^- o^c$ mutations used in our experiments it can no longer be concluded that the mutations L 8 and L 37 are located between the operator and the *z* gene. In fact it raises the possibility that the promoter is located between the *i* gene and the operator. More intensive mapping against the deletion strains isolated in the above manner should provide an answer to this question.

## Reversion Studies on the Nature of L 8 and L 37

A study of revertants from the mutants L 8 and L 37 has enabled us to define something of their genetic nature. If the lesion were a deletion we should not expect reversion to occur at the same site as the original mutation (true back mutation). In addition we should not expect N-methyl-N'-nitro-nitrosoguanidine (NG) to increase back mutations since there is evidence that this mutagen causes single base substitutions (Whitfield, Martin, and Ames, 1966; Brammar, Berger, and Yanofsky, 1967).

There is of course no reason to assume that all the revertants we isolate are due to changes at the original site. In addition to true back mutations we could expect to recover revertants due to suppressor mutations which affect the expression of the *lac* genes. Such mutations could be within the *lac* operon itself. Alternatively they could be outside the operon in other genes determining the mechanisms of transcription and translation.

The technique we have used to isolate revertant clones from the mutants L8 and L37 is to spread the bacteria on solid medium containing a carbon source, permitting only growth of strains with levels of *lac* gene expression higher than that of the mutants.

It is not possible to select revertants of L8 and L37 on minimal medium containing lactose as a sole carbon source since the activity of the *lac* genes in these mutants is enough to permit growth. We have shown previously that melibiose which at 42° employs the β-galactoside permease for its uptake is also unsuitable for our purposes. This is because it selects constitutive mutants which with their raised basal level of the *lac* permease can accumulate the sugar well enough for growth (SCAIFE and BECKWITH, 1966).

However we have found that use of the sugar raffinose (SCHAEFLER, 1967) does allow selection of revertants of our mutant switch an increased rate of expression of the *lac* enzymes. In addition to raffinose the selective medium also contains IPTG, a strong inducer of the *lac* operon, so that constitutive mutants are not favoured. The trisaccharide raffinose, which is the (1–2)-β-D-fructofuranoside of melibiose, employs the β-galactoside permease for its uptake like melibiose. It does not induce the *lac* operon. Raffinose utilization is apparently very inefficient since strains with a low permease activity cannot grow on the sugar and even those with a normal *y* gene grow very slowly indeed. On minimal medium with raffinose and IPTG revertant colonies appear after 10 to 15 days.

### a) High Level Revertants of L8 and L37

Spontaneous revertants found on raffinose plates occur at a high frequency (1 in $5 \times 10^7$). Amongst these revertants a large number have full activity for all *lac* enzymes (see Table 1). Moreover the β-galactosidase activity of these revertants grown in the absence of inducer is at the same low level as that of the wild type, implying that their regulation system is normal. These revertants therefore have the phenotype expected for true back mutants. To show that this is so we need genetic evidence indicating that the original mutation has been lost in the process of reversion. We present in Table 2 the results of crosses in which we tried to recover the original mutation from revertants of this type. The method was to mate each revertant (which is an $Sm^s$ Hfr) against an $F^- Sm^r z^- y^+$ recipient and select recombinants able to grow on minimal lactose containing streptomycin. As we have seen an L8 recombinant can grow on this medium as can the full wild type. We therefore screened these recombinants to see whether we could detect any with the L8 phenotype (see Table 2) due to a crossover event separating the original lesion from an outside mutation harboured in the revertant donor. None of the full level revertants tested yielded recombinants with the L8 phenotype. An

Table 1. *Revertants of L 8 and L 37*

| Strain | Phenotype | | | Number of revertants isolated | | | |
|---|---|---|---|---|---|---|---|
| | Percent wild type β-galactosidase activity (induced) | Growth on melibiose at 42°C | | Spont- aneous | NG | UV | Total |
| Wild type | 100 | + | | — | — | — | |
| L8, L37 | 6 | — | | — | — | — | |
| revertants Full | 100 | + | | 27 | 48 | 124 | 199 |
| revertants intermediate | 13 – 30 | + | | 7 | 27 | 3 | 37 |
| revertants Low | 6 | + | | 8 | 9 | 5 | 22 |
| Total tested | | | | 42 | 84 | 132 | 258 |

Both cultures of CA 8003 (HfrH Sm$^s$ B$_1^-$ Lac$_{L8}$) and CA 8005 (HfrH Sm$^s$ B$_1^-$ Lac$_{L37}$) were washed and resuspended in M 63 medium before plating (5 × 10$^8$ cells) on minimal medium containing raffinose (0.2%) (Pfansteihl Co.) and iso-propyl-β-D-thiogalactoside (10$^{-3}$ M) and incubated at 37°C for 10 – 15 days.

Revertants were induced by NG (N-methyl-N'-nitro-N-nitrosoguanidine, Aldrich Chemical Co.) on identical plates with 1 crystal of the mutagen placed in the center. UV revertants were recovered on the same medium from cultures irradiated to 5 – 10% survival[1].

Roughly equal numbers of each revertant class were derived from CA 8003 and CA 8005.

Table 2. *Genetic analysis of revertants from L 8 and L 37*

| Revertant type | Number tested | Number with a detectable outside mutation |
|---|---|---|
| Full | 174 | 0 |
| Intermediate | 37 | 0 |
| Low | 22 | 22 |

Young broth cultures of each of the (HfrH Sm$^s$) revertants were mated against X 9003 (F$^-$ z$^-$ M$_{15}$ y$^+$ B$_1^-$ Sm$^r$; M 15 is a small deletion near the operator end of the z-gene). Recombinants were selected on minimal medium containing streptomycin and lactose as the sole carbon source. Twenty purified recombinants from each cross were tested for the L8/L37 phenotype on melibiose minimal medium (containing streptomycin) and lactose-tetrazolium indicator medium. Control crosses against a pro$_{c-}$ Sm$^r$ recipient show that this test would give a negative result only if the second mutation were at or very near (closer than pro$_{c-}$) to the site of L8 or L37. Ninety percent of the recombinants from each low-level revertant had the L8/L37 phenotype.

---

[1] The proportion of revertants amongst the survivors was 50 times higher than in the untreated control.

outside mutation is easily detectable by this test (see section c below). Our evidence so far is, therefore, consistent with these revertants being true back mutations. Moreover the frequency of this class of revertant is increased (at least a hundred fold) by the mutagen NG which causes single base changes (WHITFIELD et al., 1966; BRAMMAR et al., 1967). On the basis of these results we conclude that the mutations can revert by a single base-change to yield a strain indistinguishable from wild type, suggesting that the original mutation was not a deletion but a single base change.

### b) Intermediate Level Revertants of L8 and L37

The raffinose medium does not limit revertant selection to those with the full wild type level of the *lac* enzymes. We have a series of different (intermediate) revertants (Table 1) with $\beta$-galactosidase activities ranging upwards from 13 % of the normal level. Genetic analysis of the type described in the previous section has also been applied to these revertants. It will be seen (Table 2) that no outside mutation was detected in any of the intermediate revertants indicating that the second mutational event occurred at or near to the primary lesion.

The properties of the intermediate revertants provide corroborative evidence in favour of our conclusion that the original mutations L8 and L37 lie outside the structural gene for $\beta$-galactosidase. This conclusion is suggested by a comparison of the $\beta$-galactosidase and transacetylase activities in the revertant strains (Table 3). It will be seen from Table 3

Table 3. *The specific activity of $\beta$-galactosidase in revertants of L8 and L37*

| Strain | | % of fully induced wild type | |
|---|---|---|---|
| | | $\beta$-gz | Transacetylase |
| CA8000 (lac⁺) | | 100 | 100 |
| CA8005 (L37) | | 5 | 3 |
| Intermediate revertants | S-11 | 23.2 | 25.2 |
| | UV-82 | 20.7 | 18 |
| | UV-89 | 14.2 | 9.6 |
| | N-20 | 13.4 | 10.8 |
| | UV-63 | 21.1 | 24 |
| | N-11 | 28.0 | 21 |
| Full revertants | N-4 | 100 | 100 |
| | UV-2 | 100 | 100 |

Induced glycerol cultures in the exponential phase were concentrated in 0.1 M phosphate buffer (pH 7) ($10^{10}$ cells/ml). Appropriately diluted aliquots of each suspension were assayed for their $\beta$-galactosidase (see Table 1) and thiogalactoside transacetylase activities (W. EPSTEIN, in press). Spontaneous, NG and UV induced revertants are designated: S, N and UV respectively.

that the activities of the two enzymes increase strictly in parallel. Thus in none of the revertants recovered has the specific activity of $\beta$-galactosidase been observed to be different, as might be expected if the original lesion were located in the $z$ gene.

The class of intermediate revertants in which the altered site has thus been shown to be in the *lac* region could throw some light on the nature of the genetic site affected in the mutants L8 and L37. For example, it is of great interest to know how sensitive is the efficiency of *lac* operon expression to substitution of different base pairs at the site corresponding to the mutations. This approach of course requires further intensive mapping to ensure that the intermediate revertants are true back mutants.

### c) Outside Mutations Permitting Growth on Raffinose

Amongst the revertants isolated from raffinose medium a third class of strain has been identified. They carry a second mutation well outside the *lac* operon.

The strains of this type have a $\beta$-galactosidase level identical with that of the L8 or L37 parent strain although they are now able to grow on melibiose at 42°C. The evidence that they have a second mutation outside the *lac* operon is presented in Table 2. All the revertants of this type, when mated against the $z^- y^+$ Sm$^r$ recipient yield recombinants (about 90 %) with the original L8 or L37 phenotype. It is concluded that the mutation permitting growth on raffinose has left the L8 mutation intact, and that in recombination the two mutations are readily separated.

The effect of this second mutation must be to permit more efficient utilization of raffinose as a carbon source. Since the second mutation also permits growth of the strain on melibiose at 42°C it is quite possible that the gene involved is that determining the $\alpha$-galactoside permease (Prestige and Pardee, 1965). Certainly the *lac* permease is not required in the new mutants because they continue to grow on melibiose even after a large deletion covering the whole *lac* region has been introduced into the strain.

No true outside suppressors of the mutations in L8 or L37 have been yet identified amongst the revertants.

### Discussion

There are now four mutations known, which affect the maximal level of expression of the *lac* operon. Three of them, including L8 and L37, reduce the maximal level to 5% of the wild type. A fourth, L1, has only 1–2% of the full wild type activity.

Suppressor studies reported elsewhere suggest that two of these mutants, L8 and L37, are not missense $z^-$ mutations (Scaife and Beckwith, 1966). The genetic and reversion studies presented here support this conclusion. They lead us to suppose that the lesion of these mutants is located outside the structural gene for $\beta$-galactosidase. There is therefore good reason to suppose that the mutants have an alteration in an element governing maximal expression of the whole operon. It is not yet clear whether the element thus defined governs transcription (the promoter) or translation of the *lac* operon.

The frequency at which these mutations arise after uv mutagenesis is extremely high. L1, L8 and L37 were recovered from a set of only 30 *lac*⁻ mutants isolated after irradiation of a *lac*⁺ inducible strain. The fourth, L29, was isolated after uv treatment of a different strain. It was one of only two *lac*⁻ mutants recovered in this isolation. Our results thus suggest that the site affected by these mutations is extremely sensitive to uv irradiation.

The evidence that at least two of the mutants (L8 and L37) can revert to wild type suggests that at most very few base pairs are affected by the mutational event. In fact the response of the mutants to nitroso-guanidine suggests that a single base pair was altered. In addition we know that L1, L8 and L37 must be very close together since it has not been possible to detect recombination between them (less than .0002%) (SCAIFE and BECKWITH, unpublished results). Taken together these results indicate that the uv sensitive site may be very limited in extent. Mutants of the same phenotype isolated after treatment with different mutagenic agents may therefore give us a clearer idea of the size of the element govering *lac* operon expression.

On the basis of the present evidence we cannot say whether the element defined by these mutations maps to the left or to the right of the operator. However, it is worth pointing out that should the mutations be located between the operator and the *i* gene we would be obliged to visualize regulation of the *lac* operon in a new way. The existence of a site between *i* and *o* governing the maximum expression of the operon would imply that transcription is initiated beyond the operator. This would raise the possibility that the role of the repressor is to block the passage of RNA polymerase into the structural genes of the *lac* operon.

## Acknowledgement

This work was supported by Grant No. 13017 from the National Institute of General Medical Sciences and by Grant No. GB5763 from the National Science Foundation. It was undertaken while one of us (J.S.) was on leave of absence from the Medical Research Council, Microbial Genetics Research Unit, London, England. One of us (R.R.A.) is on leave of absence from the International Laboratory of Genetics and Biophysics, Naples, Italy.

## References

BECKWITH, J. R.: Science **156**, 597 (1967).
—, and E. R. SIGNER: J. molec. Biol. **19**, 254 (1966).

Brammar, W. J., H. Berger, and C. Yanofsky: Proc. natl. Acad. Sci. (Wash.) 58, 1499 (1967).

Epstein, W.: J. molec. Biol. (in press).

Franklin, N., W. F. Dove, and C. Yanofsky: Biochem. biophys. Res. Commun. 18, 910 (1965).

Gilbert, W., and B. Muller-Hill: Proc. natl. Acad. Sci. (Wash.) (1967) (in press).

Jacob, F., and J. Monod: J. molec. Biol. 3, 318—356 (1961); Biochem. biophys. Res. Commun. 18, 693 (1965).

— A. Ullman, and J. Monod: C. R. Acad. Sci. (Paris) 258, 3125 (1964).

Prestige, L. S., and A. B. Pardee: Biochim. biophys. Acta (Amst.) 100, 591 (1965).

Scaife, J., and J. R. Beckwith: Cold Spr. Harb. Symp. quant. Biol. 31, 403 (1964).

Schaefler, S.: Bact. Proc. 1967, 54.

Whitfield, H. J., R. G. Martin, and B. N. Ames: J. molec. Biol. 21, 335 (1966).

Yanofsky, C., and E. S. Lennox: Virology 8, 425 (1959).

Malamy, M.: Cold Spr. Harb. Symp. quant. Biol. 31, 189 (1966).

## Discussion

*Bertani:* Are the intermediate revertants mappable within the lactose region?

*Beckwith:* They are mappable within the *lac* region and so far are not separable from the original $p^-$ mutation.

*Starlinger:* You explained the lower induction ratio of $p^-$ by a minimum level of transcription independent of repression. Do you think the following explanation also possible that both $p$ mutations and repression effect the same step and that at a low level of $p$ function the repression in no longer rate-limiting?

*Beckwith:* Yes.

*Winkler:* Do you believe that promoter sites exist also in phage genomes? I am asking because bacteria infected with certain mutants of phage $\lambda$ produce more phage-specific exonuclease than cells infected with the wild type of $\lambda$ (Radding).

*Beckwith:* This possibility has been suggested.

*Neidhardt:* What is the response of $p^-$ mutants to conditions that cause catabolite repression in the wild type?

*Beckwith:* Magasanik's group has shown that these mutants are normally sensitive to catabolite repression.

*Havender:* Would you please explain again why the $p$ region is distinguished from the $o$ region?

*Beckwith:* Because $o^c$ mutants don't affect $p$ function and the $p^-$ mutants don't affect $o$.

*Matthaei:* Would you like to comment on the evidence available favoring the idea that a special initiator-tRNA is involved in the processing promotion?

*Beckwith:* The frequency with which these mutations are picked up suggests that this site may be rather large, larger than an initiation codon. In addition, the intermediate levels in revertant suggests that this site can exist in several different states.

*Vielmetter:* Did you construct $o^c\, p^-$ and how do they respond?

*Beckwith:* Yes. They are constitutive but still for the low 6% level.

*Trautner:* Did you observe recombination between your two independently isolated $p^-$ mutants?

*Beckwith:* The techniques have not been too good for detecting recombinants. Not so far.

# Regulation of Pyruvate Dehydrogenase Synthesis: Substrate Induction

U. Henning, J. Dietrich, K. N. Murray, and Gisela Deppe

Max-Planck-Institut für Biologie, Tübingen, Germany

From an evolutionary point of view, it appears unlikely that the detailed mechanism of regulation of enzyme synthesis will be the same in each case. Gene products with regulatory functions could not, of course, have evolved without the existence of the enzyme systems to be regulated. The simplest assumption, then, would be that most such systems have evolved their own regulatory devices. Two lines of evidence appear to demonstrate that this is so. Firstly, a number of recent findings proves that the "regulatory pathways" can differ widely in different microorganisms.

Two examples may suffice for illustration. In *E. coli*, tryptophan (or a derivative thereof) effects coordinate repression of all enzymes leading from chorismate to tryptophan (Yanofsky, 1960; Ito and Crawford, 1965; for other ref. cf. Yanofsky, 1967). Crawford and Gunsalus (1966) have shown that in *Pseudomonas putida*, tryptophan synthetase is induced by its substrate, indole glycerolphosphate, while the other four enzymes of the pathway may be repressed by tryptophan. Similarly, the studies of Ornston and Stanier on the degradation of benzoate and p-hydroxybenzoate to β-ketoadipate in *Pseudomonas* species and in *Moraxella calcoacetica* (1966) have revealed that in these organisms the same sequence of enzymatic reactions will lead to β-ketoadipate. Yet, the system of control for the synthesis of these enzymes in *Pseudomonas* differs entirely from that in *Moraxella* (cf. Cánovas, Ornston and Stanier, 1967).

Secondly, as can be seen from, e.g., a comparison of the control systems operating in the inducible syntheses of the lactose (cf. Jacob, 1966) and the arabinose enzymes (Englesberg, Irr, Power, and Lee, 1965; Sheppard and Englesberg, 1967) in *E. coli* it is also true that different mechanisms exist.

Some evidence described earlier, although very indirectly, appeared to allow us the speculation (Henning, Dennert, Hertel, and Shipp, 1966) that in the pyruvate dehydrogenase system of *E. coli* K 12 still another type of regulatory mechanism might exist. In this case, one of the three components of this enzyme complex appeared to function in the regulation of the synthesis of the complex. Part of the studies summarized here have led to the identification of the internal inducer of this synthesis

(Dietrich, Deppe, and Henning, in preparation), the substrate of the enzyme. While not yet providing any conclusive evidence as to its mechanism of action, this identification reduces the number of possibilities by which it may effect the induction.

## 1. Pyruvate Dehydrogenase and Pyruvate Oxidation in E. coli

E. coli can derive acetyl-CoA from pyruvate mainly by virtue of three enzyme systems: pyruvate dehydrogenase ($pdh$[1], 1.; ref. see below), pyruvate oxidase (2.) (Stumpf, 1945; Williams and Hager, 1966) and the so called "phosphoroclastic" system (3.) (Utter and Werkman, 1944; Wolfe and O'Kane, 1953). They catalyze the following overall reactions:

1. Pyruvate + $NAD^+$ + CoA → acetyl-CoA + $CO_2$ + NADH + $H^+$
2. Pyruvate + FAD-E → $FADH_2$-E + acetate + $CO_2$
3. Pyruvate + CoA $\xrightarrow{HPO_4^{--}}$ acetyl-CoA + formate

Reaction (3) is operative only under anaerobic conditions when reaction (1) is completely inhibited (Henning, 1963; Hansen and Henning, 1966). For reasons to be described elsewhere, reaction (2) does not proceed to a significant degree under all experimental conditions mentioned in this communication. Under these conditions, therefore, glycolysis can lead to acetyl-CoA only *via pdh*. Mutants which lack *pdh* activity (*ace⁻* or *pdh⁻*) require acetate for aerobic growth on glucose.

Currently available data on the composition and biochemical function of the *pdh* complex have recently been reviewed by Reed and Cox (1966). The overall reaction (1) proceeds with enzyme bound intermediates via the following sequences:

4. *Pyruvate* + TPP-E 1 → $CO_2$ + Hydroxyethyl-TPP-E 1
5. Hydroxyethyl-TPP-E 1 + Lipoyl-E 2 → S-acetyl dihydrolipoyl-E 2 + TPP-E 1
6. S-acetyl dihydrolipoyl-E 2 + *CoASH* → *Acetyl-CoA* + dihydrolipoyl-E 2
7. Dihydrolipoyl-E 2 + FAD-E 3 → Lipoyl-E 2 + reduced FAD-E 3
8. Reduced FAD-E 3 + *NAD⁺* → *NADH* + *H⁺* + FAD-E 3

where $E1$ = decarboxylase, $E2$ = transacetylase, and $E3$ = dihydrolipoamide dehydrogenase, the components of the complex. The *pdh*

---

[1] The following abbreviations are used: Loci on the E. coli chromosome: *ace*, acetate; *ara*, arabinose; *gal*, galactose; *lac*, lactose; *leu*, leucine; *pro*, proline; *suc*, succinate; T 1, 5$^{rec}$, receptor for phages T 1 and T 5. CoA, coenzyme A; E, enzyme; *icl*, isocitric lyase; *icl$^c$*, constitutive synthesis of isocitric lyase; *pdh*, pyruvate dehydrogenase; *pps*, phosphoenolpyruvate synthetase; TPP, thiamine-pyrophosphate.

contains about 12 molecules E 1 (MW = $1.83 \times 10^5$), 1 molecule E 2 (MW = $1.7 \times 10^6$), and 6 molecules E 3 (MW = $1.12 \times 10^5$).

## 2. Genetics of the Pyruvate Dehydrogenase and the Problem of the E 3 Component

The closely linked structural genes of the E 1 and E 2 components comprise the acetate (*ace*) locus which is located near the leucine locus on the *E. coli* chromosome (cf. Fig. 1; HENNING and HERZ, 1964). Among more than one hundred *ace⁻* mutants, none has so far been found which lacks only E 3 activity. Thus, the definite localization of this gene remains unknown.

Earlier tentative evidence (HENNING and HERZ, 1964) concerning the localization of the structural genes for the α-ketoglutarate dehydrogenase complex has now been substantiated (MURRAY and HENNING, in prep.): this locus lies near the *gal* operon, i.e., about 20 % of the total chromosomal length away from the *ace* locus. MUKHERJEE, MATTHEWS, HORNEY, and REED (1965) have shown that the E 3 components of the two ketoacid dehydrogenases are functionally interchangeable. More recent studies by PETTIT and REED (1967) demonstrated that the E 3 components of the two ketoacid dehydrogenases are indistinguishable with regard to several physical, chemical and immunological parameters. To explain our failure to recover E 3 gene mutants, we also entertained the possibility that the two flavoproteins are actually identical and, furthermore, that one structural E 3 gene exists for both of them. Clearly, then, a single mutation in this gene would result in a double mutant (*ace⁻, suc⁻*) which would remain undetected in the selection for *ace⁻* mutants. However, although one E 3 gene would demand that the two flavoproteins be identical, the converse is not necessarily true. Two identical enzymes could still be the products of two separate genes.

To date, all enzymatic data from various *ace⁻* mutants are compatible with the assumption that there are two E 3 genes which are linked to the *ace* and the *suc* loci, respectively. Lack of *pdh* E 3 gene mutants may reflect complementation; i.e., if a small amount of α-ketoglutarate dehydrogenase E 3 component were available, a *pdh* E 3 mutant would at least be leaky and thus be lost in the selection procedure.

Until recently a serious difficulty had precluded a number of experiments important to the study of the pyruvate dehydrogenase system: the inability to obtain stable diploidy for the *ace* locus. Since an Hfr suitable for the usual type of selection (JACOB and ADELBERG, 1959) for a substituted episome was not available, such an episome was obtained by the procedure summarized in Fig. 1. The resulting temperature sensitive $F_{TS}$ ECO5 carries the fairly large chromosomal region T 1, $5^{rec} - ace - leu$.

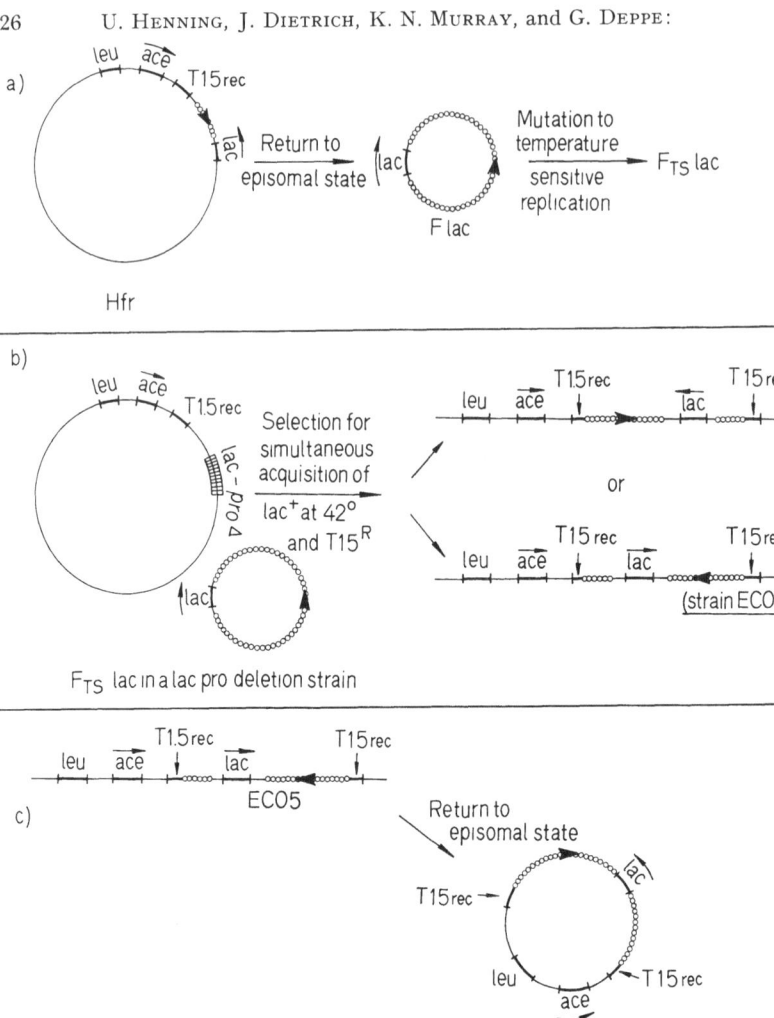

Fig. 1. Origin of the F_TS ECO5 (F_TS *lac ace leu*). A. Isolation of the substituted apisome F *lac* from which a mutant, F_TS *lac*, with a temperature sensitive replication mechanism was then isolated (Jacob, Brenner and Cuzin, 1963). B. Temperature sensitive F_TS *lac* episomes in strains lacking the whole chromosomal *lac* region can integrate at other chromosomal sites resulting in different Hfr strains with transposed *lac* regions (Cuzin and Jacob, 1964). Following the procedure of Beckwith, Signer and Epstein (1966), we selected for the integration of the F_TS *lac* into the T1,5 ^rec locus. C. Of the two types of Hfr recovered — one donating its chromosome in a counterclockwise, the other in a clockwise fashion — the latter type (ECO5) was used for the isolation of F_TS ECO5. (The resulting episome need not carry both parts of the T1,5^rec locus as assumed in C., the part to the right of the ECO5 origin may have been lost. Also, the relative positions of the ECO5 origin and the *lac* region have not been tested and may be reverse since it is unknown, where, on the F_TS *lac*, integration into the T1,5^rec locus occurred.) The arrows at the *lac* and *ace* loci indicate directions of polarisation.

## 3. The Internal Inducer for Pyruvate Dehydrogenase Synthesis

### a) The Inducing Action of Pyruvate

It had been observed earlier (cf. HENNING et al., 1966) that the largest amounts of *pdh* were obtained after growing wild type cells on pyruvate. They produce 10—15 times less enzyme after growth on acetate as sole carbon source. An inverse behavior had been found by KORNBERG (for summarizing review cf. KORNBERG, 1966) for the synthesis of the key enzyme of the glyoxylate cycle, isocitric lyase. This inverse relationship held true for almost all carbon sources tested and thus appeared to reflect a near relationship between the effectors controlling the syntheses of *pdh* and isocitric lyase. The latter enzyme is repressible, with a close derivative of pyruvate or phosphoenolpyruvate being the most likely effector (KORN-BERG, 1966). These observations led us to consider pyruvate as the internal inducer for *pdh* synthesis.

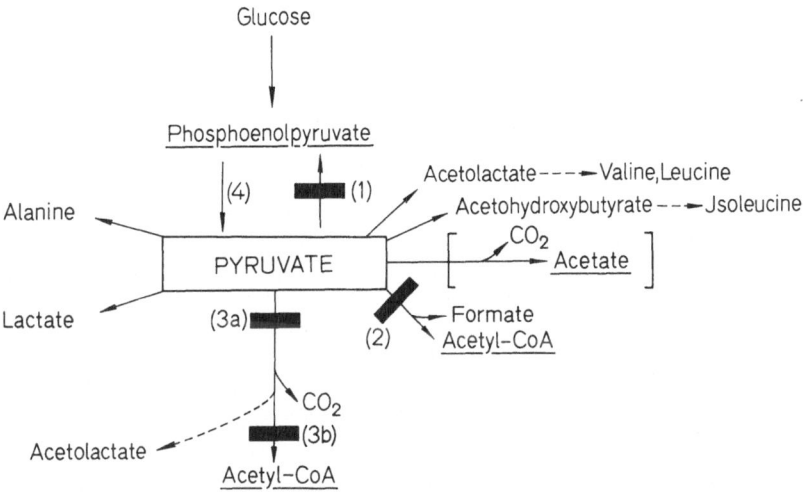

Fig. 2. The metabolic position of pyruvate in E. coli and the enzymatic blocks used in this study. For explanations of the numbers see text (Section 3a).

The central metabolic position of pyruvate (Fig. 2) appears rather insurmountable if one wishes to prove that it is the internal inducer; i.e., is acting directly and not simply serving as precursor for an effector. The following circumstances and mutants, however, made it possible to use and to test pyruvate almost as a gratutious inducer. As mentioned before, the phosphoroclastic degradation of pyruvate, which yields acetyl CoA and formate, is completely inhibited under aerobic conditions (Block 2 in Fig. 2). The pyruvate oxidase system producing acetate and $CO_2$ is not operative to any significant extent under any of the growth conditions

15*

used here (in brackets, Fig. 2). The synthesis of phosphoenolpyruvate from pyruvate is achieved only by phosphoenolpyruvate synthase ($pps$) and not by pyruvate kinase (reaction 4, Fig. 2). Therefore, mutational loss of the synthase activity ($pps^-$, block 1, Fig. 2) leads to a complete inability to grow on pyruvate (Cooper and Kornberg, 1965). Finally, the oxidation of pyruvate by $pdh$ can be eliminated by mutational loss of this enzymatic activity (blocks 3a and 3b in Fig. 2; 3a: loss of component E 1 activity, 3b: loss of component E2 activity). Aside from appearing physiologically senseless to have alanine, lactate, or the precursors of valine-leucine and of isoleucine as inducers, the precursors are excluded by the experiments described in Section 3b: thiamine pyrophosphate is required for their synthesis. Production of $pdh$ after growth on alanine or lactate is, although high, significantly lower than after growth on pyruvate.

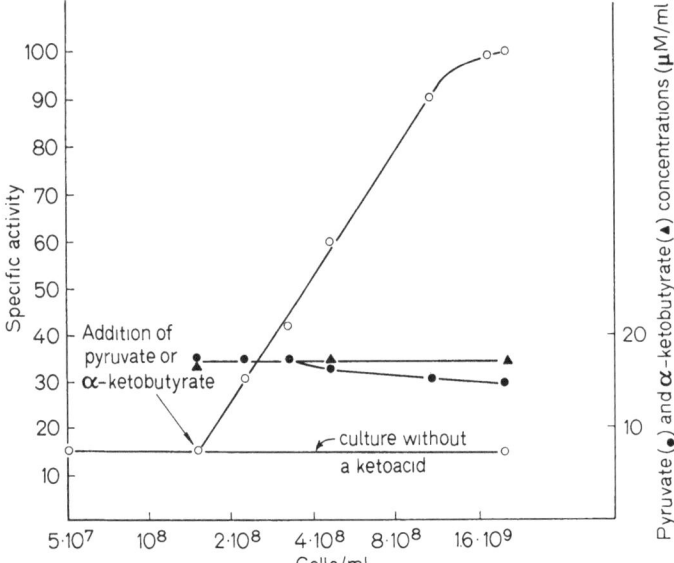

Fig. 3. Effect of pyruvate and α-ketobutyrate on the synthesis of the E 1 component of the pyruvate dehydrogenase. The triple mutant, which lacks phosphoenolpyruvate synthese activity and the E 2 component activity of pyruvate dehydrogenase, and which synthesizes isocitric lyase constitutively, was grown under vigorous aeration at 37°C with 35 mM-acetate as the sole carbon and energy source. 17.5 mM-pyruvate or α-ketobutyrate were added as indicated. The mean doubling time of all cultures during the whole experiment was 120 min. Open circles: specific activity of E 1, which was assayed as described elsewhere (Henning, Herz, and Szolyvay, 1964). In the experiment with α-ketobutyrate only the initial (at arrow) and the final spec. act. were measured, both were the same as the corresponding values in the pyruvate experiment. Addition of 2 mM-pyruvate had the same effect as 17.5 mM. Pyruvate and α-ketobutyrate concentrations in the medium were assayed with lactate dehydrogenase.

Under aerobic conditions, therefore, a $pdh^-$, $pps^-$ double mutant should be unable to metabolize pyruvate. If the inducing activity of pyruvate is to be tested with such a double mutant, two difficulties have to be overcome. Firstly, the enzyme to be induced would be inactive and yet must be measurable. Secondly, to observe a range of induced enzyme synthesis as large as possible, it would be desirable to add pyruvate to cells with a $pdh$ level as low as possible; i.e., while growing on acetate. Addition of pyruvate to cells growing on acetate, however, inhibits growth because of the inhibition of isocitric lyase by pyruvate (KORN-BERG, 1966). Thus, inhibiting the anaplerotic function of the glyoxylate cycle would also decrease protein synthesis. The first problem can easily be overcome by use of a $pdh^-$ mutant lacking only $E\,2$ activity. Such mutants synthesize enzymatically active $E\,1$ which is governed by the same control as the whole complex. The second problem can be removed by the use of mutants which synthesize isocitric lyase constitutively ($icl^c$, KORNBERG, 1966). The large amounts of isocitric lyase produced by such strains are not inhibited by pyruvate to such a degree that the cell growth on acetate is affected. Consequently, the triple mutant $pps^-$, $pdh^-$ ($= E\,2^-$) and $icl^c$ was constructed and the experiments shown in Fig. 3 were performed (DIETRICH, DEPPE, and HENNING, in prep.). After the addition of pyruvate to cells growing on acetate, the differential rate of E 1 synthesis increased immediately while the pyruvate concentration remained constant for about 2 generations. Later, a small, but significant, decrease of the pyruvate concentration was observed. It then was found that the pyruvate analogue, $\alpha$-ketobutyrate, exerted the same inducing activity. As shown in Fig. 3, $\alpha$-ketobutyrate did not dissappear from the medium before E 1 synthesis was fully induced.

### b) Pyruvate as Internal Inducer

The results presented in section 3a do not yet allow pyruvate to be assigned the role of the inducing *effector* because of the following situation. The induction of enzymatically *active* E 1 has been demonstrated. This component can catalyze both reaction (4) and the formation of acetolactate (9) (BARTHOLOMÉ and HENNING, unpubl., the analogous reactions occur with $\alpha$-ketobutyrate as substrate):

4. Pyruvate + TPP-E 1 $\rightarrow$ $CO_2$ + Hydroxyethyl-TPP-E 1

9. Hydroxyethyl-TPP-E 1 + Pyruvate $\rightarrow$ acetolactate + TPP-E 1

Therefore, one might expect a typical case of product induction which has been found in several instances (e.g., HAYASHI and LIN, 1965). In a comparable situation here, hydroxyethyl-TPP or acetolactate would serve as inducing effectors. This possibility was ruled out by the following experiments[1]. Thiamine requiring strains will grow on glucose without thiamine in the presence of 0.05 % vitamin free casamino acids. (Under these conditions, the cells accumulate pyruvate and stop growing at 50 to 70 % of the cell density reached in the same medium in the presence of

Table 1. *The effects of thiamine deprivation on the synthesis of pyruvate dehydrogenase*

A thiamine requiring, pyruvate dehydrogenase wild type strain (AB 287) was grown aerobically at 37° C in the absence or presence (1 μg/ml) of thiamine on the carbon sources indicated. In all cases 0.05 % vitamin free casamino acids (Difco) were added. The cells were harvested in the second half of the growth phase; i.e., when the culture without thiamine did not continue growth. The enzyme assays were performed in the presence ($10^{-3}$ M) or absence of thiamine pyrophosphate and with crude extracts which, after centrifugation, had received no further treatments. Accumulated pyruvate was determined in the medium after harvesting the cells.

| Carbon source | | Thiamine | Specific activity of pyruvate dehydrogenase | | Pyruvate accumulated (μM/ml) |
|---|---|---|---|---|---|
| | | | without | with | |
| | | | TPP in the assay | | |
| Glucose | (20 mM) | added | 18 | 25 | <0.01 |
| Glucose | (20 mM) | omitted | <0.1 | 80 | 5 |
| Pyruvate | (50 mM) | added | — | 60 | — |
| Acetate | (35 mM) | added | 8 | 8 | <0.01 |
| Acetate | (35 mM) | omitted | <0.1 | 13 | <0.01 |

thiamine.) Table 1 shows the effect of thiamine deprivation on *pdh* synthesis. The amount of enzyme synthesized is even higher than after growth on pyruvate and, as a result of these growth conditions, the enzyme is completely inactive in the absence of thiamine pyrophosphate. When the same experiment was performed with acetate as carbon source, enzyme synthesis was not induced. This demonstrated that it is not the lack of thiamine which caused the high *pdh* synthesis. Without thiamine pyrophosphate, the pyruvate derivatives mentioned above cannot be made and, consequently, cannot induce. Thus it seems possible to accept pyruvate as the inducing effector (see Discussion for arguments concerning an indirect action of pyruvate).

Data obtained with merodiploid strains clearly demonstrate that a high intracellular pyruvate concentration is required for full induction of *pdh* synthesis (Table 2). *Ace* 10, a mutant of the genotype $E1^+E2^-E3^+$ produces uncomplexed E1. The synthesis of this component is fully

---

[1] A simpler approach to this problem appeared to be the use of E 1-missense mutants and testing the inducibility by pyruvate of the E 2 activity in such strains. They are inducible but, remarkably enough, none of the E 1-missense mutants tested is inactive in reaction (4), they must have lost reaction (5) activity.

Table 2. *Effect of carbon sources and thiamine deprivation on pyruvate dehydrogenase and E 1 syntheses in haploid and merodiploid cells*

The same wild type and the same concentrations of carbon sources and thiamine as those in Table 1 were used. All cultures were grown at 30°C in the presence of 0.05 % vitamin free casamino acids. In all cases where a range of spec. act. is given, the experiment was repeated at least twice. Pyruvate concentrations in the media were determined on all glucose grown strains after harvesting the cells; * indicates $5-20$ μM/ml pyruvate; less than 0.01 μM/ml was found in all other experiments. Note that in the heterogenote, wild type /$F_{TS}$ ECO5 *ace* 10 (5), both complete pyruvate dehydrogenase (chromosome) and its E1 component (episome) are produced. This last experiment has also been carried out with a heterogenote, *ace* 10/$F_{TS}$ ECO5 ($E 1^+ E 2^- E 3^+$/$F E 1^+ E 2^+ E 3^+$), and the same results were obtained.

| Strain (genotype) | Pyruvate dehydrogenase (*pdh*) and/or E 1 specific activities after growth on | | |
|---|---|---|---|
| | glucose + thiamine | pyruvate + thiamine | glucose without thiamine |
| 1 Wild type ($E 1^+ E 2^+ E 3^+$) | $20-30$ (*pdh*) | $50-60$ (*pdh*) | $70-90$ (*pdh*)* |
| 2 Wild type/$F_{TS}$ ECO5 ($E 1^+ E 2^+ E 3^+$ $F E 1^+ E 2^+ E 3^+$) | $20-30$ (*pdh*) | $50-60$ (*pdh*) | $170-190$ (*pdh*)* |
| 3 *Ace* 10 ($E 1^+ E 2^- E 3^+$) | $70-90$ (E 1)* | — | $70-90$ (E 1)* |
| 4 *Ace* 10/$F_{TS}$ ECO5 *ace* 10 ($E 1^+ E 2^- E 3^+$/$F E 1^+ E 2^- E 3^+$) | $170-190$ (E 1)* | — | $170-190$ (E 1)* |
| 5 Wild type/$F_{TS}$ ECO5 *ace* 10 ($E 1^+ E 2^+ E 3^+$/$F E 1^+ E 2^- E 3^+$) | $20-30$ (*pdh*) $20-30$ $^+$(E 1) | $50-60$ (*pdh*) $50-60^+$ (E 1) | $80$ (*pdh*)* $90$ $^+$(E 1) |

induced on glucose, which is not surprising because large amounts of pyruvate are accumulated. As expected, the homogenote *ace* 10/F *ace* 10 on the same carbon source shows a clear cut gene dosage effect. This is not so with the homogenote *ace*$^+$/F *ace*$^+$, even when grown on pyruvate the specific activity of *pdh* in this diploid strain is not much higher than in the haploid wild-type. On the other hand, the gene dosage effect is again evident in the heterogenote *ace*$^+$/F *ace* 10 which produces complete *pdh* (chromosome) and uncomplexed E 1 (episome). However, the amount of E 1 synthesized is affected by the amount of *pdh* produced: both E 1 and *pdh* are present in nearly equal quantities. These are lower than from those found in *ace* 10 but the same as those found in haploid or diploid wild-type.

The explanation for these results becomes even more obvious when some of these thiamine requiring strains were grown on glucose without thiamine. As seen already (Table 1), the specific activity of *pdh* in the haploid wild-type increases over the value found on pyruvate. In addition the gene dosage effect in the homogenote *ace*$^+$/F *ace*$^+$ now becomes evident, and full induction of both *pdh* and E 1 occurs in the heterogenote *ace*$^+$/F *ace* 10. In all cases pyruvate was accumulated and the enzymes in extracts were entirely inactive in the absence of thiamine pyrophosphate. Clearly, the intracellular pyruvate concentration even during growth on pyruvate as sole carbon source does not reach levels sufficient for full induction as long as active *pdh* is present. An alternative explanation for the data of this section would be the production, by *pdh*, of a repressing effector. That this is not so, has been shown before (Fig. 3).

## 4. The Mechanism of Action of Pyruvate

Has not been elucidated yet, the main handicap being the lack of regulatory mutants. [Since the identification of pyruvate as inducer, it is clear why earlier attempts to isolate constitutive mutants could not have been successful; any agent inhibiting *pdh* activity (mutants resistant to such inhibition have been searched for) automatically will lead to full induction of enzyme synthesis thereby excluding a selective advantage for consitutive mutants.] If, analogous to the control system of the *lac* operon, an *i*-type gene were involved in *pdh* synthesis one would expect to find, in the homogenote $ace^+/F\ ace^+$, transdominant $ace^-$ mutants, i.e., mutants of the $i^s$-type (Willson, Perrin, Cohn, Jacob, and Monod, 1964). Several extensive searches for such mutants were unsuccessful (which, of course, does not eliminate the existence of such a regulatory gene in our case). If, analogous to the regulatory system of the *ara* genes, a separate activator gene were operative it should have been detected. $Ace^-$ mutants corresponding to the *ara* $C^-$ mutants described by Englesberg et al. (1965) have not been found. More generally, no mutants have been isolated which shut off *pdh* synthesis except those localized in the E 1 structural gene (see below).

Thus we can conclude the following: The data reported here demonstrate that in order for pyruvate to act as inducer, the *enzymatic* activity of the *pdh* or of its E 1 component is not required. If the primary target for pyruvate is produced by a gene outside the *pdh* structural genes a mutation in this gene cannot block *pdh* synthesis.

## 5. Mutations Affecting pdh Synthesis

To date, 109 independently isolated $ace^-$ mutants have been analyzed. Of these, 19 ($\sim 17\%$) are E $2^-$, 61 (56 %) are missense E $1^-$, and 29 ($\sim 28\%$) are of the 0°-type. The latter type is characterized by the complete absence (i.e., not less than 0.1—0.2 % of induced wild-type amounts can be detected which, however, corresponds to not more than about 1 molecule *pdh* per cell) of immunologically or enzymatically active *pdh* components. Fourteen of the 0°-type strains have been identified as nonsense mutants (12 amber, 2 ochre) which are localized within the structural gene of E 1. The tentative localization of some nonsense mutational sites is shown in Fig. 4. This genetic map has to be considered with some caution because we have met several unexplained difficulties in constructing it. Except for *ace* 123 and *ace* 137 it has been impossible to obtain meaningful recombination frequencies *between* the amber mutants. All the values were much too high[1] when compared to the reasonably consistent data from the transduction in which the E 1-missense mutants, *ace* 2 and *ace* 6, were used as recipients.

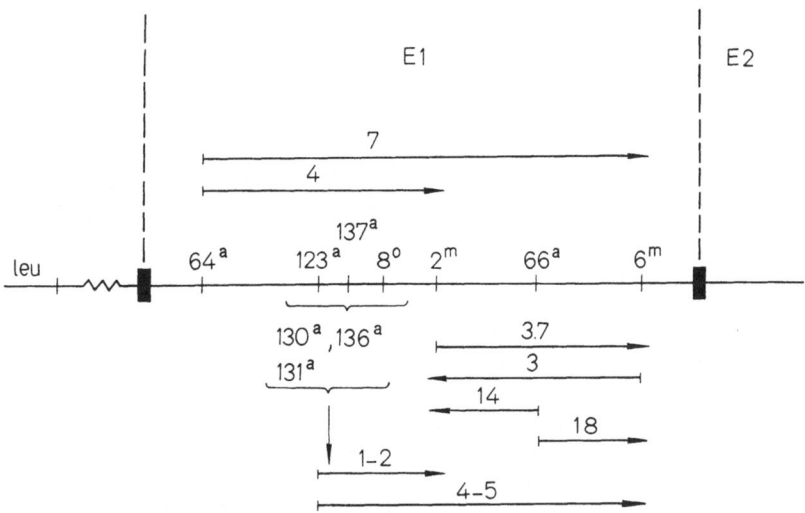

Fig. 4. Tentative localization of nonsense mutational sites in the E1 gene. *m*, missense; *a*, amber; *o*, ochre. The left and right end points of the E1 gene are set arbitrarily. Mapping was performed with phage P1 transduction. For example, the recipient, *ace* 6 (his⁻), was infected with phage grown on the donor, *ace* 2 (his⁺). The distance between mutational sites is given by the ratio *ace⁺/his⁺* recombinants. The values were corrected for the difference of single marker transduction frequencies in the *ace* and *his* regions (*ace⁺/his⁺* = 5). The arrows indicate the direction of transduction. *Ace* 66 is of intermediate polarity, all other nonsense mutants of this Fig. are of the 0°-type.

It is thus evident that 0°-type mutants in the E 1 gene are extremely frequent and, consistent with this frequency, that they are located on a segment of the E 1 gene which also appears to be unusually large when compared with the occurrence of extremely polarized, nonsense mutants

---

[1] Recently, G. R. DRAPEAU, W. J. BRAMMAR and C. YANOFSKY (J. Mol. Biol., in press) have shown that the *amber* codon may be responsible for such abnormally high recombination frequencies. Two *amber* mutants in the tryptophan synthetase A-gene (of E. coli K12) recombine much more frequently with a standard A-gene missense mutant than do missense mutants in the same codon as the two *amber* mutants. It is possible, therefore, that the frequencies of recombination between *amber* and missense mutants reported here are also too high.

in other systems (e.g., Newton, Beckwith, Zipser, and Brenner, 1965; Newton, 1966; Imamoto and Yanofsky, 1967).

In view of these data alone one might repeat our earlier speculation; namely, that the segment of the E 1 gene harboring the $0°$-type mutants specifies a product which would a) have the role of an activator, and b) yet be an integral part of the E 1 component. Two expectations resulting from such a hypothesis, however, were so far not fulfilled. Firstly, missense mutants should occur in that E 1 gene segment which also totally block *pdh* synthesis. Consequently, all $0°$-type mutants were subjected to a selection for temperature sensitive revertants not being amber or ochre suppressor mutants. Such revertants were found in only 5 strains. All of them, however, produced an enzymatically inactive E 1 component and *induced* levels of the E 2 component at high temperature. Secondly, at least some of the $0°$-type mutants may be expected to be (but need not) transrecessive to a wild type E 1 gene. Preliminary experiments with heterogenotes of the type $ace^-$ ($0°$-type)/F $E1^+E2^-E3^+$ failed to show clear complementation.

## Discussion

### 1. Pyruvate as Internal Inducer

The data presented here do not eliminate an indirect action of pyruvate. It could inhibit or stimulate some other reaction (e.g., of the citric acid cycle) the substrate or product, respectively, of which then could be the effector for induction of *pdh* synthesis. Experiments to be reported elsewhere do exclude an intermediate of the citric acid cycle as inducing or repressing effector of *pdh* synthesis. Almost all intermediates of this cycle have been used as carbon sources (including citrate with *Salmonella typhimurium*) and no pronounced differences in *pdh* synthesis have been found. The intracellular concentrations of the intermediates of glycolysis will probably be high during anaerobic growth on glucose under which conditions *pdh* synthesis is low [with reaction (3) and lactate dehydrogenase removing pyruvate]. A similar, low *pdh* synthesis is observed after aerobic growth on acetate when the intracellular concentrations of glycolytic intermediates are certainly much lower. (This appears to be clearly demonstrated by the fact that $ace^-$ mutants, while growing on acetate, do not accumulate pyruvate although they do while growing on succinate.] It appears physiologically meaningless to invoke still other effectors in addition to the intermediates of the citric acid and glyoxylate cycles and of glycolysis.

### 2. Mechanism of Action of Pyruvate

Although no evidence has been obtained as to the hypothetical role of the E 1 component in the regulation of *pdh* synthesis, it appears worthwhile to reconsider this possibility in view of the demonstration that this role cannot be an enzymatic one. In regard to the evolutionary considerations given in the introduction, it is tempting to hypothesize that regulatory gene products may (although, of course, need not) have originated from the enzymes to be regulated. If so, a protein exerting a regulatory function may not have become dissociated from its enzyme system. The question would then be, whether or not it could be at all possible to have a regulatory protein as a constituent part of the enzyme whose synthesis it is supposed to control.

Presumably, the association of all subunits into the enzyme complex could inactivate the regulatory subunit as such; i.e., formation of *pdh* would lead to the removal not only of the effector but also of the target for the effector. The action of this subunit, therfore, would have to be a stoichiometric rather than a catalytic one. Since enzyme precursors (i.e., after completed polypeptide synthesis) can exist (e.g., BILEZIKIAN, KAEMPFER, and MAGASANIK, 1967) such a situation does not seem to be an impossibility. In view of the data presented in section (5) the hypothesis of regulation at the nascent polypeptide chain advanced by GRUBER and CAMPAGNE (1965) appears less likely to apply to our case, i.e., if we wish to explain the large fraction of $0°$-type mutants occuring in a large segment of the E 1 gene.

Since at present, however, we are not even able to definitely rule out the existence of an *i*-type regulator gene further speculations are unwarranted.

## Summary

The synthesis of the pyruvate dehydrogenase complex in *E. coli* is shown to be an inducible one. The internal inducer has been identified as the substrate of the enzyme, pyruvate, which, in all probability, also acts as inducing effector. In order for the inducing action of pyruvate to be effective, neither the enzymatic activity of the pyruvate dehydrogenase nor that of its decarboxylase component is required. Some aspects of the unknown mechanism of action of pyruvate are discussed.

## Acknowledgements

We thank Mr. R. MAREK for invaluable technical assistance. J. DIETRICH thanks the Deutsche Forschungsgemeinschaft for a Predoctoral Fellowship; K. N. MURRAY, the National Institutes of Health, Bethesda, USA, for a Postdoctoral Fellowship (USPHS N⁰ 1 − F2 − GM − 31, 013 − 01). We thank Drs. J. R. BECKWITH, H. L. KORNBERG and L. J. REED for communicating unpublished results and we greatly appreciate having received the $icl^e$ and $pps^-$ mutants from Dr. H. L. KORNBERG and strain ECO $(proB\ lac)_{XIII}/F_{TS}lac$ (B.), (Fig. 1.) from Dr. J. R. BECKWITH.

## References

BILEZIKIAN, J. P., R. O. R. KAEMPFER, and B. MAGASANIK: J. molec. Biol. **27**, 495 (1967).

BECKWITH, J. R., E. R. SIGNER, and W. EPSTEIN: Cold Spr. Harb. Symp. quant. Biol. **31**, 393 (1966).

CÁNOVAS, J. L., L. N. ORNSTON, and R. Y. STANIER: Science **156**, 1695 (1967).

COOPER, R. A., and H. L. KORNBERG: Biochim. biophys. Acta (Amst.) **104**, 618 (1956).

CRAWFORD, I. P., and I. C. GUNSALUS: Proc. nat. Acad. Sci. (Wash.) **56**, 717 (1966).

CUZIN, F., and F. JACOB: C. R. Acad. Sci. (Paris) **258**, 1350 (1964).

ENGLESBERG, E., J. IRR, J. POWER, and N. LEE: J. Bact. **90**, 946 (1965).

GRUBER, M., and R. N. CAMPAGNE: Proc. kon. ned. Akad., Serie C, **68**, 1 (1965).

HANSEN, R. G., and U. HENNING: Biochim. biophys. Acta (Amst.) **122**, 355 (1966).

HAYASHI, S.-I., and E. C. C. LIN: J. molec. Biol. **14**, 515 (1965).

HENNING, U.: Biochem. Z. **337**, 490 (1963).

− G. DENNERT, R. HERTEL, and W. S. SHIPP: Cold Spr. Harb. Symp. quant. Biol. **31**, 227 (1966).

−, and C. HERZ: Z. Vererbungsl. **95**, 260 (1964).

− − and K. SZOLYVAY: Z. Vererbungsl. **95**, 236 (1964).

Imamoto, F., and C. Yanofsky: J. molec. Biol. 28, 1 (1967).

Ito, J., and I. P. Crawford: Genetics 52, 1303 (1965).

Jacob, F.: Science 152, 1470 (1966).

—, and E. A. Adelberg: C. R. Acad. Sci. (Paris) 249, 189 (1959).

— S. Brenner, and F. Cuzin: Cold Spr. Harb. Symp. quant. Biol. 28, 329 (1963).

Kornberg, H. L.: Biochem. J. 99, 1 (1966).

Mukherjee, B. B., J. Matthews, D. L. Horney, and L. J. Reed: J. biol. Chem. 240, PC 2268 (1965).

Newton, A.: Cold Spr. Harb. Symp. quant. Biol. 31, 181 (1966).

— J. R. Beckwith, D. Zipser, and S. Brenner: J. molec. Biol. 14, 290 (1965).

Ornston, L. N.: J. biol. Chem. 241, 3787, 3795, 3800 (1966).

—, and R. Y. Stanier: J. biol. Chem. 241, 3776 (1966).

Pettit, F. H., and L. J. Reed: Proc. nat. Acad. Sci. (Wash.) in press.

Reed, L. J., and D. J. Cox: Ann. Rev. Biochem. 35, 57 (1966).

Sheppard, D. E., and E. Englesberg: J. molec. Biol. 25, 443 (1967).

Stumpf, P. K.: J. biol. Chem. 159, 529 (1945).

Utter, M. F., and W. C. Werkman: Arch. Biochem. 5, 413 (1944).

Williams, F. R., and L. P. Hager: Arch. Biochem. 116, 168 (1966).

Willson, C., D. Perrin, M. Cohn, F. Jacob, and J. Monod: J. molec. Biol. 8, 582 (1964).

Wolfe, R. S., and D. J. Okane: J. biol. Chem. 205, 755 (1953).

Yanofsky, C.: Bact. Rev. 24, 221 (1960).

— Harvey Lect. 61, 145 (1967).

# Transcription in Amino Acid Starved Bacteria

N. O. KJELDGAARD, J. FORCHHAMMER, and D. W. MORRIS

Department of Molecular Biology, University of Aarhus
Århus, Denmark
and
University Institute of Microbiology, University of Copenhagen
Copenhagen, Denmark

In stringent bacterial strains, starvation of a required amino acid results in a reduction of the rate of net RNA synthesis to about 10 %of that of an unstarved culture.

This effect generally is assumed to be an expression of the normal physiological mechanism of RNA control, which in growing cells regulates the rate of RNA synthesis according to the growth medium of the culture (cf. MAALØE and KJELDGAARD, 1966).

Based mainly on the reduced rate of net RNA synthesis during amino acid starvation, several attempts have been made to formulate a general hypothesis for RNA regulation.

The models are built upon a) a regulation at the level of the substrates for the polymerase (GALLANT and CASHEL, 1967), b) a regulation at the level of the polymerase activity (KURLAND and MAALØE, 1962), c) a regulation at the level of the DNA template (STENT, 1964, 1966, 1967).

However, in their strict form, all of these models require that there is a coordinate effect of the regulation on the synthesis of rRNA, tRNA and mRNA.

It is known that the rates of rRNA and tRNA formation are decreased coordinately during amino acid starvation of a stringent strain.

To evaluate the models, it is obviously of importance to know if the rate of synthesis of mRNA shows a similar decrease.

The available information about this subject is rather controversial and for technical reasons difficult to interpret (HAYASHI and SPIEGELMAN, 1961; STERN et al., 1964; EDLIN, 1965; FRIESEN, 1966).

We shall present here the results of three different approaches in an attempt to resolve the question of the influence of amino acid starvation upon the rate of messenger formation in stringent cells.

All of our results indicate that this rate is unaffected by amino acid starvation.

The level of the messenger RNA in the cells is defined by an equilibrium between synthesis and exponential decay. A mechanism for the regulation of the level of messenger might affect either of these functions, and a detailed knowledge of both the rates of synthesis and decay, therefore, is essential for our evaluation of models of control.

The rate of decay can be rather easily measured by arresting new messenger formation and following the disappearance of the pre-existing pool.

We have as yet no means of interfering with this process of decay. Measurements of the rate of messenger synthesis, therefore, cannot be obtained independent of the decay and we have to content ourselves with estimates of the initial rates of synthesis under conditions where the concentration of messenger and consequently the decay is negligible.

For this very reason messenger synthesis is often followed by pulse-labelling experiments.

Quantitative interpretations of the results of such pulse experiments, however, are meaningful only if we know, not only the specific activity of the material outside the cells, but also the specific activity of the pool material inside the cells. Here we are in trouble when dealing with amino acid starved, stringent cultures. The situation is schematically illustrated in Fig. 1.

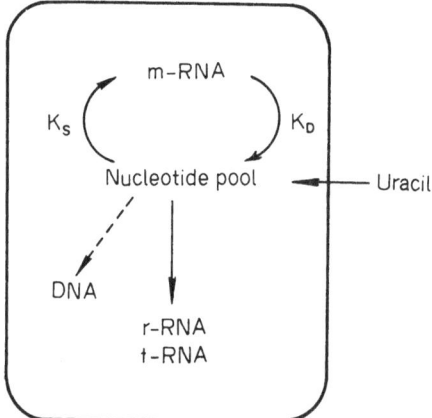

Fig. 1. Flow diagram for the incorporation of uracil into RNA

Since the mRNA at equilibrium concentration is synthesized from and degraded into nucleotides at the same overall rate, the drainage of pool materials is little influenced by the actual rate of turnover of the messenger-nucleotide cycle, but is determined mainly by the flux of nucleotides into the stable RNA species. When the rate of rRNA and tRNA synthesis decreases during amino acid starvation, the drainage is similarly diminished, followed by a slow uptake of labelled precursors from the medium into the pool (EDLIN and NEUHARD, 1967). Under these conditions the amount of label incorporated into mRNA during a pulse cannot be taken as a measure of the rate of messenger formation unless the specific activity of the nucleotide pool is taken into consideration. This is most easily done by depleting the nucleotide pool, including the messenger fraction, before the culture is subjected to a pulse.

(A) Starvation of a uracil auxotroph of the required pyrimidine results in a breakdown of the messenger fraction and a depletion of the pool of uracil-ribonucleotides (FORCHHAMMER and KJELDGAARD, 1967; NIERLICH, 1966; JAN NEUHARD, personal communication).

Using such depleted cells we have obtained a picture of the messenger formation during amino acid starvation by comparing the uptake of [$^{14}$C] uracil in cells prestarved of uracil to the uptake in control cells with a normal nucleotide pool.

The actual course of this experiment with the *E. coli* TAUbar strain is shown in Fig. 2.

This stringent strain with requirements for thymine, uracil, methionine, arginine, proline and tryptophan was grown exponentially in a

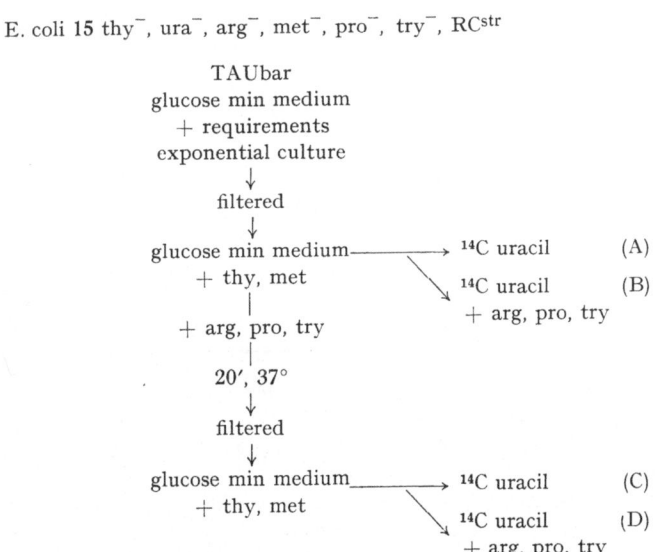

Fig. 2. Diagrammatic presentation of an experiment involving the incorporation of [$^{14}$C] uracil with or without starvation of required amino acids

glucose-minimal medium at 37°, filtered, washed and resuspended in a glucose medium containing only thymine and methionine.

Two aliquots were withdrawn, one receiving [¹⁴C] uracil, arginine, proline and tryptophan, the other [¹⁴C] uracil alone. The main culture was

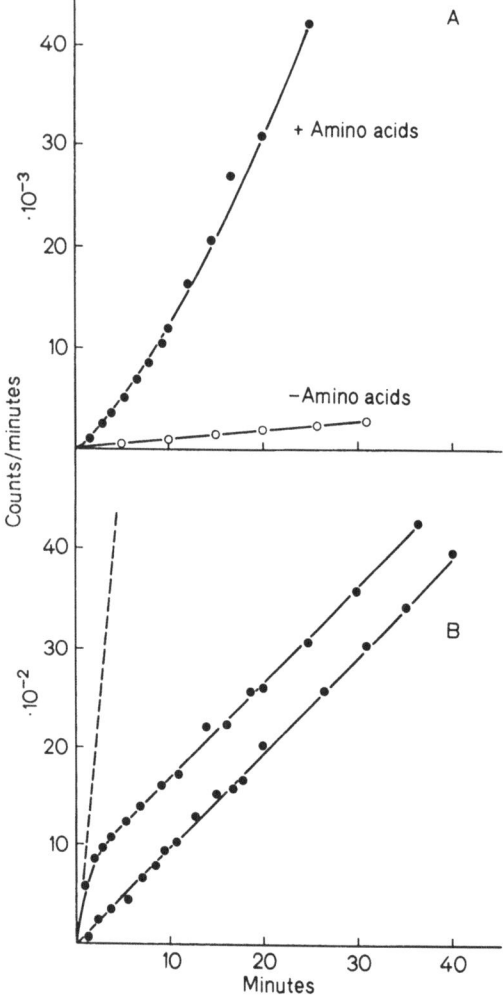

Fig. 3. A. Incorporation of [¹⁴C] uracil in a stringent culture with or without required amino acids (Cultures A and B of Fig. 2). B. Incorporation of [¹⁴C] uracil in amino acid starved, stringent cultures, with and without a period of pre-star-vation of uracil. The top curve shows the incorporation in the pre-starved culture (culture C, Fig. 2); the bottom curve gives the incorporation in a culture, containing a normal uracil pool (culture A, Fig. 2). The dashed line shows the incorporation in a control culture supplemented with all the amino acids (culture D, Fig. 2). The uracil uptake was followed by TCA precipitation and collection of the cells on membrane filters

supplemented with the three missing amino acids and further incubated at 37° for 20 min. The culture was then filtered and washed once more and resuspended in glucose medium plus thymine and methionine. Again two aliquots were given [14C] uracil with or without the three required amino acids. All radioactive cultures were incubated at 37°, and the incorporation followed by precipitation with cold TCA.

The two curves in Fig. 3 A show the uracil incorporation of the culture before uracil starvation and reveal an almost linear time course for the amino acid starved culture at a rate of about 10 % of that of the fully supplemented culture.

If amino acid starvation has no effect on the rate of mRNA synthesis, we would expect to see after depletion of the nucleotide pool an initial burst of uracil incorporation for about 1—2 min at 37°, corresponding to the resynthesis of the messenger fraction. As soon as the concentration of this fraction has attained the equilibrium value the uracil uptake would be expected to stabilize at a rate characteristic of the slow synthesis of rRNA and tRNA.

As shown in Fig. 3 B, this, indeed, is observed. The bottom curve is the amino acid starved culture of Fig. 3 A, magnified by a factor 10, whereas the top curve gives the incorporation during amino acid starvation into the culture, pre-starved of uracil. The steep, dashed line represents the incorporation into a fully supplemented culture.

Comparison of the sucrose-gradient sedimentation profiles of the radioactive material taken after a two minute pulse of labelled uracil showed nearly identical results for the two amino acid starved cultures, except for about a ten-fold higher level of radioactivity in the culture, prestarved of uracil.

We would like to suggest from these experiments that messenger is formed during amino acid starvation at a rather normal rate.

(B) Having seen the results of the labelling experiments we endeavoured to look more specifically for messenger synthesis, utilizing the unique property of mRNA to stimulate the incorporation of amino acids in a subcellular protein-forming system (FORCHHAMMER and KJELDGAARD, 1967).

The experimental design was similar to that described before, and also in this case we depleted the messenger fraction by uracil starvation of the TAUbar strain.

The "stimulating activity" of the RNA decreased during 66 min of starvation at 37° to about 10 % of that obtained from an exponentially growing culture (Fig. 4).

At this stage the culture was filtered, washed and resuspended at 25° in uracil-containing medium, either with all the required amino acids or lacking argine, proline, and tryptophan. At frequent intervals after the re-addition of uracil, samples of the cultures were harvested and the RNA extracted. To follow the resynthesis of mRNA the "stimulating activity" of the RNA preparations was measured.

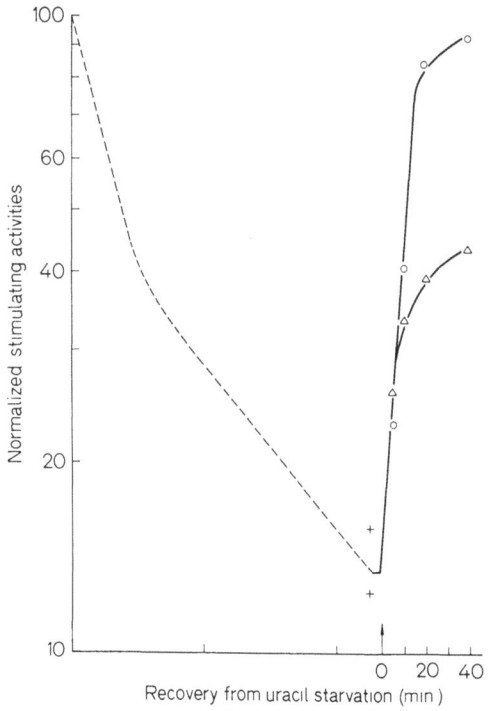

Fig. 4. Messenger resynthesis after uracil starvation. A culture of *E. coli* TAUbar was starved of uracil, RNA extracted from samples of the culture and the "stimulating activity" measured by adding aliquots of the RNA to a subcellular protein-forming system. After 66 min of starvation at 37°, uracil was readded to the culture either in the presence (O) or absence (△) of required amino acids. The resynthesis of messenger was followed by measurements of the "stimulating activity" of RNA isolated at different times after supplementing with uracil

As seen in Fig. 4, the initial rates of resynthesis of messenger were almost identical in the amino acid starved and the fully supplemented cultures. Again this suggests that messenger synthesis is independent of an amino acid control.

In the amino acid starved culture messenger formation levels off at a much lower value than in the case of the control culture.

In several experiments of this type, even after prolonged incubation, the level has never exceeded 60 % of that of an exponentially growing culture. The cause of this is not known, and we are rather free to speculate in terms of ribosomes as controlling elements according to the Stent model (1966), e.g. that the low level of messenger might be related to the availability of ribosomes in an appropriate state, ready for the release of messenger.

(C) To be more specific, we have attempted to estimate the rate of formation of $\beta$-galactosidase messenger during amino acid starvation (MORRIS and KJELDGAARD, 1967).

The measurements are based on the method devised by KEPES (1963). The capacity of the cells to form residual amounts of enzyme after removal of the inducer is taken as a measure of the amount of messenger present in the cells. Although we have no definitive proof that this capacity represents the expression of the messenger molecules present at the time of inducer removal, there is rather strong circumstantial evidence for this (NAKADA and MAGASANIK, 1964).

A stringent and a relaxed strain of *E. coli* K 12 were used in these experiments; both requiring thiamine, threonine, leucine, arginine and histidine; and isogenic except for the RC locus (FIIL and FRIESEN, 1968). A summary of the experiments is given in Fig. 5.

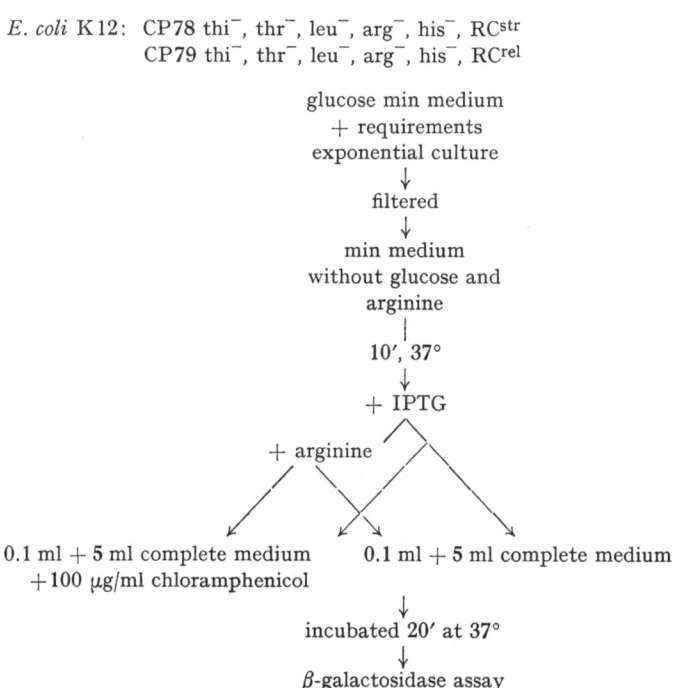

*E. coli* K12: CP78 thi⁻, thr⁻, leu⁻, arg⁻, his⁻, RCstr
CP79 thi⁻, thr⁻, leu⁻, arg⁻, his⁻, RCrel

glucose min medium
+ requirements
exponential culture
↓
filtered
↓
min medium
without glucose and
arginine
|
10′, 37°
↓
+ IPTG

+ arginine

0.1 ml + 5 ml complete medium        0.1 ml + 5 ml complete medium
+100 µg/ml chloramphenicol

incubated 20′ at 37°
↓
$\beta$-galactosidase assay

Fig. 5. Diagrammatic presentation of a $\beta$-galactosidase induction experiment

16*

As we want to study the induction of $\beta$-galactosidase under conditions of amino acid starvation, the effect of catabolite repression has to be overcome (Nakada and Magasanik, 1964). This was done by removing the energy source (glucose) at the time of arginine starvation. After 10 min of starvation, isopropyl-thio-galactoside (IPTG) was added to a concentration of $10^{-4}$ M and arginine was added back to half of the culture. At intervals, duplicate samples of 0.1 ml were withdrawn and diluted 50-fold either into a complete medium to arrest induction, or into a medium containing 100 µg/ml of chloramphenicol (CM) to stop all further protein synthesis. The diluted cultures were then incubated for 20 min at 37° before assaying for $\beta$-galactosidase activity.

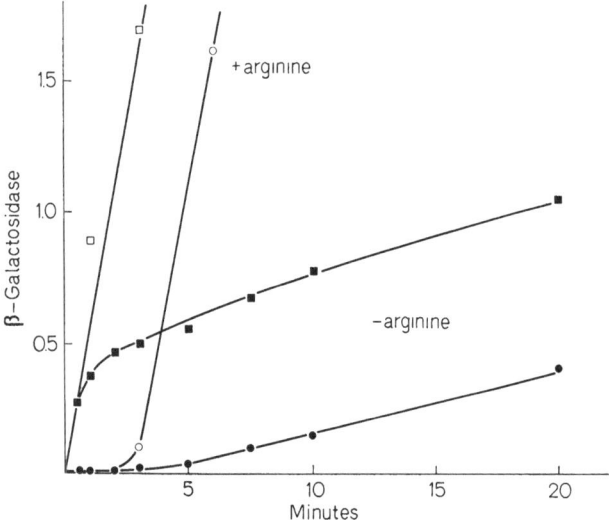

Fig. 6. The effect of arginine starvation on the synthesis of $\beta$-galactosidase and its capacity in a stringent strain of *E. coli*. The experiments were performed as indicated in Fig. 5. The assay for $\beta$-galactosidase was a conventional hydrolysis of o-nitro-phenol-galactoside and is given as increase per min in optical density at 420 mµ

The results of such an experiment are shown in Fig. 6. The circles give the values for the CM tubes, i.e. the amount of enzyme present in the cells at the time of dilution. The squares represent the values for the samples diluted into growth medium, where the messenger present at the time of dilution is allowed to be translated into enzyme.

The difference between the two curves, defined as the capacity, is taken as a measure of the amount of $\beta$-galactosidase messenger present in the cells.

The capacity curves for two starvation experiments involving the stringent and the relaxed strains are presented in Fig. 7.

It is clear from these experiments that the initial rate of $\beta$-galactosidase messenger synthesis is only slightly affected by the arginine starvation in the stringent strain and not at all in the relaxed strain. This again is a confirmation of the notion that messenger synthesis is more or less unaffected by starvation of a required amino acid.

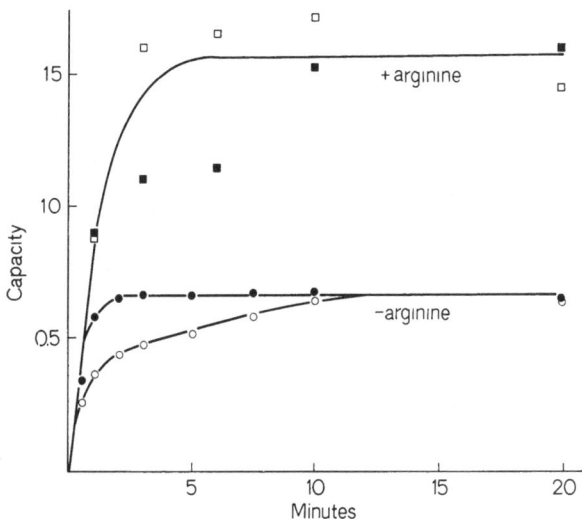

Fig. 7. Capacity for $\beta$-galactosidase formation of a stringent ($\square$, $\bigcirc$) and a relaxed ($\blacksquare$, $\bullet$) strain in the presence or absence of arginine. The capacity is defined as the difference between the amount of enzyme measured in chloramphenicol samples and the samples, incubated in complete medium (Fig. 5)

In both the stringent and the relaxed strains we found that, even after prolonged induction, the capacity of the arginine-starved cultures reached a level of about 50 % of the non-starved cultures.

Since the stringent and the relaxed strains both show this behaviour, it is not necessary here to go into further discussion of this observation.

Due to catabolite repression the presence of glucose prevents the induction of $\beta$-galactosidase.

By adding glucose to an induced, amino acid starved culture we found that the production of $\beta$-galactosidase messenger was rapidly arrested and the capacity for enzyme formation decayed (Fig. 8).

The capacity, measured at different times after glucose addition to the arginine-starved cultures, showed an exponential decay with a normal messenger halflife of about 1 min at 37° in the stringent, as well as the relaxed strain (Fig. 9).

We conclude from these results that neither the rate of messenger synthesis nor the rate of decay are seriously influenced by starvation of a required amino acid.

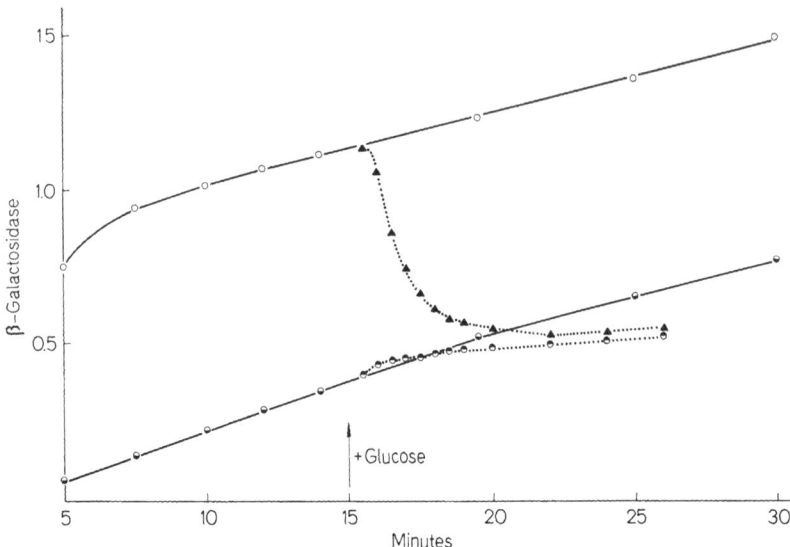

Fig. 8. The effect of glucose addition on the synthesis of $\beta$-galactosidase and its capacity in an arginine-starved, stringent culture

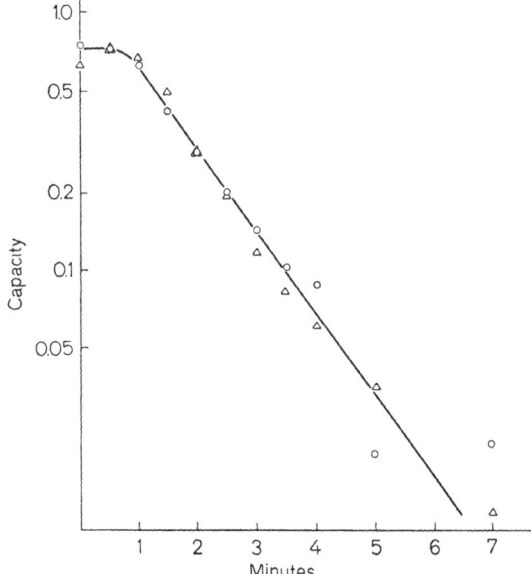

Fig. 9. The decay of $\beta$-galactosidase capacity after addition of glucose to arginine-starved stringent (O) and relaxed ($\triangle$) strains

    Taken alone, none of the experimental approaches presented above yield sufficient information; however, the combined evidence from these experiments permits us to conclude that in stringent strains, amino acid starvation does *not* result in a coordinate regulation of the synthesis of mRNA and the stable species of RNA.

This result is incompatible with the theories that propose a common regulation of RNA formation, at least in their simplest forms.

Of the three models presented earlier, the one by STENT (1966, 1967) can best be modified to take this fact into account.

Since we have other indications which support the idea that ribosomes are involved in the regulation of messenger formation (FORCHHAMMER and KJELDGAARD, in preparation), it is only necessary to modify the original notion of an intimate link between protein synthesis and messenger release.

It is known that amino acid starvation results in a decrease in the concentration of the amino acid charged tRNA molecules in question (MORRIS and DeMOSS, 1965). Our results, therefore, suggest that the translational movement of the ribosomes and a concommittant release of the messenger from the DNA template continue, even if amino acid charged tRNA molecules are not available, and net protein synthesis is arrested.

This work was supported in part by a grant from the U.S. Public Health Service (AI 04914). One of us (D.W.M.) holds a postdoctoral fellowship from the U.S. Public Health Service.

## References

EDLIN, G.: J. molec. Biol. **12**, 356 (1965).

—, and J. NEUHARD: J. molec. Biol. **24**, 225 (1967).

FIIL, N., and J. D. FRIESEN: J. Bact. **95**, 729 (1968).

FORCHHAMMER, J., and N. O. KJELDGAARD: J. molec. Biol. **24**, 459 (1967).

FRIESEN, J. D.: J. molec. Biol. **20**, 559 (1966).

GALLANT, J., and M. CASHEL: J. molec. Biol. **25**, 545 (1967).

HAYASHI, M., and S. SPIEGELMAN: Proc. nat. Acad. Sci. (Wash.) **47**, 1564 (1961).

KEPES, A.: Biochim. biophys. Acta (Amst.) **76**, 293 (1963).

KURLAND, C. G., and O. MAALØE: J. molec. Biol. **4**, 193 (1962).

MAALØE, O., and N. O. KJELDGAARD: Control of macromolecular synthesis. New York: W. A. Benjamin Inc. 1966.

MORRIS, D. W., and J. A. DeMOSS: J. Bact. **90**, 1624 (1964).

—, and N. O. KJELDGAARD: J. molec. Biol. **31**, 145 (1967).

NIERLICH, D.: Abstr. Amer. Soc. Microbiol. 66th Ann. Meet. 1966.

NAKADA, D., and B. MAGASANIK: J. molec. Biol. **8**, 105 (1964).

STENT, G.: Science **144**, 816 (1964).

— Proc. roy. Soc. B **164**, 181 (1966).

— Organizational biosynthesis (H. J. VOGEL, J. O. LAMPEN, and V. BRYSON, eds.) New York: Academic Press 1967.

STERN, J. L., M. SEKIGUCHI, H. D. BARNER, and S. S. COHEN: J. molec. Biol. **8**, 629 (1964).

# Hemoglobins of Chironomus tentans and C. pallidivittatus: Biochemical and Cytological Studies

By

HERBERT TICHY

Max-Planck-Institut für Biologie, Abteilung Beermann, Tübingen

Hemoglobin, the respiratory protein of the vertebrates, occurs among the insects only in the larvae of the Chironomides, the non-biting midges. In contrast to the hemoglobin of vertebrates, Chironomus hemoglobin is dissolved in the body fluid of the larvae. The concentration varies from 1 to 5 % depending on the larval stage.

SVEDBERG [1] determined the molecular weight of Chironomus hemoglobin and found it to be 31,400. This corresponds to a dimer molecule. With the aid of starch gel electrophoresis BRAUN [2] has shown that a number of different hemoglobins are present in the lymph of Chironomus thummi. According to the same author [3], some of the hemoglobins of Chironomus thummi can reversibly aggregate depending on pH. The presence of several different hemoglobin chains in the body fluid of Chironomus raises a number of questions of genetic and evolutionary interest.

While examining additional species of Chironomids (Fig. 1), it was found that a few of the hemoglobin bands are nearly identical in all species whereas others vary considerably so that each species has its own hemoglobin pattern. The two species, Chironomus tentans (*tent.*) and Chironomus pallidivittatus (*pall.*) are especially useful for a study of the genetic problems involved here because they can be hybridized and $F_2$ progeny can be obtained. Larvae of $15-20$ mm belonging to the IV larval instar were examined. 2 µl hemolymph, containing $17-20$ µg hemoglobin, were bled from one animal and immediately used for electrophoresis. An electrophoresis apparatus has been developed [4] by means of which a total of 49 samples in 10 % polyacrylamide gel slabs of 1 mm thickness could be tested. The best buffer for the separation of the hemoglobins was the 0.1 m TRIS EDTA boric acid buffer of ARONSSON and GRÖNWALL with a pH of 8.8. The other conditions employed were 400 V 125 mA for two hours and, in order to avoid denaturation during the run, the gel slabs were cooled with tap water of 8– 10°C. The hemoglobin bands were identified by their color which varies from light yellow to red according to their concentration in the transparent gel. Subsequently, the slabs were stained with Amidoblack 10B.

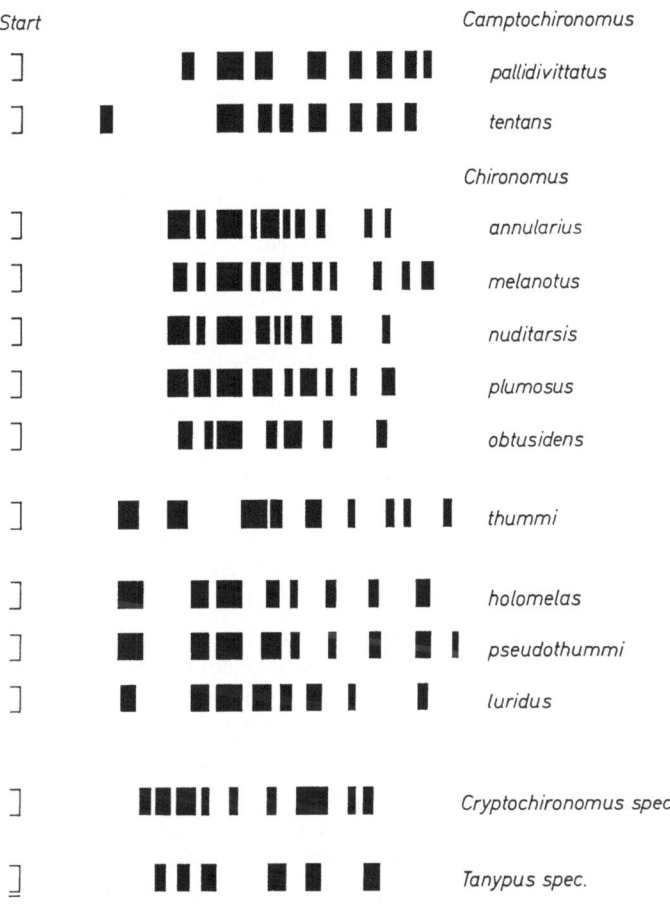

Fig. 1. Hemoglobin pattern of different Chironomus species

The differences in the hemoglobin patterns between the two species are shown in Fig. 2. The bands were numbered according to their electrophoretic mobility, no. I designating the slowest of those bands which appear constantly. There is one still slower migrating band ("VB"), which appears only in late fourth instar larvae shortly before the prepupal stage. The primary question which is raised by these results concerns the nature of the bands themselves.

250                                    H. Tichy:

Does each of the electrophoretically obtained bands represent a single
protein chain, or do they result from the aggregation of a few monomers
to form different dimer molecules?

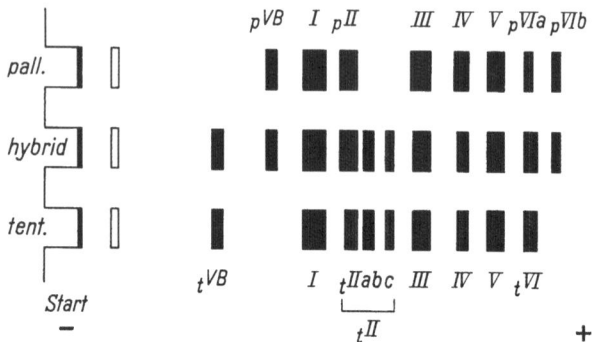

Fig. 2. Hemoglobin pattern of Chironomus pallidivittatus (pall.), Chironomus
tentans (tent.) and of the hybrid of these species. The indices $p.$, $t.$, $a$, $b$, $c$ mark the
species specific hemoglobins

To answer this question it is necessary to separate the hemoglobins
according to the method described by Braun [2] and Formanek [5]. 5 g
amounts of Chironomus larvae were homogenized and the hemoglobin
was isolated through differential precipitation with $(NH_4)_2SO_4$, purified by
gel-filtration through a Sephadex G 25 column, and transferred into the
electrophoresis buffer. Hemoglobin band no. I was separated on a DEAE
Sephadex A 50 column. Both the total purified hemoglobin and the isolated
hemoglobin fraction no. I, when tested in an analytical ultracentrifuge,
were found to have a molecular weight of 15,000 ± 1000. This means that
each of the electrophoretically obtained bands contains only a single protein chain. The existence of 8 to 10 different hemoglobin chains requires
the presence of the same numbers of globin cistrons in the Chironomus
genome. This raises the interesting question of whether these genes are
distributed randomly over the chromosomes or, are all localized in the
same region.

Since the larvae of Chironomus possess well developed giant chromosomes in the cells of their salivary glands, this question may be answered
with reasonable precision by conventional cytological techniques.

Chromosome maps for both of the species involved in the present
experiments have been published by Beermann [6]. It has been shown
that the homologous chromosomes from the two species differ only by
rearrangements of the chromosomal bands. For example, the left arm of
the I chromosome of *tent.* differs from *pall.* by the inversion of the
chromosomal region 1 A—7 B. In the hybrid, these inverted chromosomal

regions are easily distinguishable; they show no somatic pairing. Special intra-specific inversions, such as the complex inversion of the left arm of the III. chromosome III L − 1 K in *pall*., and the nucleoli, can be used as cytological markers. *Pall*. has only one nucleolus on the right arm of the II chromosome; *tent*. has two nucleoli, one on the left arm of the II chromosome and the other in the middle region of the III. chromosome. Genetic studies with the hemoglobins of Chironomus are simplified by the fact that hemoglobin inheritance is codominant. In other words, the hemoglobin pattern of the hybrid (Fig. 2) combines the patterns of the two species.

A comparison between the chromosomal configuration and the electrophoretic pattern of the hemoglobin in each individual larva should lead to the correlation of species specific hemoglobins with specific chromosomes of either of the two species. All of the species specific hemoglobins have been found to be correlated with the left arm of the III chromosome, chromosomal region 1−11C of BEERMANN's chromosomal map. In the III chromosome crossing over can only occur in the regions which are homologous in both species, i.e., when the standard arrangements of both species are present within the regions 1−5A, and 9A−11C. If *tent*. standard is paired with *pall*. inversion IIIL−1K, cross-over can only occur in the chromosomal region from 10B to 11C. Some of the latter cross-over were found to be at the same time recombinant for hemoglobin II and hemoglobins VB and VI. These findings indicate that the gene for hemoglobin II is located between 10B and 11C, whereas the genes for hemoglobins VB and VI must be situated somewhere in the chromosomal regions 1 to 10B.

A still finer delimitation of the hemoglobin loci can be tentatively achieved by observing the degree of somatic chromosome pairing. Complete somatic pairing only occurs in homologous regions of chromosomes which are derived from the same species. Homologous regions of chromosomes from different species, on the other hand, often do not pair completely, or pairing may be absent completely. If the absence of somatic pairing within chromosomal regions of identical chromomeric sequence is taken as a criterium for heterozygosis due to the prior occurrence of interspecific cross-over, the chromosomal region for hemoglobin band II can be narrowed down to the region 10B of the III chromosome with approximately 5 small bands. By using the same criterium, hemoglobins VI and VB must be localized to the left of region 9C of chromosome III (Fig. 3). Although the localization is not very exact, these results indicate the presence of at least two separate hemoglobin regions in chromosome III.

The hemoglobin producing cells are not known, but the possibility seems to be excluded that the salivery glands produce hemoglobin. Thus, in the region 10B of the salivary gland chromosomes, where the gene for the hemoglobin II should be located, messenger RNA synthesis and puffing would not be expected. In fact, Pelling [7], who examined local distribution of RNA production and puffing along the giant chromosomes of Chironomus tentans by autoradiography, found no activity of this kind in the bands of the chromosomal region 10 B.

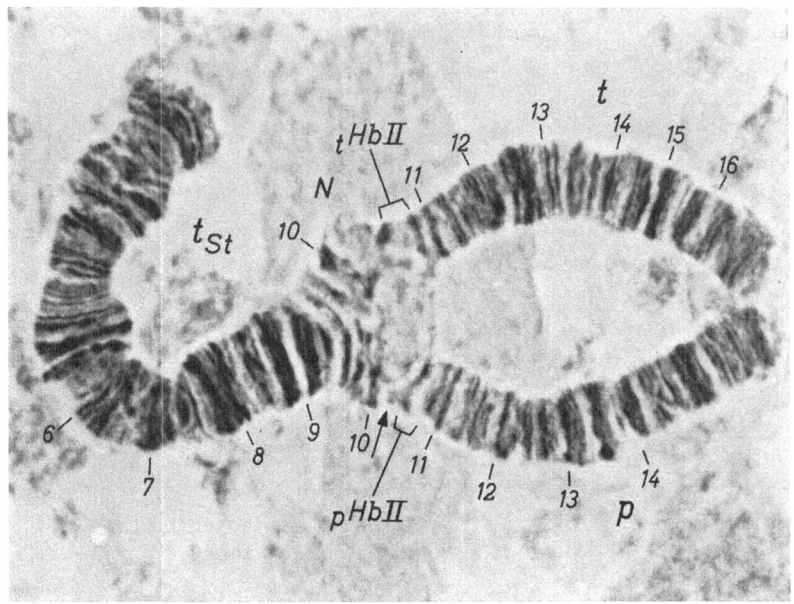

Fig. 3. The III chromosome of a hybrid larva which is homozygot tent. in the left arm and heterozygot for tent. and pall. in the right arm. $t$ = tent. $p$ = pall $tst$ = tent. standard arrangement of the left chromosome arm. $N$ = nucleolus arrow marks the crossing over fork indicates the chromosomal region where the genes for hemoglobin II are located

## References

1. Svedberg, T., u. K. O. Pederson: Die Ultrazentrifuge. Dresden: Steinkopf 1940.
2. Braun, V.: Zur vergleichenden Biochemie der respiratorischen Proteine des Blutes. Dissertation München 1964.
3. — Hoppe-Seyler's Z. physiol. Chemie (in press).
4. Tichy, H.: A multisample electrophoresis apparatus using vertical polyacryl-amide gel slabs. Anal. Biochem. 17, 320—326 (1966).
5. Formanek, H.: Gewinnung, Reindarstellung und Kristallisation des Häemo-globins der Larven von Chironomus. Diplomarbeit München 1964.
6. Beermann, W.: Cytologische Analyse eines Campto-Chironomus Artbastards. Chromosoma (Berl.) 7, 198—259 (1955a).
7. Pelling, C.: Ribonukleinsäure-Synthese der Riesenchromosomen. Chromosoma (Berl.) 15, 71—122 (1964).

# Gene Functions and Genetics of Phage S 13

Irwin Tessman and Ethel S. Tessman

Department of Biological Sciences, Purdue University
Lafayette, Indiana, USA

S 13 and $\phi$X 174 are in the same family of small icosahedral (Hall, Maclean, and I. Tessman, 1959) bacterial viruses, each containing single-stranded DNA (I. Tessman, 1958, 1959) with about 5000 nucleotides (Sinsheimer, 1959). The two viruses are nearly identical: they recombine genetically (E. S. Tessman and Shleser, 1963) and they show complementation of their gene functions (Hayashi and E. S. Tessman, unpublished), as well as being very similar in many other aspects (Zahler, 1958).

Their small size makes these viruses attractive for an attempt to identify the functions of all the phage genes and the manner of their organization and regulation. Such an analysis was begun on an exhaustive scale by E. S. Tessman (1965) using conditional lethal mutations, both temperature-sensitive and suppressible, like those discovered for phage $\lambda$ (Campbell, 1961). Such mutations are useful for identifying essential phage genes.

In the initial experiments five genes were identified, using genetic complementation as the criterion for assigning mutations to separate genes; we assume that each gene codes for a distinct polypeptide species. Since then considerably more mutants have been examined, revealing two more genes for a present total of seven (I. Tessman, Ishiwa, Kumar, and Baker, 1967). Approximately 250 mutants, induced by a large variety of methods, have been tested by complementation. No gene contains less than 14 independently isolated mutants. Therefore it seems likely that every essential gene has been identified, with the reservation that genes with abnormal genetic behavior could have escaped notice, as will become obvious later when the poor rescue phenomenon characteristic of one of the genes is discussed. Polarity effects by nonsense mutations could conceal a new gene by preventing its complementation with a neighboring gene, but approximately half of the mutants tested were temperature-sensitive and therefore should not show polarity effects.

Nonessential genes, for which lethal mutations would not normally occur, would also have excaped detection by the methods used, as was the case for phage T4 (SIMON and I. TESSMAN, 1963; HALL and I. TESSMAN, 1966; HALL, I. TESSMAN, and KARLSTRÖM, 1967). Nevertheless approximate calculations based on coding concepts indicate that the phage can code for seven to ten genes; for this reason we believe we have accounted for at least 70 % and perhaps 100 % of the phage genes (I. TESSMAN, ISHIWA, KUMAR, and BAKER, 1967).

## The Phage Map

The five genes identified originally were mapped into a linear structure, using two factor crosses (E. S. TESSMAN, 1965). With this preliminary map as a guide all seven genes have now been entirely ordered using only three-factor crosses (BAKER and I. TESSMAN, 1967). The resulting map is circular (Fig. 1).

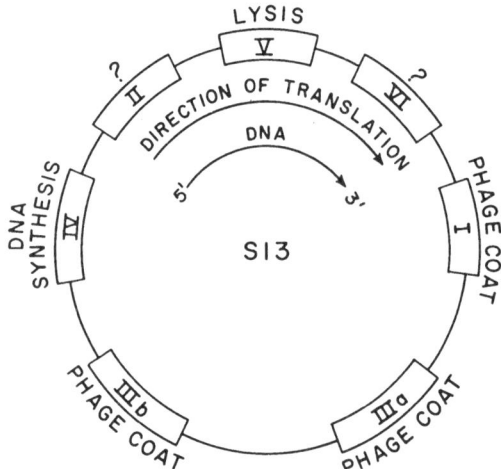

Fig. 1. The circular genetic map of phage S 13 showing the gene functions where known

A circular genetic map is noteworthy because of the implications for the relationship between the genetic map and the physical structure of the DNA, which is circular in both the single-stranded from (SS) (FIERS and SINSHEIMER, 1962) and double-stranded, replicative form (RF) (KLEINSCHMIDT, BURTON, and SINSHEIMER, 1963; CHANDLER, HAYASHI, HAYASHI, and SPIEGELMAN, 1964). The circular DNA must undergo an even number of recombination events, the minimum number being two, in order to yield a genome of unit size. But if one of the required events were always to occur in the same region the resulting map would not be circular but would be linear, as is found for phage $\lambda$ (CAMPBELL, 1963). Therefore it can be concluded that in S 13 there is no unique site for genetic recombination, as for example, a site where the DNA ring always opens; conceivably such a site might be operative in replication.

Double recombination events do appear to be the rule, as required for a circular map (BAKER and I. TESSMAN, 1967). But it is not known whether both events occur at the same or different times, and if at different times whether the second event is inevitable.

Recombination within gene IV appears to be anomalously large, exceeding by about ten fold the largest frequencies found within any other gene. This suggests that there may be a tendency for one of the recombination events to occur frequently within this gene.

There is evidence of at least two mechanisms of recombination in S 13 (I. TESSMAN, 1966). One mechanism is at least partially host controlled because it is abolished by the $rec^-$ mutation of CLARK and MARGULIES (1965) that affects bacterial recombination. Both "reckless" and "cautious" $rec^-$ bacterial strains (HOWARD-FLANDERS, THERIOT, and STEDE-FORD, 1966) have been tested and only "reckless" strains reduce S 13 recombination (I. TESSMAN, unpublished data). The indication that the phage has at least two recombination mechanisms came from the fact that there is significant residual recombination even in the "reckless" $rec^-$ strains. This residual mechanism is not stimulated by UV irradiation, while the primary mechanism in the $rec^+$ host can be enhanced 10 to 50 fold by irradiation of the phage before infection (I. TESSMAN, unpublished data).

### Direction of Translation and Orientation of the DNA with Respect to the Genetic Map

Nonsense mutations (*amber* type) in gene III a show polarity by reducing the activity of one neighboring gene, III b, but not the other neighboring gene, I (TESSMAN, KUMAR, and TESSMAN, 1967). Since this is a property of the *amber* type mutations and not the temperature-sensitive, it appears likely that the polarity results from the chain-terminating property of nonsense mutations (SARABHAI, STRETTON, BRENNER, and BOLLE, 1964; STRETTON, KAPLAN, and BRENNER, 1966); therefore the reduction in the activity of III b indicates that translation proceeds in the direction from III a to III b.

It has previously been suggested that all the phage genes are transcribed into messenger RNA (mRNA) from the strand complementary to the phage DNA strand found in the RF (HAYASHI, HAYASHI, and SPIEGEL-MAN, 1964). This would imply that all genes are translated in the same direction relative to the genetic map, so the polarity effect between genes IIIa and IIIb can be used to make the general conclusion indicated in Fig. 1 that the entire genome is translated in the clockwise direction.

A reservation must be made about this conclusion. It has already been pointed out by Summers and Szybalski (1967) that the homology between the phage mRNA and the phage DNA may not be 100 % complete, for if one gene made a small amount of messenger complementary to the phage DNA, especially at an early time after infection, it could have escaped notice. Therefore it is conceivable that perhaps one or two genes are translated in the opposite direction from that shown.

From the mRNA-DNA homology and the known direction of translation of the mRNA, the 5' to 3' orientation of the DNA nucleotides is also implied to be in the same direction in which the mRNA is translated.

The polarity effect implies that genes IIIa and IIIb are transcribed onto the same mRNA. No strong polarity effects have been observed with nonsense mutants in any of the other known genes so it cannot be concluded from the genetic evidence whether other genes are transcribed together.

## Gene Functions

The functions of five of the seven genes shown in Fig. 1 are known in a general way and will now be discussed.

*Genes I, IIIa IIIb.* A variety of experiments indicate that these three genes determine coat proteins. Experiments on the heat stability of the mature virus show that mutations in genes I, IIIa, and IIIb substantially alter the rate at which the mature virus is inactivated at $51.5°C$ (Tessman and Tessman, 1966; Baker and I. Tessman, 1967). It seems that nearly every mutation in I and IIIb alters the heat stability while only a small minority do in IIIa. No mutation in any other gene has yet been found to alter the heat stability.

Most mutants in gene I seem to alter the rate of attachment of the phage to its host so gene I must determine an attachment site. Serum blocking power (SBP), which most likely measures presence of phage attachment-site proteins, is made by *amber* type mutants in IIIb. I, IIIa, IV, and II mutants have been tested and found not to make SBP.

*Gene V.* This gene controls lysis as found for $\phi X 174$ (Hutchison and Sinsheimer, 1966) but its mechanism of action is not known. Other genes can also affect lysis; although *amber* mutants of all genes except gene IV (see below) show normal lysis, many temperature-sensitive mutants of genes I, IIIa, and IIIb (the protein coat genes) show greatly delayed lysis (E. S. Tessman, 1967).

*Gene IV.* This gene plays a role in phage DNA synthesis; mutants in this gene block synthesis of phage RF after the initial synthesis of the "early" RF (which is probably equivalent to parental RF) from the infecting parental SS DNA (E. S. TESSMAN, 1965). The kinetics of RF formation in wild type S 13 and in a gene IV mutant are illustrated in Fig. 2.

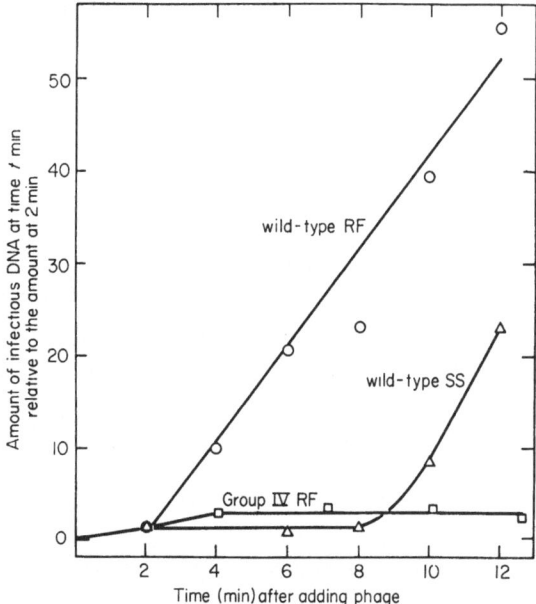

Fig. 2. Kinetics of synthesis of infectious double-stranded DNA (RF) and infectious single-stranded DNA by wild type S 13; kinetics of synthesis of double-stranded DNA by a gene IV mutant, *su* 100. The amounts of RF and SS at 2 min are set equal to 1.0

The initial amount of RF formed by a gene IV mutant is proportional to its multiplicity of infection, so it appears as if each infecting single-stranded DNA molecule is converted to a double-stranded molecule, which cannot be replicated further.

The synthesis of the gene IV product is chloramphenicol-resistant (E. S. TESSMAN, 1966); it requires at least 100 μg/ml, rather than the customary 30 μg/ml of CM (KURLAND and MAALØE, 1962) to inhibit the function of gene IV. This resistance was responsible for the original impression (DENHARDT and SINSHEIMER, 1965) that RF replication did not involve any phage-induced enzymes. CM-resistance may imply that very little gene IV product is needed or that perhaps its synthesis is different from the usual mode of protein synthesis.

Gene IV mutants are poorly rescued in complementation experiments under non-permissive conditions (E. S. TESSMAN, 1965). The burst consists almost entirely of the non-gene IV partner with a very small burst of the gene IV type. There is also poor rescue of gene IV RF in such mixed infections. The amount of mutant gene IV DNA made is proportional to the multiplicity of infection, which indicates that the infecting DNA is converted to early RF, but that the replication of this RF is very limited. Thus, a gene IV mutant is limited in its ability to use the enzyme made by a functioning wild-type gene IV; this may indicate that the enzyme does not diffuse. Another possibility is that the gene IV enzyme can act only on the genome that makes it, perhaps because its ability to function is limited to a very short time after synthesis, or because the enzyme must be in a special spatial relationship with respect to the genome. Both the CM resistance, with its implication that very little enzyme is needed, and the poor rescue support the idea of a close association of the gene IV product and the phage DNA. It is clear that mutants blocked in two different poorly rescued genes would not complement well and give the false impression of being in the same gene. This is a limitation inherent in the complementation test.

Gene IV *amber* mutants appear to be blocked in the function of phage genes other than gene IV. Gene IV mutants are the only mutants blocked in exclusion of superinfecting phage; gene IV mutants also fail to lyse the non-permissive host. Thus for S 13 there is a dependence of expression of gene function on DNA replication. This phenomenon was previously found for T 4 (WIBERG et al., 1962; LEVINTHAL, HOSODA, and SHUB, 1967) and for $\lambda$ (DOVE, 1966).

There is a definite correlation between RF synthesis and SS synthesis: RF synthesis shuts off as SS synthesis begins (Fig. 3) (E. S. TESSMAN, 1967), and if CM (30 $\mu$g/ml) is used to inhibit SS synthesis any time up to 20 min after infection at 37°C, RF synthesis will immediately resume (Fig. 4).

What is the specific role of the gene IV product? Two possibilities have been tested and rejected:

a) It may be responsible for converting the 16S double-stranded DNA [which contains at least one single-strand break, (ROTH and HAYASHI, 1966)] into the superhelical 21S form. If so, a block of gene IV should lead to a build-up of 16S RF and little or no formation of 21S. However, the contrary is found, the relative amounts of 16S and 21S RF being about the same as for wild-type phage (SHLESER, ISHIWA, MANNES, and E. S. TESSMAN, unpublished data).

b) It may be responsible for converting a non-infectious 16S or 21S RF to an infectious form. However, no accumulation of either non-infectious 16S or non-infectious 21S RF is found (SHLESER, E. S. TESSMAN, and MANNES, unpublished; SINSHEIMER, LINDQUIST, and HUTCHISON, 1967).

It remains possible that gene IV determines a DNA polymerase.

A host mutation has been found (FRIEDMAN and I. TESSMAN, un-published data) that blocks RF replication at what appears superficially to be the same stage at which the gene IV mutational block occurs (ISHIWA and I. TESSMAN, data unpublished). Thus at least two enzymes,

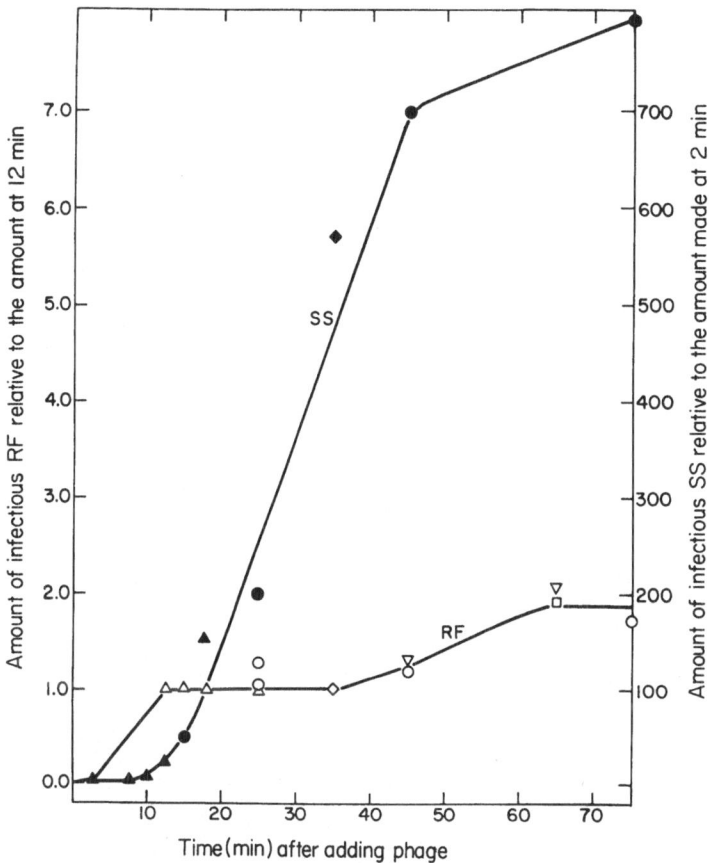

Fig. 3. Kinetics of synthesis of infectious RF and infectious SS by an S13 lysis (gene V) mutant, $su$N15. The amount of RF at 12 min (the time of the initial shut-off of RF synthesis) is set equal to 1.0 on the left-hand ordinate. The amount of SS at 2 min (the earliest time measured) is set equal to 1.0 on the right-hand ordinate

one host-induced and one phage-induced, are necessary for RF replication. Conceivably the phage induces a modification of the host enzyme. A mutant strain which blocks formation of progeny RF of $\phi$X 174 has also been described by DENHARDT, DRESSLER, and HATHAWAY (1967).

The functions of genes II and VI are unknown at present. Mutations in II or VI prevent the appearance of SS DNA. But so do mutations in all

17*

other genes except V. SS DNA never appears free, even for wild-type phage infection, but is always encapsulated (Sinsheimer, Starman, Nagler, and Guthrie, 1962); thus coat proteins are necessary for appearance of SS DNA. This may reflect involvement of the coat proteins in SS

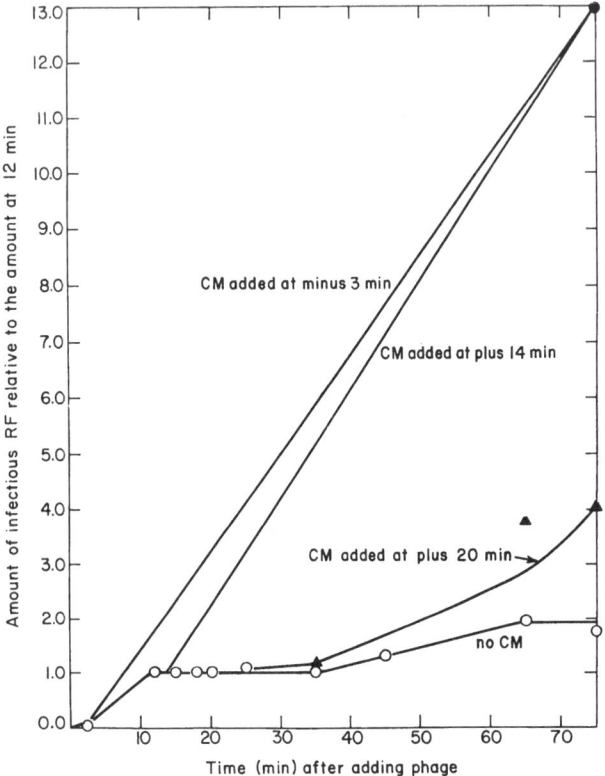

Fig. 4. Formation of infectious RF by an S 13 gene V mutant, *su* N 15, in the presence and absence of 30 μg/ml chloramphenicol (CM)

synthesis, or an indirect involvement such as the protection of the SS DNA from breakdown or the conversion to some other form (such as RF). Therefore, it cannot be said at present which phage gene, if any, directly controls SS DNA synthesis. It is even conceivable that gene IV makes both RF and SS under different conditions.

It seems that the mechanism of synthesis or maturation of SS DNA is highly specific for only one of the two complementary DNA strands. If there are any phage containing the other strand, they must comprise less than about 0.2 % of the infectious particles (I. Tessman, Poddar, and Kumar, 1964).

Other functions which seem likely to be performed by phage-specified proteins are the shut-off of host DNA synthesis (LINDQVIST and SINS-HEIMER, 1967) and shut-off of RF synthesis (E. S. TESSMAN, 1966). Either or both of these functions may be under the control of genes II or VI.

## Acknowledgements

Some of the research described was supported in part by NSF grant GB-2748 to I. TESSMAN and NIH grant AI-3903 to E. S. TESSMAN.

## References

BAKER, R., and I. TESSMAN: Proc. nat. Acad. Sci. (Wash.) **58**, 1438 (1967).
CAMPBELL, A.: Virology **14**, 22 (1961).
— Virology **20**, 344 (1963).
CHANDLER, B., M. HAYASHI, M. N. HAYASHI, and S. SPIEGELMAN: Science **143**, 47 (1964).
DENHARDT, D. T., D. H. DRESSLER, and A. HATHAWAY: Proc. nat. Acad. Sci. (Wash.) **57**, 813 (1967).
—, and R. L. SINSHEIMER: J. molec. Biol. **12**, 647 (1965).
DOVE, W. F.: J. molec. Biol. **19**, 187 (1966).
FIERS, W., and R. L. SINSHEIMER: J. molec. Biol. **5**, 424 (1962).
HALL, C. E., E. C. MACLEAN, and I. TESSMAN: J. molec. Biol. **1**, 192 (1959).
HALL, D. H., and I. TESSMAN: Virology **29**, 339 (1966).
— —, and O. KARLSTRÖM: Virology **31**, 442 (1967).
HAYASHI, M., M. N. HAYASHI, and S. SPIEGELMAN: Proc. nat. Acad. Sci. (Wash.) **51**, 351 (1964).
HOWARD-FLANDERS, P., L. THERIOT, and J. B. STEDEFORD: Abstracts of the tenth annual meeting of the biophysical society, p. 69. 1966.
HUTCHISON, C. A., and R. L. SINSHEIMER: J. molec. Biol. **18**, 429 (1966).
KLEINSCHMIDT, A. K., A. BURTON, and R. L. SINSHEIMER: Science **142**, 961 (1963).
KURLAND, C. G., and O. MAALØE: J. molec. Biol. **4**, 193 (1962).
LEVINTHAL, C., J. HOSODA, and E. SHUB: The molecular biology of viruses, p. 71. (ed. by J. S. COLTER, and W. PARANCHYCH). New York: Academic Press 1967.
SARABHAI, A. S., A. O. W. STRETTON, S. BRENNER, and A. BOLLE: Nature (Lond.) **201**, 13 (1964).
STRETTON, A. O., S. KAPLAN, and S. BRENNER: Cold Spr. Harb. Symp. quant. Biol. **31**, 173 (1966).
SIMON, E. H., and I. TESSMAN: Proc. nat. Acad. Sci. (Wash.) **50**, 526 (1963).
SINSHEIMER, R. L.: J. molec. Biol. **1**, 43 (1959).
— C. A. HUTCHISON, and B. H. LINDQVIST: The molecular biology of viruses, p. 175 (ed. by J. S. COLTER, and W. PARANCHYCH). New York: Academic Press 1967.
— B. STARMAN, C. NAGLER, and S. GUTHRIE: J. molec. Biol. **4**, 142 (1962).
SUMMERS, W. C., and W. SZYBALSKI: Virology **34**, 9 (1968).
TESSMAN, E. S.: Virology **25**, 303 (1965).
— J. molec. Biol. **17**, 218 (1966).
— The molecular biology of viruses, p. 193 (ed. by J. S. COLTER, and W. PARANCHYCH) New York: Academic Press 1967.
—, and R. SHLESER: Virology **19**, 239 (1963).
TESSMAN, I.: Programs and abstracts of the biophysical society, p. 42. 1958.
— Virology **7**, 263 (1959).

Tessman, I.: Biochem. biophys. Res. Commun. **22**, 169 (1966).
— H. Ishiwa, S. Kumar, and R. Baker: Science **156**, 824 (1967).
— S. Kumar, and E. S. Tessman: Science **158**. 267 (1967).
— R. K. Poddar, and S. Kumar: J. molec. Biol. **9**, 352 (1964).
—, and E. S. Tessman: Proc. nat. Acad. Sci. (Wash.) **55**, 1459 (1966).
Wiberg, J. S., M.-L. Dirksen, R. H. Epstein, S. E., Luria, and J. M. Buchanan:
    Proc. nat. Acad. Sci. (Wash.) **48**, 293 (1962).
Zahler, S. A.: J. Bact. **75**, 310 (1958).

## Discussion

*Wittmann:* What is known about the difference between the three coat proteins and what is their relative amounts?

*Tessman:* The gene I product is important for phage adsorption and the gene III b product is inferred to be needed in large amount because nonsense mutations in III a can block III b from functioning effectively. But we have not purified proteins corresponding to the three coat proteins genes.

*Hofschneider:* Do gene No. IV, II and VI which are supposed to control DNA synthesis interfere with coat-protein synthesis?

*Tessman:* The effect of a mutation in gene VI on coat protein synthesis has not yet been examined. But we believe genes II and IV have some effect on coat protein synthesis because if these genes are not functioning the formation of serum blocking power is prevented.

*Zillig:* Is the single-stranded daughter DNA made from the RF or on it?

*Tessman:* I believe the answer to this is unclear at present.

# Distribution of Thymidilate Incorporation into the DNA of X 174-Infected Triphosphate Permeable Cells

By

H. HOFFMANN-BERLING

Max-Planck-Institut für Medizinische Forschung, Heidelberg

Investigation of the replication of bacteriophage $\phi X 174$ ($\phi X$) has shown that infecting single stranded DNA enters a double stranded replicative form (RF). This molecule undergoes repeated semiconservative replication and further RF appears (SINSHEIMER, STARMAN, NAGLER, and GUTHRIE, 1962). In an infected culture RF-DNA increases at a constant rate, and it has been suggested that only one RF-molecule in a cell, probably that containing the parental DNA-strand, provides the template for the synthesis of further RF (DENHARDT and SINSHEIMER, 1965b). At a rate of $1-2$ RF-molecules synthesized per cell per minute labeled DNA-precursors flow rapidly through replicating DNA into mature RF (LINDQVIST and SINSHEIMER, 1967) which according to DENHARDT and SINSHEIMER needs not replicate further. Preferential labeling of $\phi X$-DNA molecules engaged in the process of replication accordingly is difficult.

Experiments to be described have employed cells after extraction with ether. This procedure presumably removes lipids from the cell membrane and was intended to open cells to desoxyribonucleoside triphosphates. These are immediate precursors of *in vitro* synthesized DNA and do not penetrate into normal cells. One purpose was to achieve control over the DNA-precursors available and the rate of DNA-synthesis in the cells and to facilitate pulse labeling of DNA by removing non-labeled precursors from leaky cells.

BUTTIN and KORNBERG (1966) have studied triphosphate permeable cells prepared by incubating (non-infected) cells with EDTA in a Trisbuffer and besides extensive breakdown of RNA have observed endonuclease I dependent uptake of nucleotide label into the DNA of such cells. Mutants producing $0.1\%$ of normal endonuclease have been found to be normally efficient hosts for DNA-phages (DÜRWALD and HOFFMANN-BERLING, 1968) including $\phi X$, suggesting that wild type activity of endonuclease I is not required for a replication of viral DNA. Endonuclease I deficient mutants were used for the following experiments and were handled in the presence of sufficient magnesium to prevent ribosomal disintegration in the cells and release of latent RNase I from ribosomes. This enzyme efficiently activates RNA-inhibited endonuclease I in *E. coli* extracts.

Bacterial suspensions, if treated with ether, lost viability from $3 \times 10^{10}$ to less than $10^4$ colony formers per ml and most suspensions became sterile. In a modified DNA-polymerase assay mixture, extracted non-infected cells incorporated label from tritiated TTP (thymidine tri-phosphate) into an acid precipitable form at rates which varied from preparation to preparation and which were especially low if mechanical rupture of cells during extraction and washing from ether had been avoided. Ultrasonication or pancreas DNase present in the assay mixtures increased the initial rate of label uptake (Fig. 1) in accordance with findings from other laboratories (for references see LEHMAN, 1967) that fragmented DNA has enhanced priming activity for DNA-polymerising enzymes. Presence of the other three common desoxyribonucleoside tri-phosphates besides TTP was required for a maximal rate of label uptake;

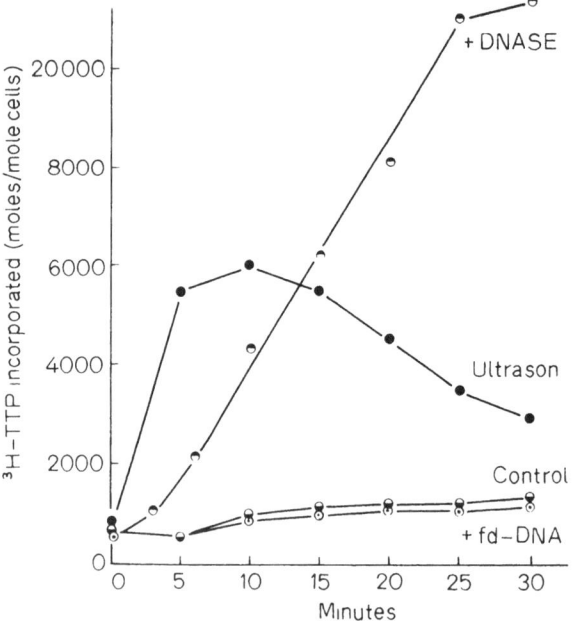

Fig. 1. Thymidilate incorporation into noninfected cells. Exponential cells were labeled and incubated as described in Fig. 2. No addition ( ⊖ ); fd-DNA (100 phage equivalents/ml added ( ○ ); 0.1 μg/ml pancreas DNase (Worthington) added ( ◓ ); 10 sec ultrasonic treatment of the assay mixture prior to incubation ( ● )

thymidine did not replace TTP and, in a fourfold excess, did not compete with TTP for incorporation. Low speed band centrifugation of DNase treated, labeled cells sedimented 75—85 % of the acid precipitable radioactivity with the cells while added single stranded or double stranded DNA remained unadsorbed near the starting point of the run. We conclude that at least part of the ether treated cells was permeable to triphosphates and, in addition, to DNase and that, under the conditions maintained, this enzyme promoted uptake of label into the cells rather than into DNA released from the cells. Penetration of an enzyme as DNase into cells is an interesting phenomenon from an operational point of view.

Single stranded DNA from phage, although excellent as a primer of extracted DNA-polymerase (MITRA, REICHARD, INMAN, BERTSCH, and KORNBERG, 1967), did not prime in our system (Fig. 1) suggesting exclusion of added DNA from the DNA-synthesising centers of ether extracted cells.

$\phi$X-infected cells, in similar tests (omitting ultrasonics or DNase), incorporated label at a several times higher rate than non-infected cells, conditions for rapid labeling being as described for non-infected cells (Fig. 2). Again 80 % of the precipitable radioactivity remained with sedimented cells suggesting insignificant rupture of cells and insignificant release of priming DNA-fragments from cells.

The diagrams given in Figs. 3—7 give a survey of the distribution of thymidilate label in the DNA of ether-extracted, $\phi$X-infected cells. All these preparations had been derived from cultures after 5 min of starvation-synchronized $\phi$X-infection when synthesis of infectious single stranded DNA had not yet begun. Two kinds of experiments are presented, involving either short term labeling (Fig. 4—6; 400—500 molecules thymidilate incorporated per cell) or long term labeling of cells (Fig. 7) with incorporation of a ten times higher amount of thymidilate per cell (protein content of an ether extracted cell: about $1.2 \times 10^{-13}$ g; Lowry method). For reasons not to be discussed here given amounts of incorporated thymidilate are minimal estimates.

Phenol-extracted DNA from labeled $\phi$X-infected cells showed up to four distinct radioactive product fractions in a low salt sucrose density gradient. These included products at s > 30, 21 s, 15 s and near 10 s. Products at 10 s or lower will be designated as the "10 s-fraction" of radioactivity, disregarding heterodispersity. Given s-values refer to single-stranded fd-DNA, assuming 11 s as the sedimentation rate of such DNA under the conditions employed. This rate is not well defined, however.

Radioactive products at s > 30 (in Fig. 3—7 at the bottom of the tubes) were seen also in noninfected cells and presumably represented *E. coli* DNA which had taken up label. Other radioactive products appeared

to be constituents specific of $\phi$X-infected cells. Radioactivity at 21 s was
prominent especially after a long term labeling procedure (Fig. 7B).
Products in this region sedimented in different ionic media as a double

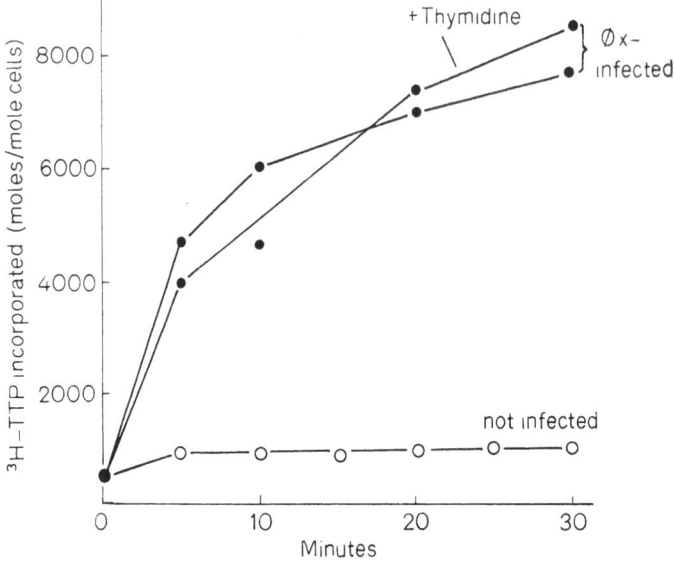

Fig. 2. Thymidilate incorporation into $\phi$X-infected cells. Cells were centrifuged
after 5 min of starvation-synchronized $\phi$X-infection (DENHARDT and SINSHEIMER,
1965a) resuspended to give $3 \times 10^{10}$ cells/ml. in a solution containing (in μmoles/ml)
KCl 80; Tris-HCl (pH 7.8) 40; Mg-acetate 8 ; EGTA 2.5; sucrose 500; sperminc-
HCl 0.2. After shaking with an equal volume of ether for 5 min in the cold the cells
were washed and resuspended in the same medium and in a $6 \times 10^{10}$ cells/ml containing suspension were stored in the cold. Incubation mixtures (35°C) contained in
a solution as described $6 \times 10^9$ cells/ml (thawed once) and as further additions (in
μmoles/ml) dATP, dGTP, dCTP 0.02 each; $^3$H-TTP (specific activity 0.48 C/mmole)
0.004; ATP 1; NAD 0.0005. At the times indicated 0.1 ml samples removed from
the mixture were precipitated and washed with 5 % TCA (containing 1 mg/ml
hydrolysed RNA) on membrane filters (Membranfilter-Gesellschaft, Göttingen).
Counting of the dried filters was in a Packard liquid scintillation spectrometer.
Filled symbols: infected cells with or without 0.016 μmoles/ml cold thymidine
added. Open symboles: noninfected cells

stranded DNA (Fig. 4) with an s-value close to RF I-DNA of replicating
$\phi$X (for references concerning RF I see JAENISCH, HOFSCHNEIDER, and
PREUSS, 1966). There were furthermore analogies with RF I-DNA in the
reactions of 21 s-product. These included 1. unaltered sedimentation properties of heated (5 min to 98°C) and chilled 21 s-product and resistance
of heated or alkali treated material to single strand specific exonuclease I
(Fig. 5). Both effects indicate non-denaturability or easy renaturation of

21 s-product. There was, however, in the gradient media employed, an increase of the sedimentation rate to 24 s after alkali treatment and reneutralisation of 21 s-product (Fig. 5) suggesting a permanent structural change in the treated molecule. This effect has not been observed after a heating procedure and has not been reported for authentic RF I. 2. Pancreas DNase sufficient for an 80 % infectivity loss of $\phi$X-DNA in a 1 µg/ml. DNA containing assay created 15 s-product out of 21 s-product with no products sedimenting at intermediate rates (Fig. 3); 15 s-product thus obtained was a denaturable DNA as shown by band sedimentation of alkali treated and reneutralised product and digestion with exonuclease I (experiments not shown).

According to present knowledge, non-denaturable DNA can derive its property either from a topological link between base paired circles of DNA (as is the case with authentic RF I) or, in a non-ring molecule, from covalent connections between complementary nucleotide sequences.

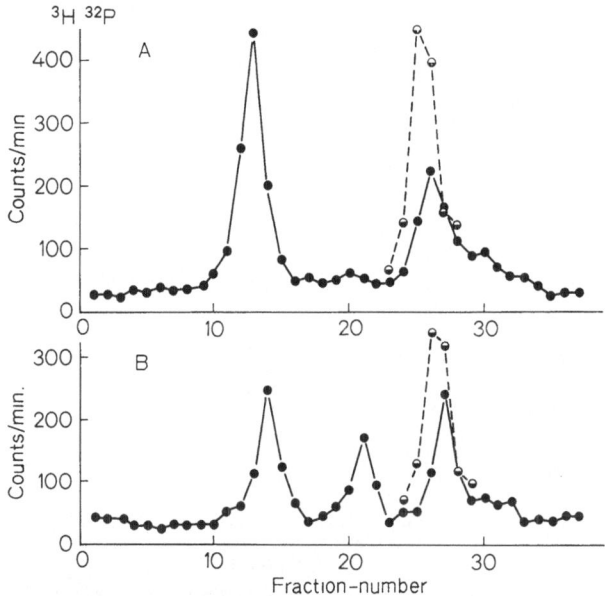

Fig. 3. Distribution of incorporated thymidilate in $\phi$X-infected cells. Cells were prepared and labeled as described in Fig. 2 (total volume 6 ml; 15 min 35°C). Centrifuged cells were disintegrated by maintaining for 1 h at 0°C and subsequently freezing in 3 ml buffered salt containing per ml. 5 µmoles EDTA, 200 µg lysozyme and 10µg pancreas RNase. The DNA was then extracted with phenol, dialysed and subjected to band sedimentation (Spinco SW 41 rotor; 40000 rpm.; 290 min; 20°C) in 12.2 ml neutral 5 – 20 % sucrose containing 0.005 M phosphate; 0.0002 M EDTA. A. control; B. preparation treated prior to sedimentation with 0.0001 µg/ml pancreas DNase; 6 µmoles/ml Mg-acetate (30 min 35°C). Solid line in this and the following figures: [3H]-label; dotted line: [32P]fd-DNA added as a reference

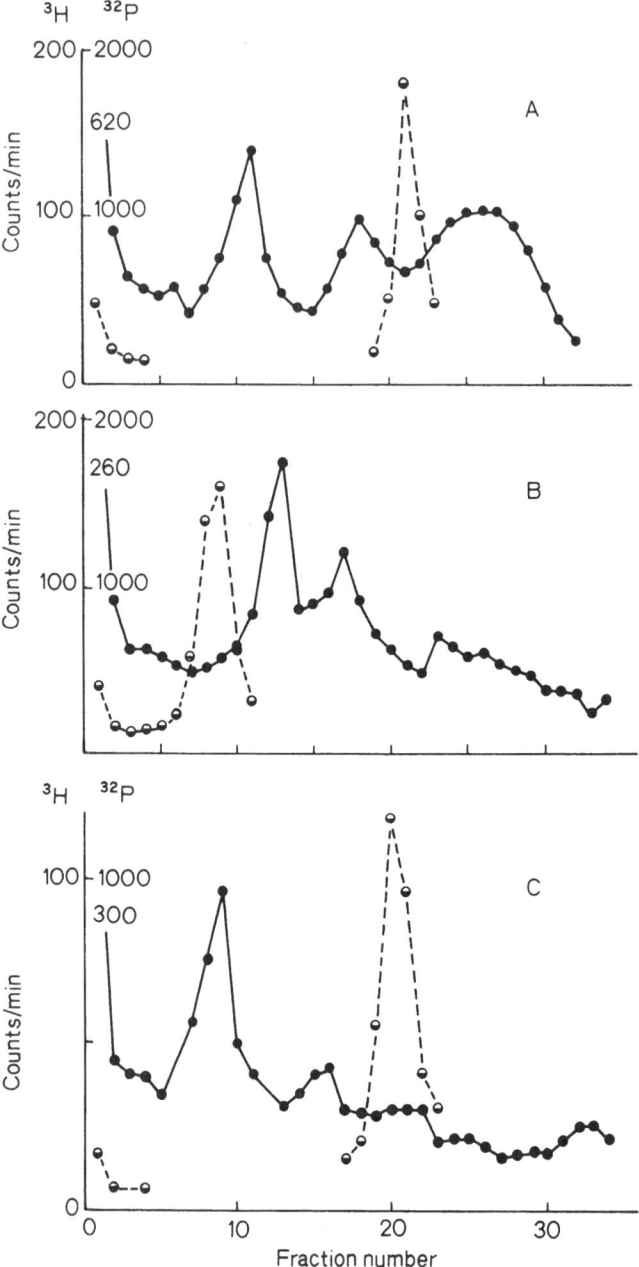

Fig. 4. Cells infected for 5 min as described except for the presence of 30 μg/ml chloramphenicol in the growth medium. Labeling with ³H-TTP (0.004 μmoles/ml; 0.85 C/mmole) was for 5 min (400 – 500 molecules thymidilate incorporated per cell); other conditions as in Fig. 2. Labeled cells were extracted with phenol. A. Gradient in 0.005 M phosphate; B. same as A with 0.2 M NaCl present in the gradient; C. DNA digested with 10 units/ml exonuclease I for 30 min at 37°C prior to sedimentation. Gradient in 0.005 M phosphate

DNA of the latter kind is known to result from extensive replication of DNA-template with DNA-polymerase *in vitro* (confer MITRA et al., 1967). Denaturability established by careful treatment with DNase is the reaction expected for intertwined rings rather than for a linear DNA containing crosslinks or paired loops of DNA.

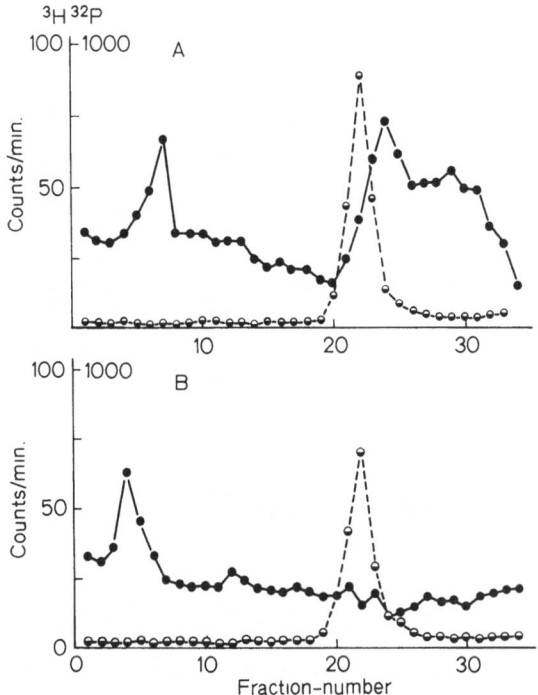

Fig. 5. Same deproteinised material as in Fig. 4. Gradients in 0.005 M phosphate. A. after treating with 0.2 N NaOH (10 min 37°C) and reneutralisation. B. after alkali treatment and exposure to exonuclease I

Circular DNA produced in a subcellular system suggests survival in the system of active polynucleotide joining enzyme (GELLERT, 1967; OLIVERA and LEHMAN, 1967a) which requires NAD as a cofactor for activity (OLIVERA and LEHMAN, 1967b; ZIMMERMAN, LITTLE, OSHINSKY, and GELLERT, 1967). In the experiments described incubation mixtures contained $5 \times 10^{-7}$ M NAD which according to the data given in the literature should have saturated enzyme accessible to cofactor.

Authentic RF I, besides being a duplex of separately closed circles, is thought to exist in a tightly twisted state which is released on endonucleolytic interruption in one of the strands. The pronounced influence of low DNase on the sedimentation rate of 21-product suggests, but does not prove, a structure in this product similar to that of authentic RF I.

The products at 15 s and 10 s will be considered together. Sensitivity to exonuclease I of undenatured 10 s-product (Fig. 4) together with a strong dependence of the sedimentation rate on the ionic environment (Fig. 4) clearly indicated a predominance of single stranded DNA in this fraction. Single strands appeared whether or not the cells had been infected in chloramphenicol [this agent is known to prevent the appearance of infectious single stranded DNA in $\phi$X-infected living cells (SINS-HEIMER et al., 1962)]. Products at 15 s, in the same tests, gave ambiguous results. Being partially digestible by exonuclease as a non-denatured product this DNA had sedimentation properties under conditions of high or low salt concentration and after application of alkali as if essentially double stranded (Figs. 4 and 5).

The situation became clearer when it was found that the distribution of label between 10 s and 15 s varied according to the mode of deproteinisation of the cells. It can be seen from Fig. 6 that pronase (BERNS and THOMAS, 1965) with subsequent addition of sodium dodecylsulfate (SDS), omitting phenol, increased the radioactivity at 15 s and that little or no radioactivity remained at slower positions; 15 s-product from a pronase-SDS treated preparation had sedimentation properties and the easy denaturability reported for 15 s-product from phenol extracted specimens. It appeared from these experiments that prior to phenolisation single stranded DNA had existed in a pronase-SDS resistant link connected to some larger molecule, probably a double stranded DNA, and that it had been freed from this DNA by shaking with phenol.

Single stranded DNA (as far as revealed by infectivity tests) is a product exclusively of late $\phi$X-infection. Such DNA does not appear in chloramphenicol-treated cells, as already mentioned, and once made is thought to be excluded from replication in the cell of its origin (SINS-HEIMER et al., 1962). The appearance of single strands in cells which prior to ether extraction presumably made no such DNA poses an unexpected problem. A straight forward interpretation of the results would be in terms of a degradative process, conceivable being exonucleolytic breakdown of double strands with release of single strands. Arguments against this interpretation are 1. the results of pulse-chase experiments (Fig. 7) which indicate that on incubation of cells label can migrate from 10 s to 21 s, i.e. into a double stranded product with $\phi$X-specific properties; 2. the fact that the single strands are rapidly labeled DNA. In calculating the ratio of the radioactivity at 10 s relative to that at 21 s one finds a value of 5 for the short term labeling experiment given in Fig. 4 and a ratio of 0.7 for the long term labeling experiment of Fig. 3. 3. A further argument against general breakdown of DNA under the conditions of our experiments comes from a study of parental labeled $\phi$X-DNA in infected cells. Such DNA sedimented with essentially unaltered profile after extraction of the cells with ether and incubation (for 15 min) with triphosphates and NAD omitted from the mixture.

In a certain way the situation reminds to that observed in studying RNA-polymerase from RNA-phage infected cells. In this case newly

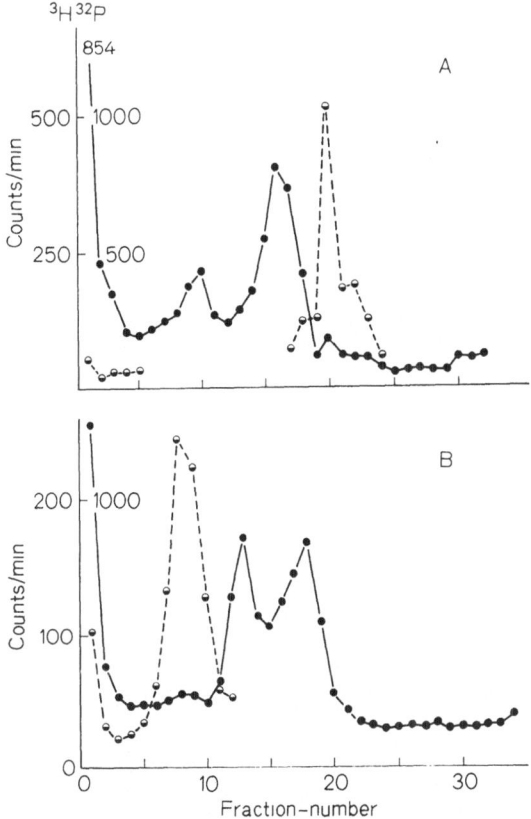

Fig. 6. Same labeled material as in Fig. 4. Deproteinsation was effected by incubating for 60 min at 37°C with 500 μg/ml pronase; 0.01 μmole/ml EDTA. The mixture was then made 1 % in SDS and centrifuged. A. gradient in 0.005 M phosphate; B. as A, with 0.2 M NaCl in the gradient. Gradients had received different amounts of radioactivity

synthesized RNA reacts as if single stranded but undergoes base pairing and formation of double stranded RNA after a protein denaturing treatment of the enzyme-RNA complex (BORST and WEISSMANN, 1965). In fact, recent experiments indicate that pronase-SDS tends to increase the extent of pairing of newly synthesised DNA in a 15 s-complex, although some new DNA can still be removed from the pretreated complex as a single strand. Provided that the labeled DNA at 15 s has the structure of a natural product, its study may reveal additional insight into the mode of $\phi$X-DNA replication.

The results presented have been obtained in collaboration with Miss HILDEGARD DÜRWALD and Dr. H. P. VOSBERG.

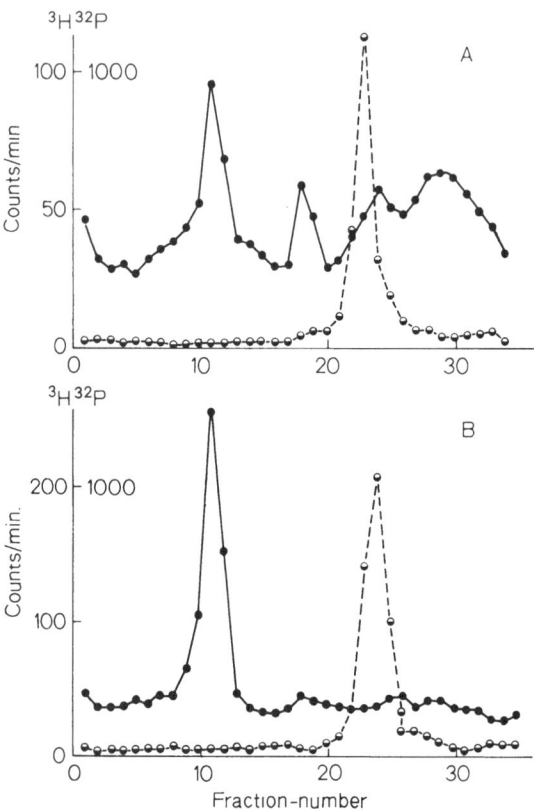

Fig. 7. Pulse-chase experiment. Same cell preparation as in Fig. 4—6. Labeling was for 5 min with 0.006 μmoles/ml. ³H-TTP (0.85 C/mmole). Phenol extraction. Gradients in 0.005 M phosphate. A. cells extracted immediately after labeling; B. labeled cells incubated for further 15 min with 0.5 μmole/ml cold TTP added and then extracted

## References

BERNS, K. J., and C. A. THOMAS: J. molec. Biol. 11, 476 (1965).

BORST, P., and C. WEISSMANN: Proc. nat. Acad. Sci. (Wash.) 54, 962 (1965).

BUTTIN, G., and A. KORNBERG: J. biol. Chem. 241, 5419 (1966).

DENHARDT, D. T., and R. L. SINSHEIMER: J. molec. Biol. 12, 641 (1965a).

— — J. molec. Biol. 12, 642 (1965b).

DÜRWALD, H., and H. HOFFMANN-BERLING: J. molec. Biol. 34, 331 (1968).

GELLERT, M.: Proc. nat. Acad. Sci. (Wash.) 57, 148 (1967).

JAENISCH, R., P. H. HOFSCHNEIDER, and A. PREUSS: J. molec. Biol. 21, 501 (1966).

LEHMAN, I. R.: Ann. Rev. Biochem. **36**, 645 (1967).
LINDQVIST, B., and R. L. SINSHEIMER: Fed. Proc. **26**, 449 (1967).
MITRA, S., P. REICHARD, R. B. INMAN, L. L. BERTSCH, and A. KORNBERG: J. molec. Biol. **24**, 429 (1967).
OLIVERA, B. M., and I. R. LEHMAN: Proc. nat. Acad. Sci. (Wash.) **57**, 1426 (1967 a).
— — Proc. nat. Acad. Sci. (Wash.) **57**, 1700 (1967 b).
SINSHEIMER, R. L., B. STARMAN, C. NAGLER, and S. GUTHRIE: J. molec. Biol. **4**, 142 (1962).
ZIMMERMAN, S. B., I. W. LITTLE, C. K. OSHINSKY, and M. GELLERT: Proc. nat. Acad. Sci. (Wash.) **57**, 1841 (1967).

## Discussion

*Scholtissek:* In order to accumulate highly labelled deoxynucleoside triophosphates in *E. coli*, it should be tried to incubate the bacteria at 4°C with labelled deoxynucleosides. At least in an animal cell at 4°C RNA- and DNA-synthesis is completely stopped, while the phosphorylation of nucleosides and deoxynucleosides up to the triphosphate level occurs at a resonable rate (SCHOLTISSEK, BBA, 1967). By this way the treatment with EDTA could be avoided.

*Hoffmann-Berling:* Thank you for your suggestion! Please note that no EDTA has been used in the present experiments (in order to avoid activation of latent residual endonuclease I by RNase I released from disintegrating ribosomes; EGTA which has been used does not complex with magnesium and leaves ribosomal particles intact). One purpose of the ether extraction procedure was to remove endogenous DNA-precursors from the sells. This would not have been achieved by proceeding as you suggested.

*Schuster:* How do the ether-treated cells look like? Do they still have their walls?

*Hoffmann-Berling:* Rodshaped and more globular cells have retained their characteristic shape after ether extraction. Whether this means that the cell has been retained cannot be said; but from the well known stability of the basis structure of the wall towards organic solvents one would expect that part of the wall structure still exists.

# Primary Structure of fd Coat Protein. Comparative Studies with Proteins of Phages fl, M13 and ZJ-2

By

GERHARD BRAUNITZER, FRANK ASBECK, KONRAD BEYREUTHER*, HEINZ KÖHLER, and GERHARD VON WETTSTEIN

Max-Planck-Institut für Biochemie, München

At present four small rod shaped DNA-containing phages, which infect E. coli, are known. These are: fd, M 13, fl and ZJ-2 [1]. They all have a molecular weight of $11 \times 10^6$ and a DNA-content of 12 % [2]. The molecular dimensions are about 8,000 by 50 Angström [2]. These phages are all specific for "male" E. coli K 12 F$^+$- and Hfr-strains. By virtue of the special biological properties gram amounts of these non lysogenic viruses can be obtained. Using the procedures first described by HOFF-MANN-BERLING [3], we are able to isolate 100 mg of lyophilised phage from one litre of culture. By free flow electrophoresis using the method of HANNING [4] the intact phage can be readily separated from non viral proteins.

I will speak about the structure of the coat protein of fd phage.

Chemical and enzymic methods have been used exclusivly for molecular weight determinations of the subunit of fd coat protein. For a long time these methods were not very successful. Only using a specially prepared protein were we able by the hydrazinolysis method of BRAUN and SCHROEDER [5] to obtain the correct value of 5,900 (uncorrected) for molecular weight (Fig. 1). From this result and on the basis of amino acid analysis it is now clear, that the peptide chain consists of only 49 amino acid residues.

This result is extremly surprising, since this peptide chain is only one third the size of the coat protein subunit of TMV, also a rod shaped virus [6].

Amino acid analysis indicated in addition, that histidine, arginine and cysteine were absent. The four valine residues were recovered only after 400 h hydrolysis.

---

* Auszug K. BEYREUTHER, Dissertation, Naturwiss. Fak., Universität München, in preparation.

End group determination by the methods of STARK and SMYTH [7] and Edman strip-method in the modification of SCHROEDER [8] gave alanine as N-terminal amino acid. Carboxypeptidase digestion and hydrozinolysis both gave serine as C-terminal amino acid.

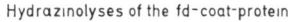

Hydrazinolyses of the fd-coat-protein

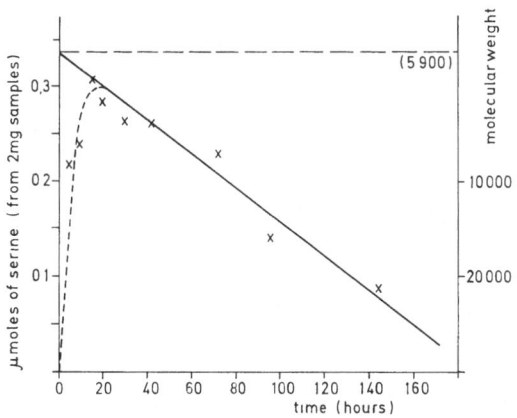

Fig. 1. The removal of serine in μmoles is plotted against time. From extrapolation to zero time a molecular weight of 5900 is obtained. This value is not corrected for the decomposition of serine which occurs during hydrazinolysis. The hydrazinolysis is carried out using 2 ml distilled hydrazine, 20 mg IRC-50 (H+-form) at 80°C

The coat protein is resistant to enzymic cleavage. The best results were obtained with pepsin in 10 % formic acid. By this method peptides were obtained in good yields only from the N- and C-terminal part of the peptide chain. The insoluble core after enzymic cleavage was found to be to very heterogenous and made the arrangement of peptides from the enzymic cleavage along the entire sequence of the chain impossible. At this stage of the investigation, cleavage with hydrochloric acid, as had been used by SANGER [9], for insulin, enabled a number of peptides from the insoluble heterogenous core to be obtained. As a result of this treatment in the couse of which the cleaving agent causes denaturation, we were able for the first time to isolate all overlapping peptides from the entire peptide chain (Fig. 2 and 3). In spite of this great difficulty was encountered in the purification of the HCl-core. This core peptide showed a great tendency to aggregate: The peptide was eluted with the front on chromatography using G-50 Sephadex (Fig. 4) and analysis showed it to contain only 11 mostly nonpolar amino acid residues. This is the first time a reliable analytical result has been obtained for this range. The result of HCl cleavage confirm at the peptide level those obtained by hydrazinolysis and amino acid analysis, that the original chain contains 49 amino acid residues.

18*

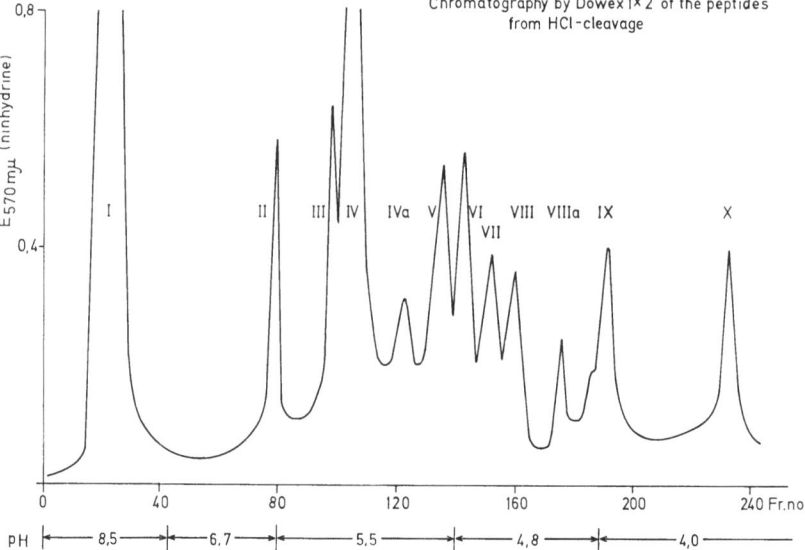

Fig. 2. Chromatographie on Dowex 1 x 2 of the peptides from HCl-cleavage. Column dimensions: $50 \times 2$ cm. Equilibration and elution with volatile pyridine-, picolin- and lutidine-acetat buffers [13]. All peptides obtained from this experiment are detailed in Fig. 3

Today a better breakdown of the peptide chain may be obtained by subjecting the chemically modified protein to cleavage by thermolysin [10]. Such a chemically modified coat protein is denatured and does not show a Cotton effect at 233 mμ in ORD (Fig. 5). The fingerprints of the products of thermolysin cleavage are remarkably simple.

Peptides of the cleavage with HCl

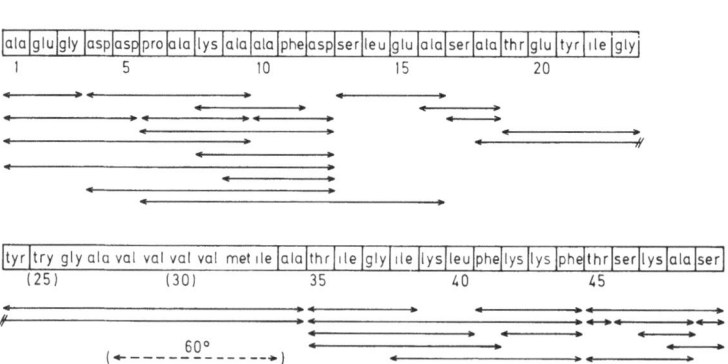

Fig. 3. Peptides obtained from cleavage with HCl are shown in their arrangement in the peptide chain

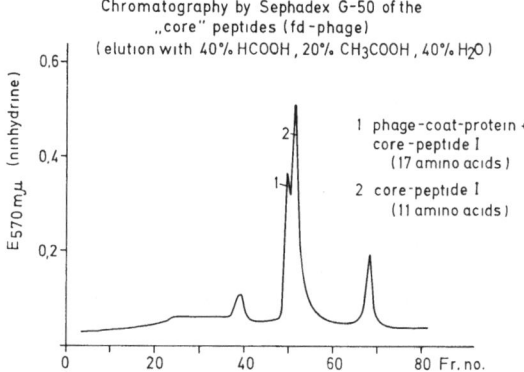

Fig. 4. Chromatographie of the core-peptides from HCl-cleavage of fd-phage on Sephadex G 50. Elution with 40 % HCOOH, 20 % CH₃COOH, 40 % H₂O. Column dimensions: 150 × 2 cm

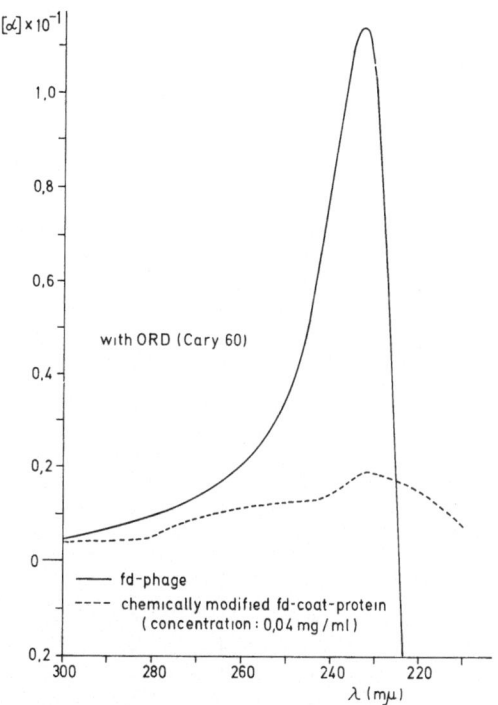

Fig. 5. Optical rotatory dispersion using a Cary 60 (1 cm cell) of fd phage (—) and of the chem. modified fd coatprotein (-----). The concentration of the solutions was 0.04 mg/ml. The negative optical rotation in degrees is shown. The small Cotton effect at 233 mμ of the chem. modified coat-protein is due to the non quantitative chemical modification

If we consider the structure (Fig. 6), it is remarkable, that it can be classified into three parts in regard to the distribution of the side chains [11]. This may have some morphological significance. The N-terminal part is acidic and contains only one lysine and all of the aspartic acid and glutamic acid residues of the protein. The glutamic residue in position 15

The three parts of the peptide chain

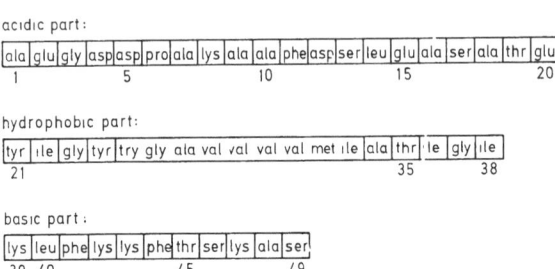

Fig. 6. The complete peptide chain can be shown to consist of three parts on the basis of the nature of the amino acid side chains. The N-terminal part may be responsible for the solubility of the native phage in aquous systems. The middle part, which has a high tendency to aggregate (Fig. 4) forms a "coat" around the DNA. The C-terminal section is bound eventually to the DNA

Table 1. *Comparison of amino acid composition of fd, M13, fl and ZJ-2 coat proteins. The results are expressed on the basis of the results for fd (methionine = 1)*

Amino acid composition of the coat proteins of the bacteriophages fd, M13, fl and ZJ-2

| Phage | fd | M 13 | f 1 | ZJ-2 |
|---|---|---|---|---|
| Amino acid: | | | | |
| Lysine | 5 | 5 | 5 | 5 |
| Aspartic acid | 3 | 3 | 3 | 3 |
| Threonine | 3 | 3 | 3 | 3 |
| Serine | 4 | 4 | 4 | 4 |
| Glutamic acid | 3 | 3 | 4 | 3 |
| Proline | 1 | 1 | 1 | 1 |
| Glycine | 4 | 4 | 4 | 4 |
| Alanine | 9 | 9 | 8—9 | 10 |
| Valine | 4 | 4 | 4 | 4 |
| Methionine | 1 | 1 | 1 | 1 |
| Isoleucine | 4 | 4 | 4 | 4 |
| Leucine | 2 | 2 | 2 | 2 |
| Tyrosine | 2 | 2 | 2 | 2 |
| Phenylalanine | 3 | 3 | 3 | 3 |
| Trypthophan | 1 | 1 | 1 | 1 |
| Total | 49 | 49 | (49) | 49 |

is amidated. The only proline residue is found in position 6. The hydrophobic middle part of the protein contains all of the 4 valine residues, the methionine and the tryptophan. In the C-terminal section we find four of the five lysine residues.

The comparison of the amino acid analyses of the four DNA-containing rod shaped phages (Table 1) show, that the differences between these viruses are extremly small. From these analysis it is apparent, that these viruses differ in at least in one amino acid residue.

To summarize: The coat protein of native fd phage contains 1,900 peptide chains. The molecular weight of one chain is 5,169 [7]. It appears therefore, that the coat protein subunit of the phage fd is the smallest known protein coded by one gene, which has a clearly defined function. This fact introduces several very interesting protein chemical aspects. It is hoped, that soon an apparatus [12] will be available, which will enable us to carry out automatic sequence analysis of a large part if not of the entire protein. With such a small protein interesting synthetic possibilities arise. The chemically synthesized complete protein and intermediates could be employed in investigations into reconstitution with the phage DNA and it brings nearer the possibility of synthesizing a complete viral particle.

## Acknowledgements

Our thanks are due to Professor A. BUTENANDT and to the Deutsche Forschungsgemeinschaft for their generous support of this work, to Dr. H. HOFFMANN-BERLING, Dr. N. ZINDER, Dr. PH. H. HOFSCHNEIDER and Dr. D. KAY for gifts of phages and to Mrs. K. STEINHOFF, Miss. J. VAHRENKAMP and Mr. A. STANGL for their excellent technical assistance.

## References

1. BRAUNITZER, G., F. ASBECK, K. BEYREUTHER, H. KÖHLER u. G. v. WETTSTEIN: Hoppe-Seyler's Z. physiol. Chem. **348**, 725 (1967).
2. HOFFMANN-BERLING, H., H. C. KAERNER, and R. KNIPPERS: Adcanc. Virus Res. **12**, 329 (1966).
3. — H. DÜRWALD, and J. BEULKE: Z. Naturforsch. **18**b, 893 (1963).
4. HANNIG, K.: Z. analyt. Chem. **181**, 244 (1960); — BRAUNITZER, G., G. HOBOM, and K. HANNIG: Hoppe-Seiler's Z. physiol. Chem. **338**, 276, 278 (1964).
5. BRAUN, V., and W. A. SCHROEDER: Arch. Biochem. **118**, 241 (1967).
6. ANDERER, F. A., H. UHLIG, F. WEBER, and G. SCHRAMM: Nature (Lond.) **186**, 922 (1960).
7. STARK, G. R., and D. G. SMYTH: J. biol. Chem. **238**, 214 (1963).
8. SCHROEDER, W. A., J. R. SHELTON, and J. B. SHELTON: Anal. Biochem. **2**, 87 (1961).
9. SANGER, F., and H. TUPPY: Biochem. J. **49**, 463, 481 (1951); — SANGER, F., and E. O. P. THOMPSON: Biochem. J. **353**, 366 (1953).
10. MATSUBARA, H., R. SASAKI, A. SINGER, and T. H. JUKES: Arch. Biochem, **115**, 324 (1966).
11. BRAUNITZER, G., F. ASBECK, K. BEYREUTHER, H. KÖHLER u. G. v. WETTSTEIN: Hoppe-Seyler's Z. physiol. Chem. **348**, 1689 (1967).

12. EDMAN, P., and E. BEGG: J. Biochem. **1**, 80 (1967).
13. RUDLOFF, V., and G. BRAUNITZER: Hoppe-Seyler's Z. physiol. Chem. **323**, 129 (1961).

## Note added in proof

The amino acid sequence of fd phage coat protein is now completly known ([11] and K. BEYREUTHER, Dissertation, Naturwiss. Fak., Universität München, in preparation):

```
1                       5                                  10
Ala – Glu – Gly – Asp – Asp  – Pro – Ala – Lys – Ala – Ala
                        15                                 20
Phe – Asp – Ser – Leu – Glu – Ala – Ser – Ala – Thr – Glu
                        25                                 30
Tyr – Ile  – Gly – Tyr – Ala – Try – Met – Val – Val – Val
                        35                                 40
Ile  – Val – Gly – Ala – Thr – Ile  – Gly – Ile  – Lys – Leu
                        45                      49
Phe – Lys – Lys – Phe – Thr – Ser – Lys – Ala – Ser
```

## Discussion

*Hoffmann-Berling:* What are the solvents used to proceed with the insoluble peptide?

*Beyreuther:* The HCl-core 1 and 2 are very soluble in a solution of 40 % formic acid, 20 % acidic acid and 40 % water. The core $Val_4$, $Met_1$, $Ile_1$ could be dissolved in 10 % pyridine solutions.

*Hofschneider:* As we both know in mass-cultures of fd-type phages the danger of contamination is very high. Do you have controls showing that the four different fd-type phages have not been mixed up by contamination during propagation in your laboratory.

*Beyreuther:* The amino acid analyses of the four DNA-phages were made initially with material which was in the case of M 13 a gift from your own group, in the case of ZJ 2 a gift from KAY (lyophilised phage material) and only for f 1 have we used for the analysis material prepared in our laboratory (f 1 infected bacteria from ZINDER).

*Wittmann:* After this paper by BRAUNITZER et al, giving information on the amino acid differences between strains of the DNA containing bacteriophage it is of interest to compare the protein coats of various strains of two RNA viruses, namely an RNA bacteriophage and tobacco mosaic virus.

### A. RNA bacteriophages:

The amino acid sequences of fr and $f_2$, two strains of an RNA bacteriophage, are compared in Table 1. Although the number of amino acids within both coat proteins is identical there is a considerable number (18 %) of positions with different amino acids. 60 % of these amino acid differences between fr and $f_2$ can be "explained" by alteration of only one nucleocide per codon.

According to the present knowledge the RNA bacteriophages whose coat proteins have been studied can be divided into the following three groups:

1. Four strains, namely MS2, M 12, R 17 and $f_2$, are almost identical. As LIN et al: J. molec. Biol. **24**, 1 (1967) have recently found MS2 and M 12 are identical in the amino acid composition of their protein coats. R 17 and $f_2$ differ from MS2 and M 12 by only one amino acid replacement.

2. As shown in Fig. 1 fr differs from $f_2$ in 23 of the 129 amino acid positions. This is the same degree of relationship as between the two TMV strains vulgare (V) and dahlemense (D) whose amino acid sequences are given in Table 2.

3. The RNA bacteriophage $Q\beta$ isolated in Japan is very different from $f_2$ in its protein coat (W. KONIGSBERG, personal communication) and therefore also from MS 2, M 12, R 17 and fr.

*B. Tobacco mosaic virus:*

The comparison of the four TMV strains whose amino acid sequences were studied (Fig. 2) leads to the following results:

1. In contrast to the other TMV strains and phytopathogenic viruses whose N-terminal end is acetylated the strain U 2 begins with a non-acetylated amino acid.

2. Strain H has two amino acids less than the other three strains and the 300 TMV mutants whose protein coats have been analyzed. The deletion within the coat protein of H is in the region between positions 145 — 149.

3. In only 35 % of the 158 positions occur the same amino acid at the same position (e.g. tyrosine in position 2, isoleucine in 4, etc.) in all four TMV strains. The number of amino acids which have so far been found to be invariable in TMV is reduced from 36 % to 30 % if amino acid replacements within TMV mutants (in addition to the TMV-strains) are also considered.

4. Two big clusters of invariable amino acids with together 18 amino acids are present in positions 87 — 94 and 113 — 122. The folding of the protein chain probably does not allow amino acid replacements in certain regions.

5. By comparison of the codons for different amino acids in a given position (e.g. serine and proline in position 1; serine, threonine and asparagine in position 3; valine, serine, methionine and proline in position 58) one can conclude that in 50 % of these positions the amino acid replacements can only be "explained" by altering more than one nucleotide per codon.

Table 1. *Amino acid sequences of RNA two bacteriophage strains*

The sequence of fr is according to WITTMANN-LIEBOLD et al. (Molec. Gen. Genet. **100**, 358 (1967) and to be published) and that of $f_2$ according to WEBER and KONIGSBERG (J. biol. Chem. **242**, 3563 (1967)]

|     | 1 | | | 5 | | | | | 10 | | |
|-----|---|---|---|---|---|---|---|---|---|---|---|
| fr  | Ala | — Ser | — Asn | — Phe | — *Glu* | — *Glu* | — Phe | — Val | — Leu | — Val | — Asn — |
| $f_2$ | Ala | — Ser | — Asn | — Phe | — *Thr* | — *Gln* | — Phe | — Val | — Leu | — Val | — Asn — |

|     | | | 15 | | | | 20 | | | | |
|-----|---|---|---|---|---|---|---|---|---|---|---|
| fr  | Asp | — Gly | — Gly | — Thr | — Gly | — *Asp* | — Val | — *Lys* | — Val | — Ala | — Pro — |
| $f_2$ | Asp | — Gly | — Gly | — Thr | — Gly | — *Asn* | — Val | — *Thr* | — Val | — Ala | — Pro — |

|     | | | 25 | | | | 30 | | | | |
|-----|---|---|---|---|---|---|---|---|---|---|---|
| fr  | Ser | — Asn | — Phe | — Ala | — Asn | — Gly | — Val | — Ala | — Glu | — Try | — Ile — |
| $f_2$ | Ser | — Asn | — Phe | — Ala | — Asn | — Gly | — Val | — Ala | — Glu | — Try | — Ile — |

|     | 35 | | | | 40 | | | | | | |
|-----|---|---|---|---|---|---|---|---|---|---|---|
| fr  | Ser | — Ser | — Asn | — Ser | — Arg | — Ser | — Gln | — Ala | — Tyr | — Lys | — Val — |
| $f_2$ | Ser | — Ser | — Asn | — Ser | — Arg | — Ser | — Gln | — Ala | — Tyr | — Lys | — Val — |

|     | 45 | | | | 50 | | | | | 55 | |
|-----|---|---|---|---|---|---|---|---|---|---|---|
| fr  | Thr | — Cys | — Ser | — Val | — Arg | — Gln | — Ser | — Ser | — Ala | — *Asn* | — Asn — |
| $f_2$ | Thr | — Cys | — Ser | — Val | — Arg | — Gln | — Ser | — Ser | — Ala | — *Gln* | — Asn — |

Table 1 (Fortetzung)

|     |     | 60  |     |     |     |     |     |     | 65  |     |
| --- | --- | --- | --- | --- | --- | --- | --- | --- | --- | --- |
| fr  | Arg – Lys – Tyr – Thr – *Val* – Lys – Val – Glu – Val – Pro – Lys – |
| f₂  | Arg – Lys – Tyr – Thr – *Ile* – Lys – Val – Glu – Val – Pro – Lys – |

|     |     | 70  |     |     |     |     |     | 75  |     |     |
| --- | --- | --- | --- | --- | --- | --- | --- | --- | --- | --- |
| fr  | Val – Ala – Thr – Gln – *Val* – *Gln* – Gly – Gly – Val – Glu – Leu – |
| f₂  | Val – Ala – Thr – Gln – *Thr* – *Val* – Gly – Gly – Val – Glu – Leu – |

|     |     | 80  |     |     |     |     | 85  |     |     |     |
| --- | --- | --- | --- | --- | --- | --- | --- | --- | --- | --- |
| fr  | Pro – Val – Ala – Ala – Try – Arg – Ser – Tyr – *Met* – Asn – *Met* – |
| f₂  | Pro – Val – Ala – Ala – Try – Arg – Ser – Tyr – *Leu* – Asn – *Leu* – |

|     |     | 90  |     |     |     | 95  |     |     |     |     |
| --- | --- | --- | --- | --- | --- | --- | --- | --- | --- | --- |
| fr  | Glu – Leu – Thr – Ile – Pro – *Val* – Phe – Ala – Thr – *Asx* – *Asp* – |
| f₂  | Glu – Leu – Thr – Ile – Pro – *Ile* – Phe – Ala – Thr – *Asn* – *Ser* – |

|     | 100 |     |     |     | 105 |     |     |     |     | 110 |
| --- | --- | --- | --- | --- | --- | --- | --- | --- | --- | --- |
| fr  | Asp – Cys – *Ala* – Leu – Ile – Val – Lys – Ala – *Leu* – Gln – Gly – |
| f₂  | Asp – Cys – *Glu* – Leu – Ile – Val – Lys – Ala – *Met* – Gln – Gly – |

|     |     | 115 |     |     |     | 120 |     |     |     |     |
| --- | --- | --- | --- | --- | --- | --- | --- | --- | --- | --- |
| fr  | *Thr* – *Phe* – Lys – *Thr* – Gly – *Ile* – *Ala* – Pro – *Asn* – *Thr* – Ala – |
| f₂  | *Leu* – *Leu* – Lys – *Asp* – Gly – *Asn* – Pro – Ile – *Pro* – *Ser* – Ala – |

|     |     | 125 |     |     |     | 129 |     |     |
| --- | --- | --- | --- | --- | --- | --- | --- | --- |
| fr  | Ile – Ala – Ala – Asn – Ser – Gly – Ile – Tyr |
| f₂  | Ile – Ala – Ala – Asn – Ser – Gly – Ile – Tyr |

Table 2. *Amino acid sequences of four TMV strains*

The sequences for the four strains is according to: Anderer, Wittmann-Liebold and Wittmann [Z. Naturforsch. **20**b, 1203 (1965)] for strain vulgare (V). Wittmann-Liebold and Wittmann [Z. Vererbungsl. **94**, 427 (1963)] for strain dahlemense (D). Wittmann [Z. Naturforsch. **20**b, 1213 (1965)] and Rentschler [Molec. Gen. Genet. **100**, 96 (1967)] for strain U 2. Wittmann-Liebold et al. [Molec. Gen. Genet. **100**, 358 (1967) and to be published] and Funatsu (to be published) for strain H.

|     | 1   | 2   | 3   | 4   | 5   | 6   | 7   | 8   | 9   | 10  |
| --- | --- | --- | --- | --- | --- | --- | --- | --- | --- | --- |
| V   | Acetyl – Ser – Tyr – Ser – Ile – Thr – Thr – Pro – Ser – Gln – Phe – |
| D   | Acetyl – Ser – Tyr – Ser – Ile – Thr – Ser – Pro – Ser – Gln – Phe – |
| U2  | Pro – Tyr – Thr – Ile – Asn – Ser – Pro – Ser – Gln – Phe – |
| H   | Acetyl – Ser – Tyr – Asn – Ile – Thr – Asn – Ser – Asn – Gln – Tyr – |

|     | 11  | 12  | 13  | 14  | 15  | 16  | 17  | 18  | 19  | 20  | 21  |
| --- | --- | --- | --- | --- | --- | --- | --- | --- | --- | --- | --- |
| V   | Val – Phe – Leu – Ser – Ser – Ala – Try – Ala – Asp – Pro – Ile – |
| D   | Val – Phe – Leu – Ser – Ser – Val – Try – Ala – Asp – Pro – Ile – |
| U 2 | Val – Tyr – Leu – Ser – Ser – Ala – Tyr – Ala – Asp – Pro – Val – |
| H   | Gln – Tyr – Phe – Ala – Ala – Val – Try – Ala – Glu – Pro – Thr – |

|     | 22  | 23  | 24  | 25  | 26  | 27  | 28  | 29  | 30  | 31  | 32  |
| --- | --- | --- | --- | --- | --- | --- | --- | --- | --- | --- | --- |
| V   | Glu – Leu – Ile – Asn – Leu – Cys – Thr – Asn – Ala – Leu – Gly – |
| D   | Glu – Leu – Leu – Asn – Val – Cys – Thr – Ser – Ser – Leu – Gly – |
| U 2 | Glu – Leu – Ile – Asn – Leu – Cys – Thr – Asn – Ala – Leu – Gly – |
| H   | Pro – Met – Leu – Asn – Gln – Cys – Val – Ser – Ala – Leu – Ser – |

Table 2 (Fortsetzung)

| | 33 | 34 | 35 | 36 | 37 | 38 | 39 | 40 | 41 | 42 | 43 |
|---|---|---|---|---|---|---|---|---|---|---|---|
| V | Asn | Gln | Phe | Gln | Thr | Gln | Gln | Ala | Arg | Thr | Val |
| D | Asn | Gln | Phe | Gln | Thr | Gln | Gln | Ala | Arg | Thr | Thr |
| U2 | Asn | Gln | Phe | Gln | Thr | Gln | Gln | Ala | Arg | Thr | Thr |
| H | Gln | Ser | Tyr | Gln | Thr | Gln | Ala | Gly | Arg | Asp | Thr |

| | 44 | 45 | 46 | 47 | 48 | 49 | 50 | 51 | 52 | 53 | 54 |
|---|---|---|---|---|---|---|---|---|---|---|---|
| V | Val | Gln | Arg | Gln | Phe | Ser | Gln | Val | Try | Lys | Pro |
| D | Val | Gln | Gln | Gln | Phe | Ser | Glu | Val | Try | Lys | Pro |
| U2 | Val | Gln | Gln | Gln | Phe | Ala | Asp | Ala | Try | Lys | Pro |
| H | Val | Arg | Gln | Gln | Phe | Ala | Asn | Leu | Leu | Ser | Thr |

| | 55 | 56 | 57 | 58 | 59 | 60 | 61 | 62 | 63 | 64 | 65 |
|---|---|---|---|---|---|---|---|---|---|---|---|
| V | Ser | Pro | Gln | Val | Thr | Val | Arg | Phe | Pro | Asp | Ser |
| D | Phe | Pro | Gln | Ser | Thr | Val | Arg | Phe | Pro | Asp | Asp |
| U2 | Ser | Pro | Val | Met | Thr | Val | Arg | Phe | Pro | Ala | Ser |
| H | Ile | Val | Ala | Pro | Asn | Gln | Arg | Phe | Pro | Asp | Thr |

| | 66 | 67 | 68 | 69 | 70 | 71 | 72 | 73 | 74 | 75 | 76 |
|---|---|---|---|---|---|---|---|---|---|---|---|
| V | Asp | Phe | Lys | Val | Tyr | Arg | Tyr | Asn | Ala | Val | Leu |
| D | Val | Tyr | Lys | Val | Tyr | Arg | Tyr | Asn | Ala | Val | Leu |
| U2 | Asp | Phe | Tyr | Val | Tyr | Arg | Tyr | Asn | Ser | Thr | Leu |
| H | Gly | Phe | Arg | Val | Tyr | Val | Asn | Ser | Ala | Val | Ile |

| | 77 | 78 | 79 | 80 | 81 | 82 | 83 | 84 | 85 | 86 | 87 |
|---|---|---|---|---|---|---|---|---|---|---|---|
| V | Asp | Pro | Leu | Val | Thr | Ala | Leu | Leu | Gly | Ala | Phe |
| D | Asp | Pro | Leu | Ile | Thr | Ala | Leu | Leu | Gly | Thr | Phe |
| U2 | Asp | Pro | Leu | Ile | Thr | Ala | Leu | Leu | Asn | Ser | Phe |
| H | Lys | Pro | Leu | Tyr | Glu | Ala | Leu | Met | Lys | Ser | Phe |

| | 88 | 89 | 90 | 91 | 92 | 93 | 94 | 95 | 96 | 97 | 98 |
|---|---|---|---|---|---|---|---|---|---|---|---|
| V | Asp | Thr | Arg | Asn | Arg | Ile | Ile | Glu | Val | Glu | Asn |
| D | Asp | Thr | Arg | Asn | Arg | Ile | Ile | Glu | Val | Glu | Asn |
| U2 | Asp | Thr | Arg | Asn | Arg | Ile | Ile | Glx | Val | Asx | Asx |
| H | Asp | Thr | Arg | Asn | Arg | Ile | Ile | Gln | Thr | Glu | Glu |

| | 99 | 100 | 101 | 102 | 103 | 104 | 105 | 106 | 107 | 108 | 109 |
|---|---|---|---|---|---|---|---|---|---|---|---|
| V | Gln | Ala | Asn | Pro | Thr | Thr | Ala | Glu | Thr | Leu | Asp |
| D | Gln | Gln | Ser | Pro | Thr | Thr | Ala | Glu | Thr | Leu | Asp |
| U2 | Glx | Ala | Asx | Pro | Thr | Thr | Ala | (Asx, | Thr, | Glx, | Glx, |
| H | Gln | Ser | Arg | Pro | Ser | Ala | Ser | Gln | Val | Ala | Asn |

| | 110 | 111 | 112 | 113 | 114 | 115 | 116 | 117 | 118 | 119 | 120 |
|---|---|---|---|---|---|---|---|---|---|---|---|
| V | Ala | Thr | Arg | Arg | Val | Asp | Asp | Ala | Thr | Val | Ala |
| D | Ala | Thr | Arg | Arg | Val | Asp | Asp | Ala | Thr | Val | Ala |
| U2 | Pro, | Val, | Ile) | Arg | Val | Asp | Asp | Ala | Thr | Val | Ala |
| H | Ala | Thr | Gln | Arg | Val | Asp | Asp | Ala | Thr | Val | Ala |

| | 121 | 122 | 123 | 124 | 125 | 126 | 127 | 128 | 129 | 130 | 131 |
|---|---|---|---|---|---|---|---|---|---|---|---|
| V | Ile | Arg | Ser | Ala | Ile | Asn | Asn | Leu | Ile | Val | Glu |
| D | Ile | Arg | Ser | Ala | Ile | Asn | Asn | Leu | Val | Asn | Glu |
| U2 | Ile | Arg | Ala | Ser | Ile | Asn | Asn | Leu | Ala | Asn | Glu |
| H | Ile | Arg | Ser | Gln | Ile | Gln | Leu | Leu | Leu | Asn | Glu |

Table 2 (Fortsetzung)

| | 132 | 133 | 134 | 135 | 136 | 137 | 138 | 139 | 140 | 141 | 142 |
|---|---|---|---|---|---|---|---|---|---|---|---|
| V | Leu | Ile | Arg | Gly | Thr | Gly | Ser | Tyr | Asn | Arg | Ser |
| D | Leu | Val | Arg | Gly | Thr | Gly | Leu | Tyr | Asn | Gln | Asn |
| U 2 | Leu | Val | Arg | Gly | Thr | Gly | Met | Phe | Asn | Gln | Ala |
| H | Leu | Ser | Asx | His | Gly | Gly | Tyr | Met | Asn | Arg | Ala |

| | 143 | 144 | 145 | 146 | 147 | 148 | 149 | 150 | 151 | 152 | 153 |
|---|---|---|---|---|---|---|---|---|---|---|---|
| V | Ser | Phe | Glu | Ser | Ser | Ser | Gly | Leu | Val | Try | Thr |
| D | Thr | Phe | Glu | Ser | Met | Ser | Gly | Leu | Val | Try | Thr |
| U 2 | Gly | Phe | Glu | Thr | Ala | Ser | Gly | Leu | Val | Try | Thr |
| H | Glu | Phe | Glu | ← | Ala | Ile | → | Leu | Pro | Try | Thr |

| | 154 | 155 | 156 | 157 | 158 |
|---|---|---|---|---|---|
| V | Ser | Gly | Pro | Ala | Thr |
| D | Ser | Ala | Pro | Ala | Ser |
| U 2 | Thr | Thr | Pro | Ala | Thr |
| H | Thr | Ala | Pro | Ala | Thr |

# Control Mechanisms in T 4 Infection: Studies on the Exclusion of the RNA Phage M 12 in Mixedly Infected Cells

By

S. HATTMAN and P. H. HOFSCHNEIDER

Max-Planck-Institut für Biochemie, München, Goethestraße 31

## 1. Introduction

It is well known that following infection of *E. coli* cells with phages of the T-even class (T 2, T 4, T 6) the macromolecular syntheses of the host cell are rapidly stopped (COHEN, 1947, 1948; VOLKIN and ASTRACHAN, 1956; NOMURA, HALL and SPIEGELMAN, 1960). In an attempt to elucidate the mechanism by which T-even phage controls RNA and protein-synthesis determined by genetic elements other than its own, the exclusion of RNA phage M 12 was studied in cells mixedly infected with phage T 4. In particular, the following points were examined: (1) The possible interference of T 4 infection with M 12 penetration, (2) the intracellular fate of parental M 12 RNA, (3) the synthesis of replicative intermediates and single-stranded progeny RNA, (4) the formation of defective phage particles and (5) the production of phage coat protein. Results of these experiments suggested that T 4 may specifically interfere with the mRNA function of the M 12 chromosome. In a second series of experiments this hypothesis was further strengthened by studying the formation of M 12 specific polysomes and RNA polymerase. In mixed infections the production of RNA polymerase is completely blocked, but the formation of polysomes containing parental M 12 RNA is normal. These findings are interpreted to mean that T 4 interferes with translation of the M 12 RNA, possibly on the polysomal level.

## 2. Material and Methods

All materials and methods used have been or will soon be published elsewhere (HATTMAN and HOFSCHNEIDER, 1967, 1968). To avoid the inhibition of M 12 phage production due to T 4 induced lysis prior to M 12 maturation the early amber mutant T 4 am 55 has been used in some experiments.

## 3. Results

### a) Does M12 Exclusion Involve T4 Interference with M12 RNA Penetration?

It is known that following adsorption of phage f2 and prior to cell penetration, the RNA is sensitive to exogenous RNase (Loeb, 1961; Valentine and Wedel, 1965). Whereas T4 does not interfere with M12 adsorption, it is conceivable that exclusion may be established during the RNase-sensitive phase. To test this possibility, M12 was adsorbed to cells in the cold, a condition which prevents RNA injection (Valentine and Wedel, 1965). The culture was then warmed to 37°C and portions transferred at intervals into anti-M12 serum containing either RNase, T4am or medium. After a suitable interval, the samples were chilled and diluted in the cold for plating before lysis. As can be seen from Fig. 1, there was a rapid increase in resistance to the presence of exogenous RNase. By 10 min after warming to 37°C, all the potential M12 yielders were recoverable,

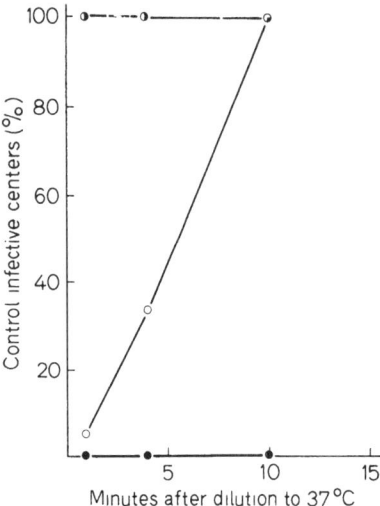

Fig. 1. Kinetics of M12 penetration as measured by RNase resistance. E. coli AB301 was grown to $4 \times 10^8$/ml in F medium. Cells and M12 were equilibrated at ice-bath temperature and mixed; after 10 min adsorption, an equal volume of warmed, fresh medium was added and aeration begun (t = 0) at 37°C. At 1.5 and 10 min, triplicate 1.0 ml samples were removed and mixed with 1.0 ml volumes of medium containing anti-M12 serum (K $\sim$ 5 to 10) supplemented either with RNase (100 μg/ml), T4am50 ($4 \times 10^9$/ml), or F medium. At 20 min, 8.0 ml cold medium was added and serial dilutions made at ice-bath temperature. Appropriate samples were plated for M12-infective centres. Note that in the control the number of infective centres remained constant, indicating that adsorption was completed in the cold and that adsorbed M12 are not affected by antiserum. ⟶●—●⟶, Control; —○—○—, RNase-treated; —●—●—, T4 am 50 superinfected

suggesting that at least one functional M 12 chromosome had penetrated each of these cells. Nevertheless, there was no appreciable increase in the proportion of escape-yielders; addition of T 4 am at 10 min still produced exclusion in well over 95 % of the cells. These results show that exclusion does not occur during penetration but at some later intracellular stage.

### b) Fate of Parental M 12 RNA

Nucleic acids were extracted from cells 15 min after infection with $^{32}$P-labelled M 12 and analysed for the amount of acid-insoluble label and by sucrose gradient sedimentation (Fig. 2). The recoveries of parental label were the same for both control and mixed infections, indicating that no extensive degradation to acid-soluble material occurred. The gradient profiles depicted in Fig. 2 (a) and (b) show that the parental RNA from the mixed infection sedimented homogeneously, even though exclusion was observed in 97 % of the cells. That the $^{32}$P-label recovered from both cultures consists of uniform intact phage RNA molecules was shown in a separate experiment. A sample of the $^{32}$P-label analysed in Fig. 2 (a) was sedimented with $^{3}$H-labelled RNA isolated from M 12-infected cells. The profiles obtained [Fig. 2 (c)] show that the $^{32}$P RNA sediments at about 27 to 28s; this is the sedimentation coefficient expected for intact M 12 RNA (DELIUS, 1966).

In other experiments, deproteinization was omitted and lysates were analysed directly in sucrose gradients. No evidence for M 12 RNA degradation was obtained. These observations still do not rule out the possibility that a very small amount of degradation of parental M 12 RNA may occur (e.g. at the end(s) of the molecule) which does not significantly affect the sedimentation properties of a resistant core.

### c) Conversion of Parental RNA to an RNase-Resistant Form

The results described above indicate that the parental M 12 RNA appears to remain intact during exclusion. To determine whether any replicative intermediates are synthesized, the conversion of parental-labelled RNA to an RNase-resistant form was studied.

Cells were infected with $^{3}$H-labelled M 12 in the presence and absence of T 4 am 50, and samples were taken at various intervals. The nucleic acids were prepared and assayed for total acid-insoluble and RNase-resistant acid-insoluble $^{3}$H-label. The results depicted in Fig. 3 show that T 4 am markedly depresses (or completely eliminates) the synthesis of any RNase-resistant structure(s) containing parental M 12 RNA. The kinetics observed in the control M 12 infection are similar to those reported for the RNA phage R 17 (ERIKSON, FENWICK, and FRANKLIN, 1964).

As in the experiment described above [section (b)], recoveries of parental label in the nucleic acid fractions from control and mixedly

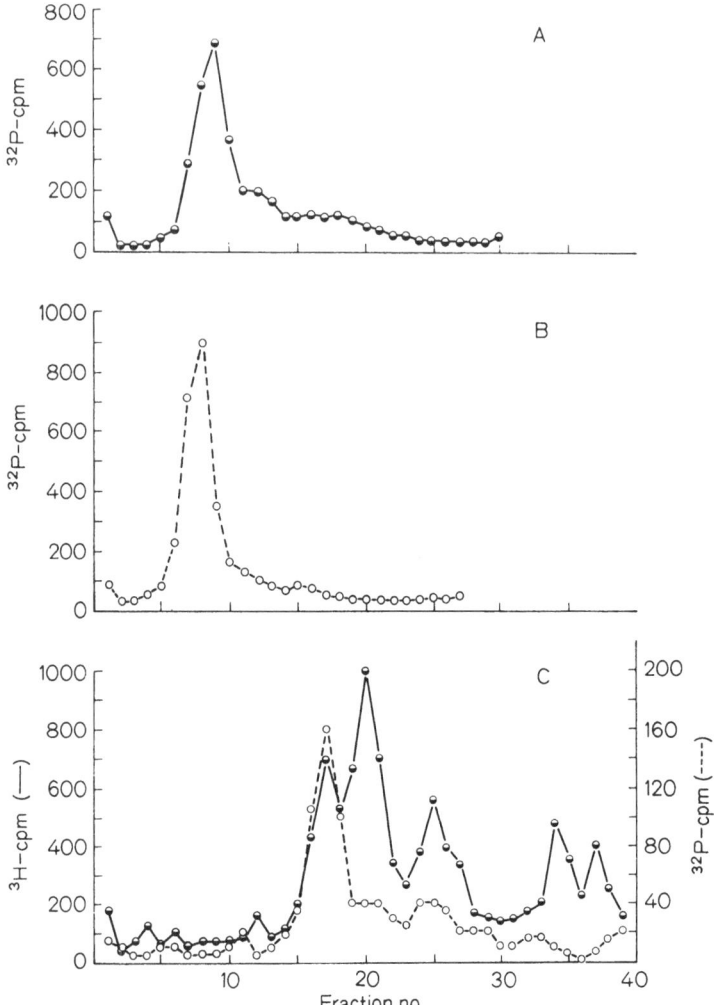

Fig. 2. Sucrose gradient analysis of nucleic acids isolated from cells 15 min after infection with $^{32}$P-labelled M 12 alone (control) or with $^{32}$P-labelled M 12 and T 4 am 55. E. coli AB 301 was grown to $2 \times 10^8$/ml in F medium and infected with $^{32}$P-labelled M 12 in the presence or absence of T 4 am 55. Included in Fig. 3 (c) is a separate centrifugation made with an artificial mixture of the $^{32}$P-labelled M 12 control nucleic acid fraction and one from M 12-infected cells incubated in the presence of [$^3$H]-uridine. The sucrose gradients were run for 14 h at 25,000 rev/min in a Spinco SW 25 rotor. (a) $^{32}$P-labelled M 12 control; (b) $^{32}$P-labelled M 12 + T 4 am 55; (c) —o—o—, $^{32}$P-labelled M 12 control; —●—●—, M 12-infected cells incubated with [$^3$H] uridine

infected cells were not significantly different. Furthermore, since the washing procedure used prior to phenol extraction would have been sufficient to remove even extracellularly adsorbed M 12 (IPPEN and

VALENTINE, 1965; VALENTINE, WEDEL, and IPPEN, 1965), the label must have been recovered from intracellular M 12 RNA.

Therefore, these observations suggest that one of the earliest observable sites of T 4 interference with M 12 replication in simultaneous mixed infection is an inhibition of the synthesis of M 12 double-stranded RNA.

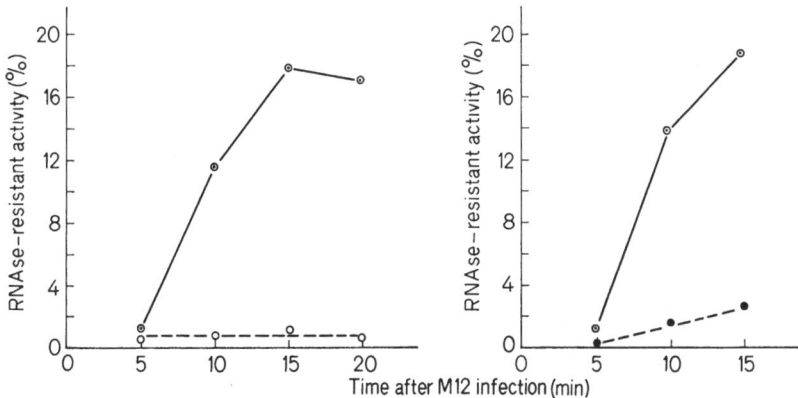

Fig. 3. Kinetics of conversion of [3]H-labelled M12 RNA to an RNase-resistant form in normal and mixed infections. E. coli 112-12 was grown to $4 \times 10^8$/ml in D medium and infected with [3]H-labelled M 12 alone or in combination with T 4 am 50. At the times indicated, samples were removed and diluted 1 : 5 into cold 0.3 M-NaCl containing 0.003 M-EDTA and kept in ice-water until all samples were collected. The cultures were washed 3 times in the cold by centrifugation in NaCl-EDTA solution. The cells were lysed with 1 % sodium dodecyl sulphate and extracted 3 times with phenol at room temperature; after the third extraction, the aqueous phase was removed and precipitated overnight in ethanol at $-20°$C. After sedimenting 30 min at 10,000 rev./min in a refrigerated Servall centrifuge, the pellet was resuspended in 0.003 M-EDTA and stored at $-20°$C. Measurement of RNase-resistant [3]H label was as follows: 1.0 ml of sample was mixed at 37°C with 1.0 ml of solution containing 4 μg RNase/ml in 0.02 M-MgCl$_2$ + 0.2 M-sodium citrate and incubated for 25 min. The reaction was stopped by adding cold trichloroacetic acid to 5 %. The acid-insoluble [3]H label was then filtered on to membranes, washed, dried and counted. 100 % in the figures would correspond to 700 cts/min. The results shown are from two independent experiments: —⊚—⊚—, [3]H-labelled M 12 control; —○—○—, [3]H-labelled M 12 + T 4 am 50 at t = 0; —•—•—, [3]H-labelled M 12 + T 4 am 50 at 2 min

## d) Synthesis of M 12 Progeny RNA

The interference with synthesis of M 12 double-stranded RNA cannot be the sole process by which exclusion of M 12 by T 4 is mediated. Exclusion is still effectively established when T 4 am superinfection occurs at times when M 12 double-stranded RNA synthesis is already in progress.

It was of interest, therefore, to determine whether M 12-specific RNA synthesis could occur in cells superinfected with T 4 am at times when

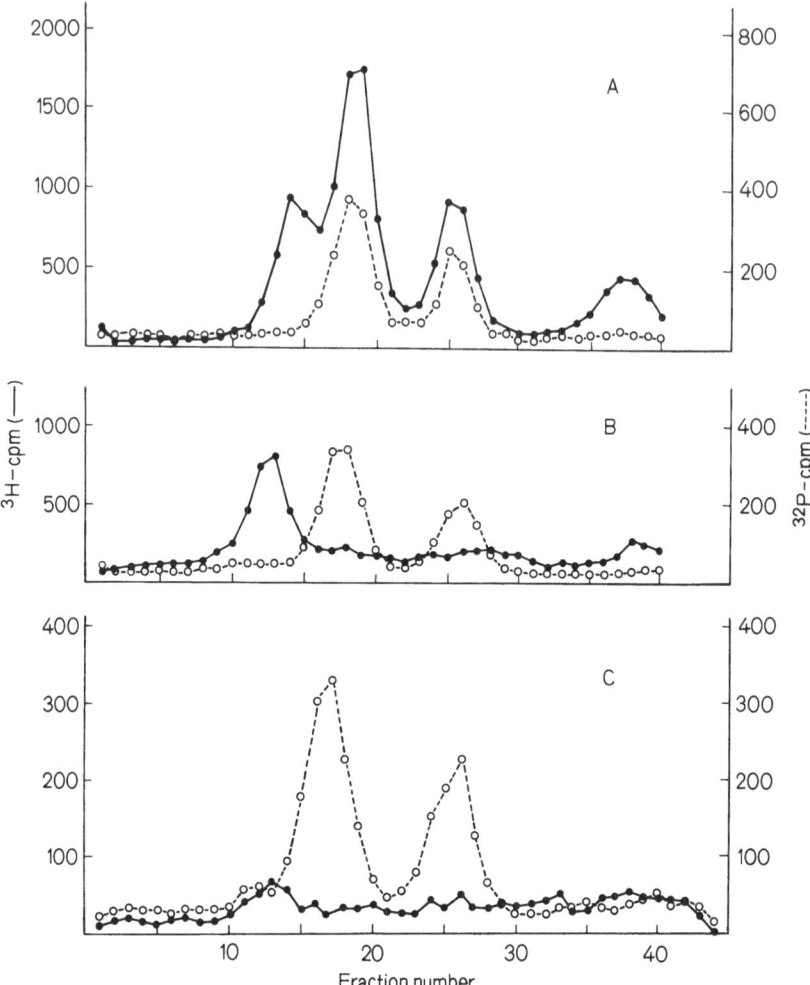

Fig. 4. Sucrose gradient analysis of RNA synthesized in M 12 and T 4 am super-infected cells. E. coli AB 301 was grown to $2 \times 10^8$/ml in F medium, infected with M 12 and divided into 3 equal portions. At 1 min, one of the cultures was super-infected with T 4 am 55, and at 21 min a second culture was superinfected. Relative to the control infection, only 0.3 % of the 1 min- and 0.7 % of the 21 min-superinfected cells produced M 12. [5-$^3$H] Uridine was added to 0.4 µC/ml at 30 min and cold uracil (100 µg/ml) was added at 45 min. At 50 min, the cells were harvested in the cold by centrifugation, washed, and nucleic acids prepared. Samples were taken for determination of incorporation of $^3$H into acid-insoluble material or for infectious RNA. Normalized to the control infection, the relative incorporation of $^3$H was about 1.7 and 15 %, and the infectivity for protoplasts was ∼ 0.4 and 25 % for the early and late superinfections, respectively. Sucrose gradients were run for 14 h at 25,000 rev/min; $^{32}$P-labelled ribosomal RNA was included in each tube as a marker. (a) 0.1 ml of the M 12 control; (b) 0.4 ml of the 21 min superinfection; (c) 0.5 ml of the 1 min superinfection. Radioactivity: $^3$H counts, —•—•—; $^{32}$P counts, —o—o—

double-stranded RNA [as well as the specific RNA-synthesizing enzyme(s)] is already present. The patterns of RNA synthesis in cells infected with M 12 alone or superinfected at various times were examined in sucrose gradients (Fig. 4). For the M 12 control infection, one sees the typical 4, 16 and 23s cell RNA peaks, plus the M 12 RNA peak at 27 to 28s [Fig. 4 (a)]. This pattern is similar to that shown with R 17 infection (PA-RANCHYCH and ELLIS, 1964) under conditions where there was no interference with cellular RNA production. With T 4 am superinfection at 1 min, an almost complete suppression of cellular and M 12 RNA synthesis occurred [Fig. 4 (c)]. On the other hand, addition of T 4 am at 21 min did not eliminate M 12-specific RNA formation [Fig. 4 (b)], although exclusion occurred in 99 % of the cells.

It should be noted that $^3$H uridine was added at 30 min; i.e. 9 min after infection with T 4 am. This eliminates the possibility that what is observed is merely residual M 12 RNA synthesis occurring during a lag in the expression of exclusion.

Assays for infectious RNA at 50 min showed infectivities of 0.4 and 25 % for the superinfections at 1 and 21 min, respectively, relative to the control. Since the amount of infectious RNA present at 20 min is only a small fraction of that synthesized from 20 to 50 min in controls, then the infectious RNA measured at 50 min represents net synthesis of M 12 RNA. From the sucrose gradient profiles shown in Fig. 4, one can determine the amount of $^3$H-label incorporated into M 12 RNA. The ratio of infectious RNA to M 12-specific tritium label was essentially the same for both the control and 21-min superinfected cultures. It is also concluded that there is no large-scale production of non-infectious M 12 single-stranded RNA in the superinfected cells.

That normal M 12 RNA is produced in the cells superinfected with T 4 am at later times lends support to the notion that parental M 12 RNA remains intact during exclusion in simultaneous mixed infection. It should be pointed out that, in the culture superinfected at 21 min, about 10 % of the RNA produced was RNase-resistant; thus, double-stranded M 12 RNA is also synthesized under these conditions.

*e) Does T 4 Superinfection Favour the Formation of Defective M 12 Phage Particles?*

In an experiment analogous to that above, T 4 am 55 was added to M 12-infected cells at 19 min and $^3$H uridine was added at 35 min. At 60 min samples were taken for either extraction with phenol or for crude lysates; the crude lysates were further clarified by two low-speed centrifugations. The nucleic acids exhibited a sedimentation pattern similar to that shown in Fig. 4. Biological assay of the lysates showed that the yield of active M 12 phage in the mixed infection was only 0.2 % of the

19*

control at 60 min; however, the yield of infectious RNA after phenol extraction was 17 % of the control. These results confirm that infectious M 12 RNA synthesis can occur under appropriate superinfection conditions, although no viable phage particles are made.

To characterize further the sedimentation properties of the RNA produced in the presence of T 4 am, the ³H-labelled crude lysates were

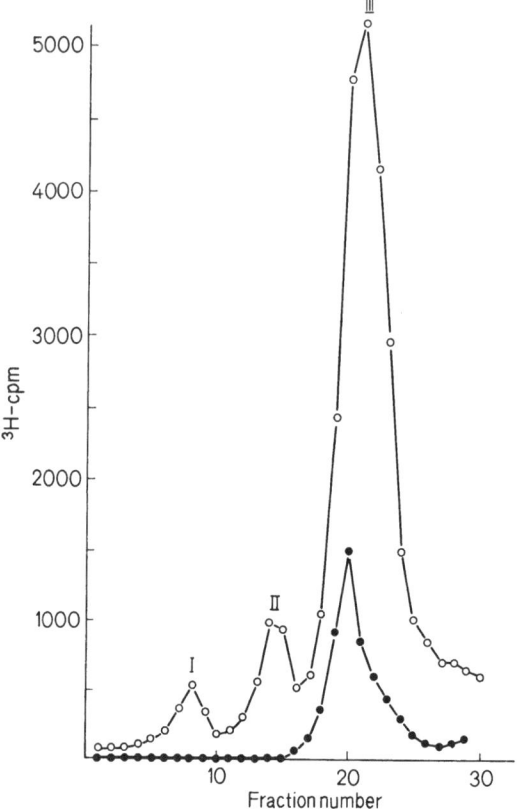

Fig. 5. Sucrose gradient analysis of crude lysates from M 12- and T 4 am-superinfected cells. E. coli AB 301 was grown and infected as described in the legend to Fig. 4. At 19 min, one-half the culture was superinfected with T 4 am 55; at 35 min, [5-³H] uridine was added to each culture to 0.6 to 0.7 μC/ml. In the superinfected culture, the M 12-yielder frequency was 6 % that of the control infection. At 60 min, samples were taken for either production of crude lysates or preparation of nucleic acids. The nucleic acids were assayed for infectious RNA, and the crude lysates assayed for acid-insoluble ³H and M 12 phage. Taking the control values as 100 %, the following values were obtained for the superinfected culture: infectious RNA = 17 %; M 12 progeny phage = 0.2 %; ³H incorporated = 18 %. Shown in the figure above are the superposed distributions of ³H label after sedimentation of the crude lysates in sucrose gradients (5¹/₂ h 25,000 rev/min). M 12 phage (peak I) was located in separate experiments using purified ³H-labelled M 12 as a marker. —○—○—, Control M 12 infection; —●—●—, superinfection

examined by sucrose gradient centrifugation. In the superinfected cells, no detectable ³H-label is found under the phage peak (I) or under the 50 s ribosome peak (II) of the control (Fig. 5). All label is localized under the peak (III) corresponding to a mixture of 28 s and 23 s RNA, as well as 30 s ribosomes, which are not separated in the short time allowed for centrifugation.

An additional experiment was done to determine whether transfer of M 12 RNA into inactive particles might occur at times up to 120 min. It again prooves that the M 12 RNA synthesized in late superinfections does not become incorporated into inactive particles of size similar to M 12 phage. It is not excluded that highly labile defective particles are formed which are dissociated by the lysis procedure employed here. An alternative explanation for the absence of M 12 particle formation is that maturation is prevented by an inhibition of coat protein synthesis.

Table 1. *Measurement of formation of M 12 coat-protein antigen in control and mixed infections*[+]

| Infection | Labelling period (min) | Total ³⁵S incorporated (cts/min/ml) | Total per cent precipitated | Per cent[++] M 12 specific label |
|---|---|---|---|---|
| **Exp. 1** | | | | |
| M 12 control | 16 to 30 | 33,900 | 9.2 | 5.9 |
| M 12 + T 4 am at 15 min | 16 to 30 | 10,760 | 4.3 | 1.0 |
| M 12 control | 26 to 40 | 37,400 | 26.0 | 22.7 |
| M 12 + T 4 am at 25 min | 26 to 40 | 12,900 | 7.1 | 3.8 |
| M 12 control | 36 to 50 | 41,900 | 32.1 | 28.8 |
| M 12 + T 4 am at 35 min | 36 to 50 | 12,940 | 8.4 | 5.1 |
| Uninfected cells | 25 to 60 | 59,580 | 3.3 | — |
| **Exp. 2** | | | | |
| M 12 control | 20 to 30 | 29,800 | 11.3 | 7.3 |
| M 12 + T 4 am at 15 min | 20 to 30 | 15,600 | 3.8 | 0.0 |
| M 12 control | 30 to 40 | 16,400 | 14.0 | 10.0 |
| M 12 + T 4 am at 25 min | 30 to 40 | 8,400 | 5.6 | 1.6 |
| M 12 control | 40 to 50 | 30,600 | 25.3 | 21.3 |
| M 12 + T 4 am at 35 min | 40 to 50 | 11,500 | 9.4 | 5.4 |
| M 12 control | 50 to 60 | 37,500 | 33.9 | 29.9 |
| M 12 + T 4 am at 45 min | 50 to 60 | 15,600 | 10.6 | 6.6 |
| Uninfected cells | 50 to 60 | 29,000 | 4.0 | — |
| T 4 am infected cells | 50 to 60 | 18,400 | 4.0 | — |

[+] E. coli AB 301 was grown to log phase in synthetic medium and infected with M 12. At the times indicated, duplicate samples were removed into tubes with T 4 am 50 (input ratio 5 to 10 phage/cell) or with medium. ³⁵S was then added to a final activity of 1 to 2 μC/ml. Preparation of lysates and serum precipitation procedure is described elsewhere (HOFSCHNEIDER and HATTMAN, 1967).

[++] The percentage M 12 specific label was calculated by subtacting the fraction of label precipitated in uninfected cell lysates from that for the test lysates.

*f) Production of M12 Phage Coat Protein*

To study whether T4 infection also affects M12-specific protein synthesis, the formation of M12 coat protein antigen was measured by ³⁵S-labelling and immuneserum precipitation. The results summarized in Table 1 show that addition of T4am50 inhibits the formation of M12 coat protein antigen at least fourfold, even when added as late as 45 min after M12. The extent of specific inhibition is probably greater than seen from Table 1, because no corrections were made for the contribution from the escape yielders to coat synthesis. In addition to this specific inhibition, the absolute level of M12 coat protein per cell is reduced by an additional factor of 2 or 3, i.e. the factor corresponding to the reduction in over-all rate of protein synthesis produced by infecting with T4am.

Similar results were obtained using ¹⁴C phenylalanine as the radioactive marker, or with T4am55 as the superinfecting phage. Additional superinfection experiments also showed that the inhibition of coat antigen synthesis is considerably greater than the reduction in synthesis of infectious M12 RNA.

*g) Formation of M12 Specific Polysomes*

It was of interest to determine first whether addition of T4am phage to M12-infected cells leads to a destruction of polysomes containing parental M12 RNA. Cells were infected with ³H-labelled M12 and superinfected with T4am55 15 min later. After an additional incubation for 10 min, lysates were prepared and contrifuged in sucrose gradients. The distribution of ³H-label is shown in Fig. 6. The total recovery of radioactivity from the superinfected culture was about 40 % of that of the control infection; however, there were no preferential losses from any particular region in the gradient. The proportions of parental M12 ³H-label in the polysomes were 45 % and 50 % for the control- and superinfected preparation, respectively.

It should be noted that a large amount of parental ³H sedimented in the 70s monosome region. Additional superinfection experiments showed that this label was not in phage particles, but rather in a RNase-sensitive form. It is not known whether this M12 RNA-ribosome complex exists *in vivo* or is a breakdown product of polysomes. In addition, 25 % of the label found in polysomes region was RNase-resistant for both control- and super-infections, whereas little if any RNase-resistant label was found in non-polysome regions.

The experiments described above indicate that following superinfection with T4am, M12-specific polysomes are not disrupted. Since we have no information as to whether polysomes containing parental M12-RNA are still functional late in the infection cycle, then the above results must be interpreted with some caution. For example, if polysomes containing parental M12-RNA had stopped functioning in protein synthesis prior to addition of T4am, then the observed resistance to disruption by T4am may only reflect some abnormal property of the polysome

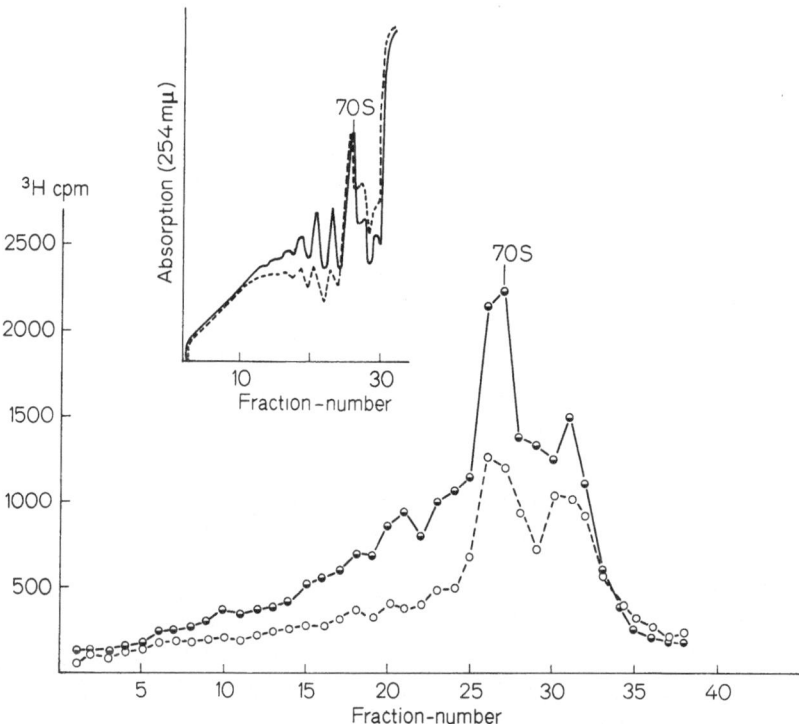

Fig 6. Sucrose gradient analysis of polysomes containing parental M12 $^3$H label from control- and super-infected cells. E. coli AB301 was grown in 100 ml synthetic medium to $1-2 \times 10^8$/ml. At time 0, the culture was split into 2-equal portions and infected with $^3$H-labelled M12 $(1-2$ phage/cell). After 15 min, T4am55 was added to one culture $(5-10$ phage/cell) and incubation continued. At 25 min, the cells were harvested and lysed. The lysates were centrifuged in sucrose gradients for 4 h at 25,000 rpm in a Spinco SW 25 rotor. The fractions were treated directly with TCA and the radioactivity determined. The curves drawn are the superposition of 2 separate gradients. —●—●— = Control infection; —o—o— = Superinfection with T4am55 at 15 min; Solid line in inset = control infection; Broken line in inset = superinfection with T4am55

complex (e.g. "sticking"). On the other hand, newly-formed (non-labelled) functional polysomes may have been degraded, a circumstance which would not have been detected by the labelling procedure employed. For these reasons, it was essential to analyze parental M12-polysomes isolated from simultaneous-mixedly infected cells at relatively short times after infection.

The results of a series of such experiments are summarized in Table 1. The fraction of total label found in the polysome region of sucrose-gradients is listed. In all these experiments, there were no marked differences in $^3$H-label distribution for control- and mixed-infections.

It is clear that there is no large-scale, consistent inhibition of the formation of parental M 12-specific polysomes in mixed-infections. Similar amounts of free 28s parental M 12 RNA were observed in the analyses of control- and mixedly-infected cells. Confirming earlier findings (see section 3b), no evidence for breakdown of parental M 12 RNA was observed despite the occurrence of exclusion. It should be noted that the absence of non specific binding of ribosomes to M 12 RNA during the preparation procedures could be demonstrated by appropriate control experiments.

### h) Formation of M 12 RNA Polymerase

The above findings indicate that parental M 12-RNA participates in polyribosome production. It was of interest, then, to determine whether these polysomes were producing any functional M 12-specific protein. Therefore, M 12 RNA-polymerase was assayed in lysates from uninfected and infected cells as described by August, Shapiro, and Eoyang (1965).

This procedure was chosen since it was desirable to assay for enzyme activity in lysates where the polysome distributions were known to be the same for control- and mixed-infections. It should be noted that in enzyme assays of MS2-infected cells, addition of purified MS2-RNA did not stimulate incorporation of nucleoside-triphosphates (Weissmann, Simon, and Ochoa, 1963). However, since T 4 am blocks M 12-RNA replication *in vivo*, it is possible that the template could be limiting and that the polymerase levels would appear much lower in lysates from mixedly-infected cells. For this reason, purified M 12 RNA (ca. 15 µg) was routinely included in the assay system.

Polymerase activity was not detectable in uninfected cells or mixedly-infected cells (at 30 to 35 min post-infection). In 3 experiments, the control infection showed specific enzyme activities of 4.6—11 units at 30 min post-infection, and in a fourth expt. 15.4 at 35 min. Artificial mixtures of lysates from control- and mixed-infections were also tested for polymerase. The enzyme activities were proportional to the amount of input control lysate. These observations suggest that the presence of T 4 am blocks the synthesis of functional M 12 RNA polymerase since no inhibitor of polymerase activity could be demonstrated in lysates from mixedly-infected cells.

## 4. Discussion

It could be shown that in M 12 infected cells: 1) late superinfection with T 4 am 55 does not eliminate M 12-RNA synthesis, while progeny phage and coat-protein antigen formation are strongly inhibited; 2) early superinfection with T 4 am completely blocks formation of any double-stranded M 12-RNA replicative intermediate, but parental RNA appears to remain intact. The two last findings reported here are that in mixed-infection between T 4 am 55 and M 12, parental M 12 RNA is capable of forming polysomes, but no RNA polymerase activity is produced.

Several observations suggest that T 4 interferes with M 12 replication by directly inhibiting M 12 protein synthesis, and that the inhibition of M 12 protein synthesis is not the consequence of primary inhibition of M 12 RNA synthesis. Firstly, when f 2-infected cells were severely restricted in their ability to synthesize RNA, either by treatment with 6-azauracil or by starvation of suitable auxotrophs for uracil or adenine, there was a normal synthesis of RNA polymerase (LODISH, COOPER, and ZINDER, 1964). Therefore, absence of polymerase production cannot be attributed to inhibition of progeny RNA formation. Secondly, infectious M 12 RNA synthesis was observed in cells superinfected with T 4 am 55 15 min after primary M 12 infection. This indicates that T 4 does not interfere *in vivo* with the functioning of M 12-RNA polymerase. In addition, when lysates of mixedly-infected cells were added to lysates of M 12-infected cells, no inhibition of the original enzyme activity was observed. Thus, there is no *in vitro* inhibitor of M 12-RNA polymerase present in lysates of mixedly-infected cells.

The details described above appear to render as unlikely that the primary effect of T 4 is on M 12-RNA synthesis. On the contrary, the simplest interpretation consistent with all the data is that T 4 blocks normal functioning of M 12 as a messenger RNA in directing protein synthesis. In support of this model is the observation that *in vitro* translation of phage R 17 RNA in extracts from T 4-infected cells is inhibited (SALSER and GESTELAND, personal communication).

Keeping in mind that T 4 does not prevent M 12 from complexing with ribosomes to form polysomes, the inhibition of M 12 RNA polymerase production may be explained by either a block in enzyme production or the formation of an incomplete enzyme. For the present, we cannot distinguish between these possibilities. In this respect, inhibition of coat-protein antigen production may be due to either a block in its synthesis or formation of an incomplete protein lacking the proper antigenic specificity.

Two mechanisms may be envisaged to explain how T 4 interferes with M 12 protein synthesis. The movement of ribosomes along the M 12 (messenger) RNA may be blocked, or there may occur a premature termination in the synthesis of M 12-specific polypeptides. Whatever factors are responsible for this inhibitory action, they must distinguish T 4 mRNA from M 12 RNA molecules. That a recognition at this level is possible may be seen by the specific effects of interferon in animal virus infected cells (see review article from Baron and Levy, 1966), as well as the destruction of host cells polysomes after poliovirus infection (Penman, Scherer, Becker, and Darnell, 1963; Scharff, Shatkin, and Levintow, 1963; Willems and Penman, 1966).

It is known that shortly after T-even phage infection of E. coli cells, a modification of leucyl-sRNA occurs (Sueoka and Kano-Sueoka, 1964; Kano-Sueoka and Sueoka, 1966). Whether such a modification may play a role in the T 4 interference of M 12 proteins synthesis remains to be elucidated. Also open to speculation is whether T 4 may modify the ribosomes, the activating enzymes or any other component of the translating maschinery. One possible approach to answering such questions would be an investigation of the *in vitro* protein synthesizing system primed by RNA-phage RNA. The various components for the *in vitro* system could be isolated from uninfected and T-phage infected cells, and then reconstructed systems tested for ability to inhibit or allow synthesis of RNA phage proteins.

## References

August, J. T., L. Shapiro, and L. Eoyang: J. molec. Biol. 11, 257 (1965).

Baron, S., and H. B. Levy: Ann. Rev. Microbiol. 20, 291 (1966).

Cohen, S. S.: Cold Spr. Harb. Symp. quant. Biol. 12, 35 (1947).

— J. biol. Chem. 174, 281 (1948).

Delius, H.: Ph. D. Thesis, Ludwig-Maximilians-Universität, München 1966.

Erikson, R. L., M. L. Fenwick, and R. M. Franklin: J. molec. Biol. 10, 519 (1964).

Hattman, S., and P. H. Hofschneider: J. molec. Biol. 29, 173 (1967).

— — Manuscript in preparation (1968).

Ippen, K. A., and R. C. Valentine: Biochem. biophys. Res. Commun. 21, 21 (1965).

Kano-Sueoka, T., and N. Sueoka: J. molec. Biol. 20, 183 (1966).

Lodish, H. F., S. Cooper, and N. D. Zinder: Virology 24, 60 (1964).

Loeb, T.: Ph. D. Thesis, The Rockefeller Institute 1961.

Nomura, M., B. D. Hall, and S. Spiegelman: J. molec. Biol. 5, 535 (1960).

Penman, S., K. Scherrer, Y. Becker, and J. E. Darnell: Proc. nat. Acad. Sci. (Wash.) 49, 654 (1963).

Scharff, M. D., A. J. Shatkin, and L. Levintow: Proc. nat. Acad. Sci. (Wash.) 50, 686 (1963).

Sueoka, N., and T. Kano-Sueoka: Proc. nat. Acad. Sci. (Wash.) 52, 1535 (1964).

VALENTINE, R. C., and H. WEDEL: Biochem. biophys. Res. Commun. 21, 106 (1965).

— —, and K. A. IPPEN: Biochem. biophys. Res. Commun. 21, 277 (1965).

VOLKIN, E., and L. ASTRACHAN: Virology 2, 149 (1956).

WEISSMANN, C., L. SIMON, and S. OCHOA: Proc. nat. Acad. Sci. (Wash.) 49, 407 (1963).

WILLEMS, M., and S. PENMAN: Virology 30, 348 (1966).

## Discussion

*Zillig:* After simultaneous infecting is the RNA going into monosomes only, or really into polysomes?

*Hofschneider:* 20—40% of the labelled parental M 12 RNA is sedimenting in sucrose gradients faster than 70 s monosome. Thus, this part of the parental RNA must be complexed in polysomes.

*Neidhardt:* The conversion of valyl-tRNA synthetase from host to phage (T 4) form begins at about 2 min and is completed by about 15 min after infection. How rapidly does T 4 interfere with M 12 protein synthesis?

*Hofschneider:* In normal infection M 12 RF appears very rapidly, at least within 5 min. In simultaneously mixed infection no RF formation is observed either late or very early in infection. Thus the slow conversion of valyl synthetase can probably not account for the fact exclusion excerted by T 4.

*Nomura:* Do you know whether function of T 4 mRNA is also reduced, that is, whether the inhibition is specific to M 12 RNA or is due to a general reduction of synthetic capacity in the cells?

*Hofschneider:* The over-all rate of protein synthesis is also reduced but to a smaller extent than that of the M 12 specific protein synthesis, the reported reduction rate for M 12 protein is already corrected for this factor. For example, if the over-all reduction is 50 % and the remaining M 12 specific synthesis 25 % of the control, the M 12 specific reduction is calculated to be 50 %.

# Defective Plant Viruses*

By

H. L. SÄNGER

Institut für Phytopathologie Justus Liebig-Universität Gießen

## 1. Introduction

During the past few years several plant viruses have been described which exhibit a fault in a biosynthetic step in viral replication. It is evident that such a fault will generally lead to the production of morphologically and (or) functionally incomplete virus. In analogy to certain bacteriophage [18, 34] such viruses are called "defective". This term implies that the molecular basis for the defectiveness are certain changes in the genetic information of the viral genome which might be caused by mutation, deletion or breakage.

Recent results indicate that functional defectiveness also plays an important role in a number of multi-component plant virus systems which are composed of particles of different size and (or) composition. In some of these systems the RNA of two (or more) defective particles is necessary for the complete expression of all viral functions. However, the mechanism of synergistic interaction between the RNA of these particles differs, depending upon whether complementation is required for early or (and) late functions. A virus which requires the RNA of two or more different components for the expression of all viral functions and for a complete cycle of viral synthesis may be called a complementing multi-component (CMC) virus system.

A previous review [113] has dealt with defective plant viruses under the aspect of latent infection and hidden viruses. This report deals with the genetical and functional aspect of defectiveness in plant viruses. Special emphasis will be placed on the complementing multi-component tobacco rattle virus system which has been studied in our laboratory during the past 3 years.

## 2. The Normal Virus

A morphologically normal or complete virus is a ribonucleoprotein particle. Tobacco mosaic virus (TMV) is the best known example of a

---

* A part of this article was presented at the symposium.

"normal" rod-shaped plant virus with a "normal" cycle of replication. It is composed of an intact and infectious RNA molecule which is wrapped in a specific way in the viral protein coat (Fig. 1, top).

The functions of the coat protein are (a) to rescue the viral genome in the infected cell in the presence of other proteins and nucleic acids by specific aggregation into virus particles and (b) to protect the viral genome from inactivation especially by nucleases during its extracellular phase. The function of the viral RNA is to guarantee the continuity of the virus. In the infected cell it induces and directs its own replication by serving as a template for the synthesis of progeny RNA. Simultaneously, viral RNA acts as a special kind of exogenous messenger RNA, which induces and directs the synthesis of virus specific proteins [e.g. RNA synthetase(s) and probably some other enzymes and virus coat protein].

The formation of a normal plant virus is consequently based on the biosynthesis of all virus-specific products which lead to the replication of infectious viral RNA and to the synthesis of a functional coat protein. Both components are synthesized independently at different sites in the cell. Normal and complete virus particles are finally formed from them by self-assembly (Fig. 1, top).

### 3. The Gene-Functions in Plant Viruses

Because of experimental limitations inherent in plant virus systems relatively little progress has been achieved in the analysis of the genomes of these viruses and their functions. This is in sharp contrast to the more detailed knowledge about the genomes and gene-functions of animal and especially of bacterial viruses. In plant viruses it is the end-product of one gene-function only, the viral coat protein, which could be investigated successfully with respect to its structure. Thus, the coat protein of TMV is a well known polypeptide consisting of 158 amino acid residues [122]. But, according to the non-overlapping triplet code the coat protein gene of TMV should theoretically consist of about 500 nucleotides which is roughly 7.5 % of the total TMV-RNA with its ca. 6500 nucleotides. From this calculation one might expect another 10—12 structural genes with certain intracellular functions for viral replication in the case of TMV [128].

The RNA of most plant viruses has a molecular weight between $1 \times 10^6$ and $3 \times 10^6$ [75]. The molecular weights of their coat protein subunits seem to range between ca. 17,500 (TMV) and ca. 52,000 (potato virus X) [75] which corresponds to polypeptide chains ranging between 158 to about 600 amino acid residues. These values indicate that similar correlations between the size of the complete genome and the size of the coat protein gene might exist in most other plant viruses. This leads to the question about the functions of the larger part of the viral genome, or, in other words, to the actual number of the structural and functional genes carried by the RNA of plant viruses.

It is known with certainty that one (or two) virus specific RNA synthetase(s) must be coded by the viral RNA. Moreover, there are indications that the viral RNA also induces and determines an acetylating enzyme in the host cell [75, 130]. Therefore, one might postulate that plant virus RNA carries at least 3 to 5 structural genes which code for virus-specific polypeptides.

There is evidence in TMV that symptomatology is not related to coat protein structure. About 100 of 150 TMV mutants tested do not exhibit any amino acid replacements in their coat protein although they are phenotypically different to common TMV with respect to their symptoms [128, 129]. It might be argued, therefore, that in most cases the changes in symptomatology are produced by mutations in genes other than the coat protein gene. These observations also suggest that several other virus-specific enzymes might exist the function of which is as yet unknown.

## 4. Defects in Plant Viruses

The variety of obvious and detectable functional defects in plant viruses is limited by the low number of functions actually known. In addition, the nature of the defects themselves (e.g. lethal mutation) and the selectivity of the experimental approach largely limit the chances for the detection and isolation of the corresponding mutants. As a matter of fact, only two types of defective plant viruses have been described (a) viruses defective in their coat protein like certain mutants of tobacco mosaic virus (TMV), tobacco necrosis virus (TNV) and (b) viruses defective in functions related to the RNA synthesis as the satellite virus (SV). In these cases it is mutation or deletion which makes a normal virus become a defective or mutant virus.

Recently, however, quite a different mechanism has been shown to produce defective virus. In the case of tobacco rattle virus (TRV) it is its multi-component character or the lack of an obligatory mutual complementation which leads to the occurence of an incomplete or defective virus. The biosynthesis of complete TRV is based on the synergistic interaction of the functionally defective RNA of an infectious long and of a non-infectious short particle in the cell. Infections caused by the infectious long particles alone give rise to "free" infectious long RNA (i.e. "proteinless" or defective TRV) while the RNA of the short particle provides the gene for the coat protein for both particles. The "short" RNA, on the other hand, appears to be defective in early functions and therefore requires the help of the "long" RNA for its own replication.

There are indications that specific interactions might also occur between different defective particles in the multi-component system of alfalfa mosaic virus (AMV) in cowpea mosaic virus (CPMV), and probably some other plant virus systems.

## 5. Defects of the Viral Coat Protein

The nucleic acid of many viruses is known to initiate infection without the viral protein coat [36, 45]. This indicates that the functions of the coat protein are of limited importance only. As a matter of fact, viruses can persistently exist as "free" nucleic acid without the accessory coat protein as was first shown in the case of defective tobacco rattle virus [27, 111]. These findings stimulated investigations in other systems and contributed to the discovery of additional defective plant virus.

### 5.1. Persistently Defective Coat Protein Mutants of Tobacco Mosaic Virus (TMV)

Nitrous acid is known to induce mutations in TMV in vitro [46, 90, 112a]. From such nitrous acid-treated TMV preparations SIEGEL et al. [114] could isolate mutants which are able to multiply in a systemic host although they produce "sterile" lesions on local lesion host plants. These mutants PM1, PM2 and later PM4 [138] were isolated after inoculation of nitrous acid-treated TMV at limit dilution to tobacco seedlings of a variety which becomes systemically infected. It appears that they are quite different from common TMV as they are defective in their capacity to induce the formation of complete virus particles. They have the following unusual features in common:

(a) Their infectious principle is unstable in leaf homogenates and little or no infectivity can be recovered from their necrotic local lesions.

(b) No infectious rodshaped virus particles can be recovered from homogenates of infected leaves.

(c) Their symptomatology in systemic hosts indicates that the infectious agent can only spread from cell to cell. Unlike virus particles it is not transported in the conducting elements of the vascular system.

It turned out that these unusual mutants were the progeny of mutated TMV particles which had been rendered defective in their ability to induce the synthesis of a functional virus coat protein. Consequently, their infectious principle remains unprotected by the normal protein coat (Fig. 1). Therefore, it does not behave like normal virus but in a manner similar to that of infectious RNA isolated from common TMV. Thus, its infectivity is largely inactivated during the procedures of transfer or purification as commonly used for normal virus. Incubation with pancreatic RNase does, in fact, largely destroy its infectivity. All attempts to recover the infectious principle from leaf homogenates by differential centrifugation have failed, and no virus-like sediment is ever obtained. When leaf

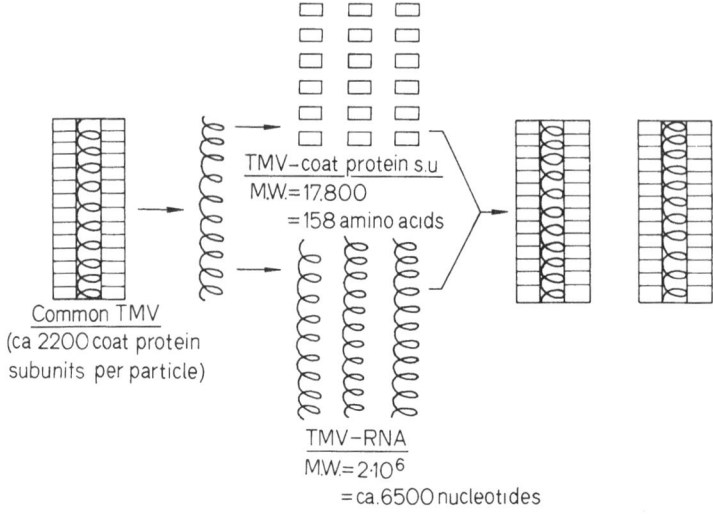

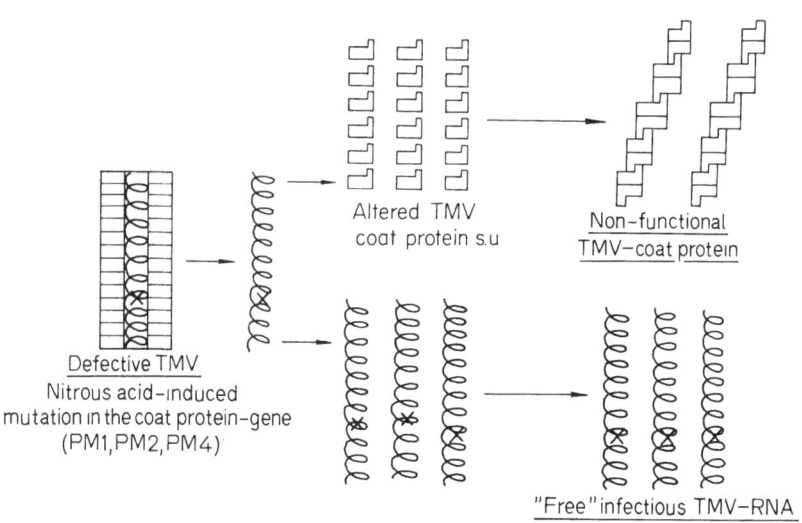

Fig. 1. Diagrammatic representation of the biosynthesis of normal and defective tobacco mosaic virus (TMV) (M.W. = molecular weight, s.u. = subunits)

homogenates are fractionated a considerable amount of infectivity is found to sediment at low forces of gravity. This indicates that a part of the infectious agent is adsorbed onto or embedded in cellular particles or debris. It is noteworthy, that some infectivity survives incubation in the presence of ribonuclease. But the lesions induced by these survivors are in no way different from lesions of the susceptible particles.

The defective mutants PM1, PM2 and PM4 induce the synthesis of a non-functional coat protein which is serologically related to the X-protein of TMV. This X-protein is normally present in TMV-infected plants [119]. It is indistinguishable from virus coat protein and is assumed to correspond to the coat protein subunits from which the viral protein coat is formed [117, 118]. Despite these similarities, considerable differences in the configuration of the non-functional coat proteins of these mutants can be observed.

*PM 1:* Defective mutant PM 1 was first believed to produce no protein at all [114]. Later, by the use of more sensitive immunological techniques [99], it was found to be present in infected cells. But the PM 1 protein unlike X-protein does not aggregate in a linear fashion but rather forms large insoluble masses in leaf homogenates [94]. These aggregates are associated with cell debris, and are therefore normally discarded in purification procedures.

*PM 2:* Unlike the X-protein of TMV, PM2-protein is defective in the sense that it does not reconstitute with TMV RNA to form intact infectious virus rods. A detailed study showed [137] that it does not aggregate *in vitro* into dense staked-disc like structures but into loose DNA-like double-helices (Fig. 1). The comparison of the amino acid composition of the PM2 protein with the one of the parent strain U1 [139] showed 2 amino acid residue replacements. This finding was confirmed by a detailed sequential analysis of the 158 amino acid polypeptide chain [131]. It revealed that a threonin is replaced by isoleucin at position 28 and that glutamic acid is replaced by aspartic acid at position 95.

*PM 4:* Homogenates from PM4 infected leaves contain a protein which aggregates reversible like the PM2 protein. But in contrast to PM1 and PM2 small amounts of short rodlike material are present [138].

In all three cases nonfunctional coat protein and uncoated infectious viral RNA are accumulating in the infected cell, and it is the defectiveness of the coat protein polypeptide chain which prevents the assembly of both products to give morphologically complete virus particles. Interestingly, all of these findings could be substantiated by a detailed study of infected living cells with the microscope [9].

### 5.2. Conditionally Defective (Temperature-Sensitive) Coat Protein Mutants of TMV

In the last few years a large number of mutants with single amino acid replacements in the 158 amino acid residue polypeptide chain of the TMV coat protein became available [42, 121, 128, 129]. It was found by Jo-KUSCH [63] that many of these mutants produce extremely low quantities of infectious virus if the infected host plants are grown at temperatures between 30 to 35°C. At normal greenhouse temperatures (20−25°C), however, they are multiplied readily. These are designated as temperature-sensitive (ts) mutants. Common TMV, on the other hand, and some of its mutants are temperature resistant. They produce high amounts of virus at 20−25°C as well as at 30−35°C. The conditional defectiveness of the majority of the ts mutants is most probably a consequence of the amino acid replacement(s) in the coat protein [63, 64]. The localization of these replacements in the polypeptide chain was achieved for some of these mutants by WITTMANN and co-workers [132]. From these investigations and the detailed studies of JOKUSCH on the behaviour of the ts mutants *in vivo* [65] and *in vitro* [66] the following conclusions can be drawn: The temperature-sensitivity of most of these mutants is solely based on amino acid replacements in the coat protein, and is not caused by the degradation of complete virus particles. At high temperatures they readily produce infectious viral RNA but their coat protein is accumulated in an insoluble, denatured and therefore non-functional form. Consequently they are unable to produce (infective or non-infective) rod-like structures at such temperatures. It is evident that the amino acid replacements must have altered the conformation of the coat protein in a way that its function is lost because of thermal instability. However, no simple rule can be given concerning the relations between temperature sensitivity and the type of amino acid replacements.

In conclusion it appears that by induced mutation TMV isolates can be produced which are persistently or conditionally defective in their coat protein polypeptide chain. Such mutants induce the synthesis of altered and non-functional coat protein. It accumulates in the infected cell because it has lost its ability to fold in the normal conformation and (or) to specifically aggregate with the viral RNA to give complete virus particles. As a result the viral RNA remains unprotected by the viral protein coat and exists in a "naked" state in the infected cell (Fig. 1). Certain amino acid replacements in the polypeptide chain can be made responsible for this loss of function. The amino acid replacements, however, are the consequence of changes in the base sequence of the coat protein gene of the viral RNA.

### 5.3. "Proteinless" Mutants of Tobacco Necrosis Virus (TNV)

TNV isolates have also been described which can be regarded to be persistently defective coat protein mutants. BABOS and KASSANIS [7] found that about 5 % of the local lesions produced by purified preparations of strains A and B of TNV contained variants which were unstable in sap and difficult to transmit mechanically. Their properties are quite similar to the ones of the defective coat protein mutants of TMV and the

defective tobacco rattle virus to be described later. Thus, defective TNV
does not produce stable (spherical) nucleoprotein particles but occurs in
the host tissue as "free" infectious RNA (Fig. 2). It has properties similar

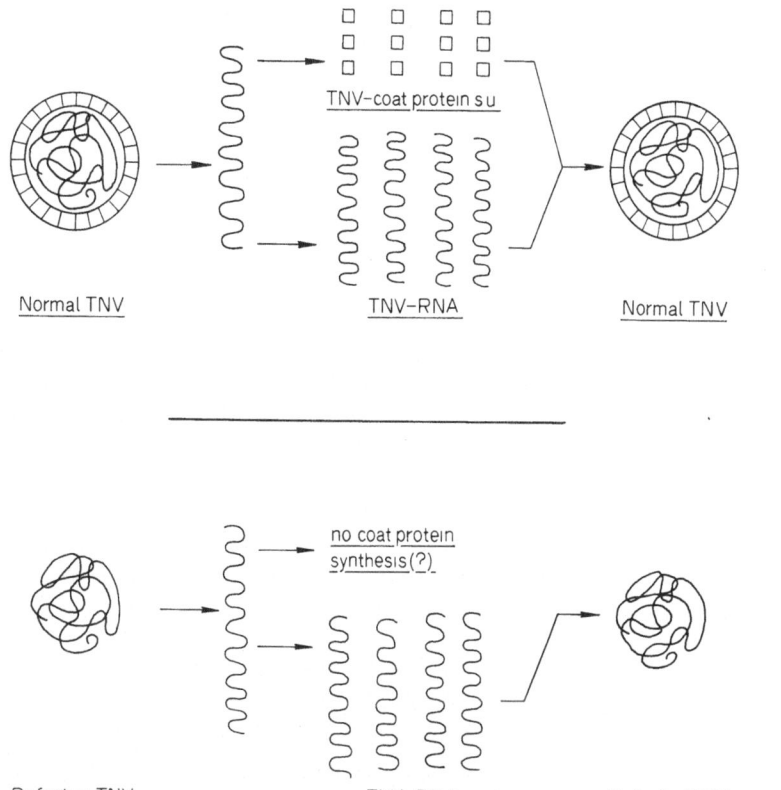

Fig. 2. Diagrammatic representation of the biosynthesis of normal and defective
tobacco necrosis virus (TNV). The original infection which produced defective TNV
was initiated by the (mutated) RNA of a complete virus particle. (s. u. = subunits)

to RNA extracted from normal virus with phenol, and in sucrose gradients
it sediments as nucleic acid and not as intact virus [72]. The infectivity of
defective TNV is destroyed when leaves are extracted in conditions to
allow inactivating enzymes to act. Accordingly any method of preparing
leaf extracts which inhibits the action of leaf ribonuclease largely prevents
the inactivation of the infectious agent of these mutants. This can be
achieved by direct extraction with phenol, by grinding the leaves in
buffers with pH values above 9, and by using bentonite as an additive
20*

[7, 72]. Although no soluble coat protein related material could be detected serologically in leaf sap [7], it might yet be present in an insoluble and denatured state as in the case of the defective protein mutant PM 1 of TMV [94]. Therefore it is uncertain at present whether defective TNV produces no coat protein at all or a denatured and therefore non-functional coat protein.

From the results described it can be concluded that defective TNV is a spontaneously occuring coat protein mutant which has been isolated from a genetically inhomogeneous population. Under natural conditions such mutants would be lost, and remain therefore undetected because they are lethal mutants.

## 6. Defects Related to the Synthesis of Viral RNA

Viruses can persistently exist with defective and nonfunctional coat protein as "free" infectious RNA. Defects, however, which occur during the synthesis of the viral RNA, interfere with the basic function of the viral genome which is to guarantee the continuity of the virus. Therefore, in most cases such defects will be lethal for the virus since they lead to the production of non-infectious RNA or non-infectious virus-like particles. At present only one stable plant virus, the satellite virus, is known which exhibits a defect in a gene-function(s) related with the replication of the viral RNA. Therefore SV cannot induce its own replication but requires a compatible "helper" virus for this function(s).

### 6.1. Satellite Virus (SV)

Certain cultures of tobacco necrosis virus (TNV) were reported to contain spherical particles of two sizes of 170 Å and 300 Å in diameter [15, 16, 70]. The significance of this was not understood until KASSANIS and NIXON [70, 71] found them to be serologically unrelated viruses. The smaller particles are multiplied detectable only in the presence of the larger particles. This association and dependency gave origin to the name "satellite virus" (SV) for the smaller particles [67]. The 300 Å particles were shown to be related to other strains of normal TNV whereas no serological relationship could be demonstrated with the 170 Å SV particles (Fig. 3). With one exception, all strains of TNV that were tested supported the multiplication of SV. Even the defective coat protein mutants of TNV, described in the preceeding section, were found to be effective in this helper function [7]. The one strain which did not activate SV is considered to be only distantly related to the viruses originally described as TNV [8, 68]. A wide range of viruses serologically unrelated to TNV have also been tested but none activated SV [69].

All studies with SV itself must start from a mixed infection of both viruses. Fortunately they can be separated from each other from purified and concentrated preparations without difficulty by centrifugation in sucrose gradients or by electrophoresis [67, 69]. Thus, highly purified preparation of SV could be obtained, which initiated detailed studies of

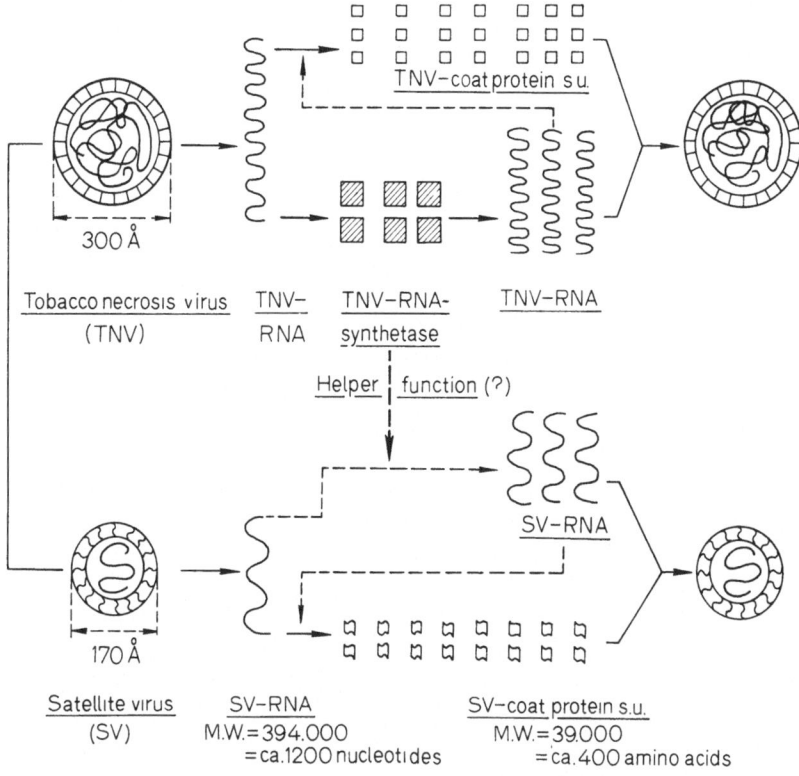

The tobacco necrosis virus–satellite virus–system

Fig. 3. Diagrammatic representation of the helper virus system of satellite virus — tobacco necrosis virus (for details see text). (M.W. = molecular weight, s. u. = subunits)

the physical and chemical properties of SV [67, 102, 103]. SV is the smallest virus discovered to date with a molecular weight of $1.79 \times 10^6$. Its RNA is a single chain molecule with a maximum molecular weight of 394,000. According to its base composition it consists of about 1200 nucleotides or 400 triplets. The minimum molecular weight of the SV coat protein was first determined by REICHMANN et al. [103] to be 25,000 and later corrected by REICHMANN [102] to 39,000 corresponding to a polypeptide chain composed of about 400 amino acid residues. According to current ideas on genetic coding the limited amount of genetic information stored in this RNA is only sufficient to code for the SV coat protein.

A more direct proof for the concept that SV-RNA only carries a mono-cistronic message was obtained by using this RNA as an exogenous messenger in a cell-free, protein-synthetizing system from E. coli [29]. It was found that the tryptic fingerprint of the SV-RNA-directed C14-labelled product is in excellent agreement with the tryptic fingerprint of the SV-coat protein. Overlapping comparisons show exact agreement in 18 out of 25 peptides and a relative freedom from non-specific peptides. This supports the conclusion that only the information for its coat protein is contained in the SV-RNA.

Moreover, in contrast to most other plant viruses, the coat protein chain of SV does not contain a acetylated aminoterminal amino acid residue [102]. Since it is very likely that the acetylating enzyme is normally determined by the viral RNA [130], one might assume that also this information is lacking in the case of SV.

From these results it can be concluded, that the RNA of SV does not seem to contain any information for early functions neccessary for viral replication. It might be postulated that it is primarily the message for the virus-specific RNA-synthetizing system (RNA-synthetase) which is lack-ing. A defect of such nature plausibly explains the association of SV with TNV and its dependency on this helper virus for replication. This leads to assume that the RNA synthetase(s) induced by TNV is also used to replicate the RNA of the defective SV (Fig. 3).

If, however, SV-RNA has been replicated it initiates and directs the synthesis of the SV-specific coat protein. Consequently morphologically complete but functionally defective SV particles are formed. Thus, the defectiveness of SV and its inability to multiply by itself is related to its unusually small RNA which evidently lacks the information for early functions of viral replication.

## 7. Defects in Complementing Multi-Component Plant Virus Systems

The occurrence of non-infectious virus related components in addition to infectious virus particles has been described for several plant viruses [50, 83, 85, 136]. It, therefore, appears that non-infectious particles are a constant feature of infection by many viruses. Any such virus which is characterized by particles of different size and composition may be called a multi-component virus system.

A prominent example of such kind is turnip yellow mosaic virus (TYMV). In addition to infectious virus plants infected with TYMV pro-duce at least three types of RNA-containing but non-infectious particles as well as hollow protein shells [83, 84, 85, 86, 115]. Findings similar in some respects to those made with TYMV have been reported for the polyhedrical wild cucumber mosaic virus [136], alfalfa mosaic virus [123, 125, 134], cowpea mosaic virus [26, 134], cherry necrotic ringspot virus

[41], bean pod mottle virus [134], tobacco streak virus [40], brome grass mosaic virus [20] and for the rod-shaped tobacco rattle virus [49, 95].

Some of the RNA-containing but non-infectious components of these systems have been rather well identified, but relatively little is known about their origin and function. In the case of TYMV, for example, it was conjectured that they might be examples of incomplete assembly with respect to the nucleic acid incorporated in them [84, 85]. In other cases these non-infectious components have been assumed to be break down or by-products of viral synthesis [20, 43, 49, 107, 115].

The biological significance of the RNA of such non-infectious components has been underestimated in the past especially because infectivity was the dominating criterium used in plant virology. Recent results, however, clearly show that the RNA of some of the non-infectious components of such multi-component systems might well have essential biological functions. In the case of tobacco rattle virus, the initiation and direction of a complete cycle of viral replication was shown to be based on the complementing interaction of the RNA of a infectious and a non-infectious virus particle. This condition would suggest, that the viral genome of TRV is not represented by one intact RNA molecule but rather by two pieces of RNA coding for early and late functions, respectively.

There are indications that in certain other multi-component systems comparable complementing interactions between the RNA of different particles might occur. Defectiveness appears to play an important role in most if not all of such systems. In some cases the interacting components carry functionally defective parts of the viral genome and only by complementation can all of the functions of the virus be expressed. In other cases, defective virus particles may occur which are only able to initiate the production of infectious RNA. Yet, this RNA remains unprotected by the viral protein coat because the complementing RNA providing the coat protein gene is lacking. Under such conditions, as in the case of tobacco rattle virus, morphologically and functionally defective viral progeny is produced.

Any virus system in which the RNA of two or more different particles is necessary for a complete cycle of viral replication may be disignated as a complementing multi-component system (CMC-system).

In the following section the pertinent results obtained with various CMC-systems will be discussed. Tobacco rattle virus will be described in detail because the interaction in this system is rather well understood. Alfalfa mosaic virus will be discussed although the actual mode of interaction between the components is not completely elucidated yet. Some of the other multi-component viruses will also be described with respect to their possible mechanisms of interaction.

## 7.1. The System of Tobacco Rattle Virus (TRV)

In retrospection defective TRV appears to be the first defective plant virus described. As early as 1956 KÖHLER [76] reported on unusual field isolates of TRV which, in contrast to normal TRV, were difficult to transmit mechanically. The existance of such aberrant isolates was confirmed repeatedly [23, 28, 107, 108]. It was found that such poorly transmissible isolates of TRV can be obtained by subculturing poorly infective local lesions produced by dilute inocula of normal TRV. In contrast, undiluted inocula produced infections of normal TRV [23, 33, 107, 108]. This "dilution effect" is a rather unique feature among plant viruses. The difficulties in transmitting poorly transmissible isolates could be related to the instability of their infectious agent which is inactivated quite rapidly in plant sap whereas normal and particulate TRV is a relatively stable virus. The actual nature of the aberrant isolates of TRV, to be called "defective TRV" (D-TRV) in the following, was demonstrated by SÄNGER and BRANDENBURG [111]. They found a) that direct phenol extracts from D-TRV tissue were highly infective in contrast to normal buffer homogenates and b) that D-TRV isolates neither induce the synthesis of detectable amounts of virus particles or virus antigen. This led to the assumption [111] that the infectious agent of D-TRV is "free" infectious viral RNA unprotected by a viral protein coat. These results were confirmed and amplified by CADMAN [27] who provided some evidence that D-TRV might be associated with the nuclei of the infected cells. Recent results indicate, however, that D-TRV does not exist in any stable complex or cell constituent from which it might be liberated during extraction but rather as "free" infectious RNA [110]. The finding, that a persistent viral infection can be based on the presence of a "naked" or defective virus i.e. proteinless or "free" infectious viral RNA [111] opened new perspectives in virology. It stimulated corresponding investigations in other systems and largely contributed to the discovery of other defective plant viruses.

According to its unusual and curious properties the "poorly transmissible" D-TRV has been subject to many speculations. Its nature has been explained by reversible, season-induced changes in the virulence of the normal virus ("winter-type" of TRV [76]), by quantitative differences in virus synthesis ("non-multiplying" (NM) form of TRV [28]), and by the inability of the viral RNA to escape from the nuclei of the cell where it is presumed to be synthetized ("non-escaping virus" [27]). It has also been assumed that "poorly transmissible" D-TRV isolates are the progeny of what is now being called defective coat protein mutants, a suggestion which would imply an unusually high mutation rate of the parent C-TRV [107, 108]. It appears, however, that none of these explanations actually apply, and that the occurrence of D-TRV is rather based on the unique feature of normal TRV, namely on its multi-component character.

In virus preparations [49,95] and leaf tissue [31] normal TRV is characterized by the occurrence of tubular particles of two different lengths which
are not in any simple integral ratio to each other. The length of the short
particles varies with the isolate ranging from about 400 to about 1050 Å.
The long particles of all isolates studied so far range from about 1800 to
2000 Å [51]. Long and short particles can be readily fractionated in sucrose
gradients. Both contain about 5% RNA, both are serologically identical
and both are always found together. However, only the long particles
and their RNA are infective and produce local lesions in plants. The short
particles are non-infective and do not influence the number of lesions
produced by the long ones [49, 107, 112]. It has been suggested that the
shorter particles might either be the degradation product of the long
particles or additional products of the virus infected cell [49, 107, 108,
112]. The normal and particulate TRV will be called "complete" TRV
(C-TRV) in the following.

There were many suggestions that the origin of D-TRV must be
related to the presence of two different particles in C-TRV preparations
[14, 23, 77]. Thus BAWDEN [14] postulated, that the short rods might lack
the part RNA that confers infectivity but contain the one able to code
for the protein. In fact, LISTER [78] found that the inocula containing
purified long particles produce predominantly lesions containing D-TRV,
whereas lesions containing C-TRV were obtained only when both long
and non-infectious short particles were involved in infections. He put
forward the hypothesis "that the RNA of the long particles of the TRV
type is deficient in the information required for some stage of the process
leading to the enrobement of the viral RNA with virus protein; possibly
the coding for the virus protein itself".

This interpretation was tested by FROST et al. [38], working with
fractions and mixtures of long and short particles of known composition.
Their results provided the quantitative support for both the findings and
hypothesis of LISTER [78]. They concluded that "each isolate of tobacco
rattle virus seems to be a system of two or more pieces of infective nucleic
acid interacting specifically in a symbiotic manner" [38]. Additional and
more direct evidence for this unique interrelationship is provided by the
author [109]. The phenomenon of an increased occurrence of D-TRV
infections after inoculation with diluted inocula of C-TRV, the so-called
"dilution effect of TRV" [23] was reinvestigated on quantitative basis.
The total number of all lesions appearing was found to be directly proportional to the amount of long particles present in the inoculum (Fig. 4).
The curve resembles a single hit dilution curve which is characteristic of
many plant viruses. If, however, the corresponding curves of the relative
numbers of C-TRV and D-TRV infections are plotted, the curves differ
because they resemble two-hit curves. Multiple hit dilution curves might

result from genetic insufficiency of many or most virus particles as
described for UV-inactivated bacteriophage [*80, 81*]. Thus two (or
more) virus particles at a site appear necessary to initiate normal

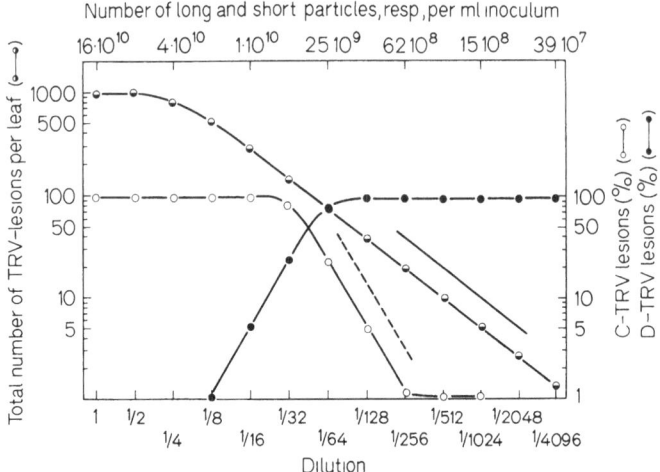

Fig. 4. Dilution curve of an artificial mixture of purified long and short particles of
tobacco rattle virus (TRV). C-TRV lesions contain virus particles and D-TRV lesions
contain "free" infectious "long" RNA. The lesions cannot be distinguished by their
appearance but have to be tested individually for the presence of the infectious
agent they contain

virus multiplication, because together they provide a complete complement of genetic units which are incomplete in individual particles. By
analogy it is concluded that C-TRV infections associated with the production of virus particles must result from some sort of cooperation among
the infectious long and non-infectious short particle present within a
given site.

D-TRV lesions, however, appear to result from the absence of this
cooperation, and are therefore initiated by the infectious long particles
alone. These findings suggest that the presence of the short particles is
necessary for the synthesis of the coat protein of both particles. If this is
so, it should be possible to complement D-TRV (= free infectious long
RNA) with either the homologous short particles or their RNA in order
to obtain the parent C-TRV which consists of both long and short particles. Appropriate experiments showed this to be the case (Fig. 5). The
RNA of the long particles, although infective, is yet defective in its
ability to initiate and direct the synthesis of a viral coat protein. The RNA
of the short particle, on the other hand, though defective for the initiation
of its own replication, complements the lacking information of the long

particles most probably by providing a functioning gene which codes for the coat protein for both particles. Direct evidence on this point was obtained by heterologous complementation using different strains of TRV with suitable genetic markers. The characteristics of this system and the pertinent results are presented diagrammatically in Fig. 6.

C-TRV-GER is a local German strain and has particles which are 1800 Å and 700 Å long and which separate characteristically in sucrose density gradients. It produces necrotic rings as primary lesions on tobacco leaves (Fig. 6A). C-TRV-USA is a single lesion isolate from the Oregon strain of TRV [3] and has particles which are 1950 Å and 1050 Å long. It produces fully necrotic spots as primary lesions and its particles also separate characteristically in sucrose gradients (Fig. 6C). Both TRV strains can be clearly distinguished serologically which indicates the existence of differences in their coat protein subunits. D-TRV cultures from both strains were isolated by single lesion transfers and propagated continuously as D-TRV [110]. The use of such cultures eliminates any interference and contamination with homologous short particles. Heterologous complementation using the "proteinless" D-TRV of one strain and the short particles of the heterologous strain produces a "mixed" C-TRV progeny. This mixed C-TRV is composed of morphologically complete long and short virus particles each of which are derived from the other TRV strain, respectively (Fig. 6E, F). The local lesions of the mixed TRV

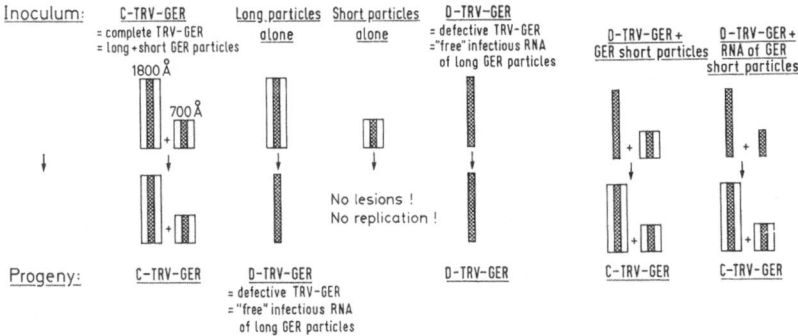

Fig. 5. The properties of a German isolate of tobacco rattle virus (TRV), and the results of the homologous complementation of defective TRV with the short particle of TRV or their RNA

as produced by the infectious long particles are indistinguishable from the ones of the corresponding D-TRV or the C-TRV from which it was isolated. However, serological tests showed that the mixed TRV appears to be indistinguishable from the strain from which the complementing short particle was derived (Table 1). These results clearly demonstrate

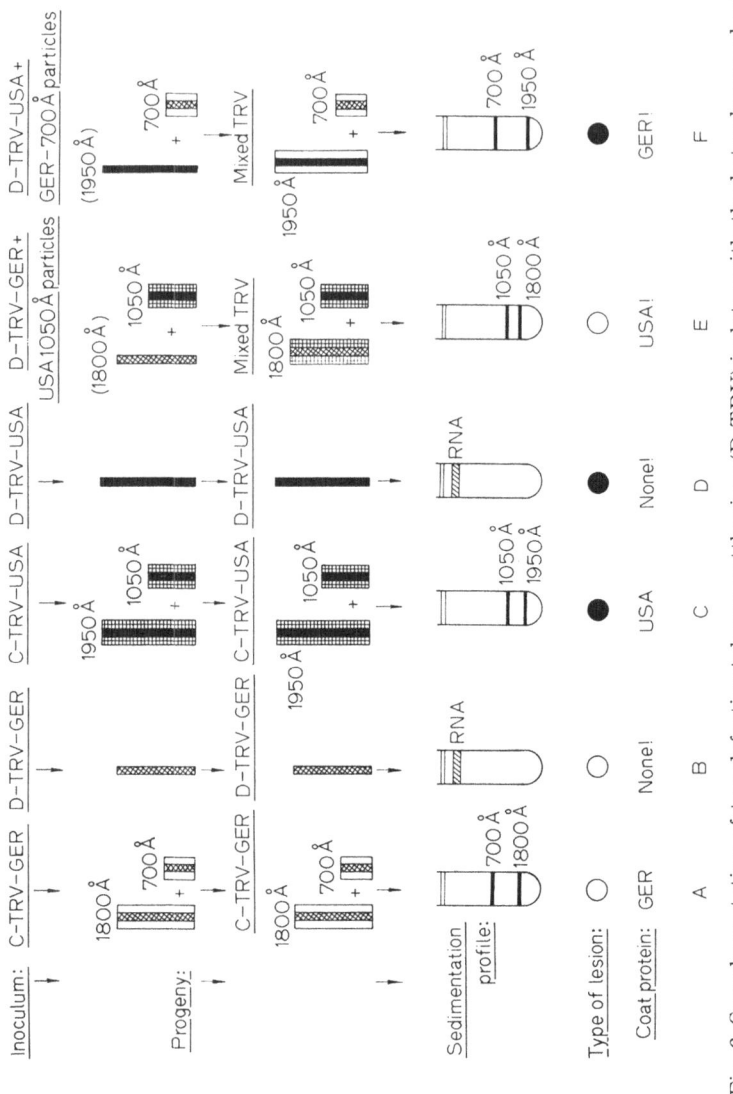

Fig. 6. Complementation of two defective tobacco rattle virus (D-TRV) isolates with the heterologous short particles. The specific markers of the two systems (particle size, type of lesion and coat protein) are re-presented diagrammatically. The sedimentation profiles of D-TRV are the ones obtained when bentonite-extracts from D-TRV infected leaves were centrifuged under the same experimental conditions as the purified parent complete TRV preparation. For further details see text

that the RNA of the short particles of C-TRV does, in fact, contain a functioning gene providing the information for both the induction and the specificity of the coat protein for both particles of the system of which they are a part.

Despite these results several features of the TRV system are not yet clearly understood and need more investigation. It is puzzling that different TRV isolates exhibit distinct differences in the lengths of their particles [51]. Isolates have been found with infectious long particles of about 1800, 1900 or 2000 Å so that the minimal infectious length of TRV appears to be about 1800 Å. The differences in lengths are more pronounced within the short particles which vary with the individual isolate. Lengths of 400, 500, 700, 800, 900, 1000 or 1050 Å were found. The length of tubular virus particles is generally determined by the length of the corresponding RNA molecule. Thus, also the RNA of these different short particles differs in length and correspondingly in the potential amount of information carried by their RNA.

Table 1. *Serological behaviour of different tobacco rattle virus systems*

| Virus isolate | Short particles used for complementation | Antiserum titer[a] | |
|---|---|---|---|
| | | C-GER antiserum | C-USA antiserum |
| C-TRV-GER | — | 512 | 4 |
| C-TRV-USA | — | 4 | 512 |
| D-TRV-GER | GER- 700A | 512 | 4 |
| D-TRV-GER | USA-1050A | 4 | 512 |
| D-TRV-USA | USA-1050A | 4 | 512 |
| D-TRV-USA | GER- 700A | 512 | 4 |

[a] Figures are the reciprocals of precipitin endpoint of the antisera as determined in the agar gel diffusion test. All antigens were highly purified and used at a concentration of 0,5 mg/ml

Preliminary analysis indicates that the RNA of the infectious 1800 Å particles has a mol wt of about $2.3 \times 10^6$ which would correspond to approximately 7100 nucleotides. Thus, the "long" RNA is of the size of the RNA of many plant viruses. The mol wts of the RNA of the non-infectious 700 Å and 1050 Å particles of the TRV systems used in this study are about $9 \times 10^5$ (ca. 2800 nucleotides) and $1.3 \times 10^6$ (ca. 4000 nucleotides), respectively. Thus even these short particles can theoretically carry enough information to code for 4 to 6 genes assuming an average of 200 amino acids per protein. Indeed, the RNA of brome grass mosaic virus (mol wt $= 1 \times 10^6$ [20]) and broad bean mottle virus (mol wt $= 1.1 \times 10^6$ [135]) are of about the same size, and both viruses are normally replicating plant viruses.

The coat protein of TRV consists of identical polypepide chains each with a molecular weight of about 24,000 and a total of about 220 amino acid residues [92, 93]. Assuming a non-overlapping triplet code an RNA molecule with about 700 nucleotides (mol wt = ca. 230,000) would be theoretically of the minimal length to carry the information for the coat protein of TRV. Thus, the presence of additional nucleotides would

indicate the short particles contain more information than that necessary
for the coat protein. Thus the shortest particles of the TRV system known
at present to specify protein synthesis is about twice as long (ca. 400 Å)
as the particle of the theoretical minimal length required for this function. It would appear, therefore, that excessive RNA of the short
particles might be found to code for several more viral functions.

Recent observations on the system of pea early browning virus which
shares all of its essential properties with the TRV-system (see the following paragraph) led LISTER [79] to speculate that the RNA of some short
particles might be capable of influencing the production of coat protein
whereas other might influence factors affecting symptomatology. But this
does evidently not apply for the 700 Å and 1050 Å particles of the two
TRV systems used in this investigation, because the symptomatology of
the "mixed" TRV is not influenced by the heterologous short particles but
solely specified by the "long" RNA.

The phylogenetic relationship of the two partners of the TRV system
and the precise nature of the defectiveness of their RNA cannot be
assessed from the above described experiments. Two interpretations are
possible with respect to the origin of the two species of RNA and they
apply for the origin of the components of any other CMC-virus system:

*Interpretation 1:* The RNA of the components represents functionally
spezialized parts of one viral genome.

*Interpretation 2:* The RNA of the components represents the genomes
of different viruses.

It appears rather difficult at the moment to clearly decide whether
interpretation 1 or 2 applies for the TRV system. But for each of these
two interpretations several distinct mechanisms can be visualized which
might have led to the developement of such a system. In the process of
natural selection any such system of complementing viral components
would acquire an advantage over a single component virus system with
a defective cycle of replication.

The high degree of specificity in most virus systems especially in
structure and function of the virus coat proteins and in the function of
virus-specific enzymes would suggest that the TRV-system developed
from one virus the genome of which specialized into two functionally
distinct pieces of RNA, responsible for early and late functions, respectively.

If, however, Interpretation 2 applies, the TRV-system would represent a helper virus system consisting of two different defective viruses.
One might argue that in the process of phylogeny the chances appear to
be relatively low for the development of such a virus system. But, if
established once its advantage over the defective parent viruses would
the guarantee its survival and further existence. If, in fact, the "long" and

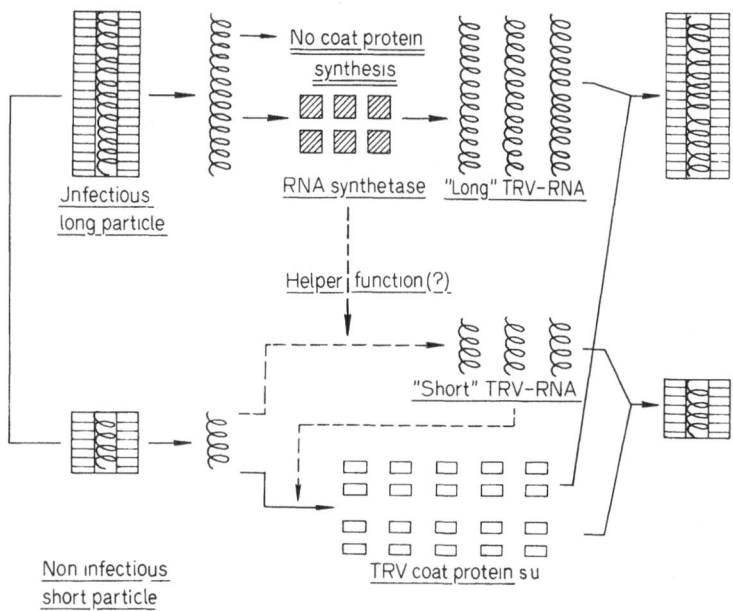

The complete tobacco rattle virus system (C-TRV)

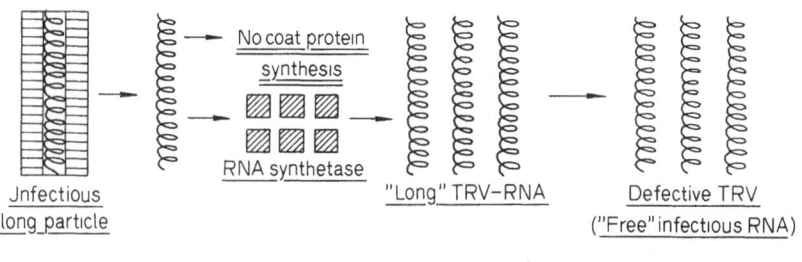

Defective tobacco rattle virus (D-TRV)

Fig. 7. Diagrammatic representation of the complementing multi-component system of tobacco rattle virus. (s. u. = subunits)

"short" RNA of TRV represent the genomes of two defective viruses, the TRV system would be different from any helper virus system described so far. In these systems as in the case of satellite virus-tobacco necrosis virus, it is an independent and normally replicating virus which helps to replicate a defective virus.

In conclusion it appears that TRV is a system of two morphologically complete but functionally defective tubular nucleoprotein particles of different length but coated with the same coat protein. The RNA of the long and infectious particle is deficient in the information necessary for the synthesis of the viral coat protein and only capable of inducing its own replication. Therefore, the progeny of the long particles is a morphologically and functionally defective TRV existing as proteinless free infectious "long" RNA unprotected by a viral protein coat. The RNA of the short particle induces and specifies the synthesis of the coat protein for both particles although it produces no lesions and does not replicate by itself. It rather appears that its replication is dependent on the helper function(s) coded by the "long" RNA. Thus the long and short virus particles of TRV can be synthesized only through the mutually complementing interaction of the RNA of both particles (Fig. 7). Because of this unique interaction the TRV system is different from any plant virus system reported so far.

### 7.2. The System of Pea Early Browning Virus (PEBV)

PEBV closely resembles tobacco rattle virus in many of its characteristics. It has been investigated intensively by Bos and VAN DER WANT [22] who demonstrated that most of its biological properties are similar to that of TRV. PEBV also has two elongated particles. The lengths of these particles is about 1000 Å and 2000 Å, and as with TRV only the long particle being infectious. Although at first no serological relationship could be detected between PEBV and TRV with low titer antisera [22], a distant relationship was later found with the use of antisera of high titer [5, 44, 82].

The differences between TRV and PEBV should not be overemphasized. It is well known, that different isolates of TRV exhibit pronounced differences in both the absolute lengths and in the ratio of the average lengths of both particles [4, 51], as well as in their serological relationship [5, 51]. Comparable differences have also been observed between different isolates of PEBV [44]. Especially the morphology of the local lesions on beans has been used to distinguish between both viruses. PEBV is known to produce large ring type lesions whereas TRV produces very small lesions. LISTER [79] found that even this marked difference is no longer a reliable criterium. Unexpectedly, certain PEBV cultures give rise to TRV-like lesions. Among the different explanations provided the most plausible one relates this difference in symptomatology to the specificity of information of the RNA of particles of different lengths.

The similarities between PEBV and TRV are still more pronounced if the properties and functions of the particles are compared. It has been indicated by LISTER [78] that the isolated long and short particles of

PEBV behave in a similar way as the ones of TRV, and that defective PEBV (D-PEBV) can be complemented to produce complete PEBV progeny by adding noninfectious short PEBV particles to the D-PEBV inoculum. Also, D-PEBV like D-TRV can be isolated by transfers of local lesions produced by highly dilute inocula [44, 79].

Since the variation within PEBV and TRV is of about the same magnitude as the differences between both viruses it appears justified to consider PEBV isolates as strains of TRV. Therefore, all the findings and problems discussed in the proceeding paragraph in connection with the CMC-system of TRV also apply for the system of PEBV.

### 7.3. The System of Alfalfa Mosaic Virus (AMV)

Next to tobacco rattle virus AMV is the system studied best with respect to the interaction of its components. The results indicate that this interaction might be rather complicated and in certain respects comparable to the one in the TRV system. But it should be noted that the actual mechanism of interaction between the AMV-components has not yet been fully elucidated and is still a matter of conjecture.

In preparations of spherical AMV 4 or 5 different components can be found all of which contain RNA. They can be separated by centrifugation in sucrose gradients and by precipitation with magnesium ions. Two components predominate, a 99s bottom component and a 68s top component. It is supposed that they all have the same coat protein [12, 62, 73]. Although it had been made plausible that the bottom component is the infectious particle of AMV [11], some uncertainty arose when it was found that infectivity is not clearly associated with one component [43]. An interesting observation might explain this disagreement. It was found that two to three times of the original infectivity is activated in the fraction of the bottom component when the non-infectious middle or a mixture of top components is added to it [134]. It is conceivable, therefore, that incomplete separation of the components in the sucrose gradient may cause a shift and broadening of the infectivity maximum. In fact, when RNA extracted from unfractionated virus was separated in sucrose gradients, maximal infectivity was found in association with the RNA of the bottom component [47]. But the distribution of infectivity in the gradient suggested an interaction between this RNA and the RNA from the middle or top component. A more detailed investigation [123] showed that it is the non-infectious RNA of top component a which activates the infectivity of the RNA of the bottom component. This RNA from top component a is of interest because in a protein-synthesizing in vitro system from E. coli it exhibited specific messenger activity. Polypeptide synthesized under its direction corresponded to the coat protein of AMV

[*100, 101*]. Therefore, this RNA must contain the gene for the coat protein of AMV.

From these observations it appears that the interaction between the components of the AMV system may be comparable to the type of inter-action in the TRV system. But it is still not clear whether the low infectivity found in the fractions of the bottom component arises from a small degree of contamination with the infectivity activating components. The same applies for the preparations of RNA from these particles. In this case, the activation of infectivity could still be caused by a mechanism which has some characteristics of multiplicity reactivation.

If, however, the bottom component lacks genetic information for the coat protein, it must be assumed that this is in some way correlated with the ability to cause local lesions. This would be in contrast to the situation in the tobacco rattle virus system where the defective "long" RNA does induces local lesions and initiates the replication of its RNA.

The possibility that in the case of AMV the non-infectious RNA of the top component *a* only stimulates the reaction of the plant in forming a visible lesions without playing a part in the actual process of virus reproduction seems to be excluded. Recent investigations [*124, 125*] showed that the activation of infectivity could also be obtained by combining the components from different isolates which are characterized by different symptoms. A phenotypically stable AMV isolate with inter-mediary symptoms is obtained. From this "mixed" virus the parent type viruses can be obtained again by "back crossings" with the corresponding homologous non-infectious particles. It is hoped that the characterization of the coat proteins of these isolates might help to support a single mechanism of interaction in this system.

*7.4. Possible Mechanisms of Interaction in other Multi-Component Plant Virus Systems*

The RNA of the non-infectious short particles of the TRV system contains the gene for the coat protein. It activates and specifies the syn-thesis of the viral coat protein and thus contributes a late function in viral multiplication. However, all early functions of viral replication are initiated by the RNA of the infectious long particle. The specialization of their RNA in only late functions would explain the observation that the short particles do not affect the quantity of lesions produced by the long particles which carry all the information for the primary process of infection.

In contrast, the RNA of the smaller and non-infectious components of all other multi-component plant virus systems investigated so far pri-marily activate infectivity. They increase the number of lesions produced by the fractions containing the larger and apparently infectious particles.

This phenomenon of infectivity-activation has been reported for the systems of cherry necrotic ringspot virus [41], bean pod mottle virus [134], cowpea mosaic virus [26, 134], alfalfa mosaic virus [123, 125, 134] and tobacco streak virus [40]. It is likely that similar phenomena might have remained undetected in some of the other multi-component systems.

Three possible mechanisms have been proposed to explain this activation of infectivity by the non-infectious components in the fractions containing infectious particles:

*Mechanism 1:* Infectivity is stimulated by a protein or a substance other than RNA not present in the infectious component. This mechanism may act by stimulating host responses [40, 41], or it might be concerned with uncoating of the virus [134].

*Mechanism 2:* In the fraction of the infectious component particles are present which contain damaged or incomplete and therefore non-infectous RNA molecules. This RNA is "rescued" in the presence of other non-infectious RNA-containing components [134]. This would imply that the lacking or defective information of such damaged viral genomes is provided by the RNA of the smaller component. This RNA should therefore carry an incomplete viral genome containing the complement of the non-functional region of the RNA of a part of the population of the larger component. Accordingly, this mechanism has some of characteristics of "multiplicity reactivation" as found with bacteriophage [80, 81], influenza viruses [13, 116] and vaccinia virus [1, 2].

*Mechanism 3:* The primary processes of infection are initiated on principle by the interaction of two (or several) functionally distinct types of RNA and the observed increase of infectivity reflects this synergistic interaction [26].

Unfortunately the methods of separating the components are still inadequate, and consequently the data published thus far do not support any single hypothesis to the exclusion of the others. It is hoped that with the improvement of these methods this uncertainty will be clarified. On the other hand, it is to be expected that the mechanisms of infectivity activation might be different in different virus systems.

Mechanism 3 is of special interest because it is based on essentially the same principle as the interaction in the tobacco rattle virus system. In any such case the expression of certain gene-activities of one viral genome would depend on the interaction of two (or more) different types of RNA. Accordingly all virus systems of this type — in analogy to higher organisms — would have in common the one general feature, that their genome exists in two (or more) functionally spezialized pieces. It is evident, that the initiation and (or) completion of an infection cycle is, therefore, based on the interaction of these different pieces of the one viral genome.

21*

## 8. Defects of Unknown Nature

Two viruses will be described in this section which exhibit defects the origin of which has not yet been fully elucidated. It is hoped that further investigations will soon lead to an understanding of their nature.

### 8.1. TMV Mutants Conditionally Defective in an Unknown Funktion(s)

In addition to the temperature-sensitive (ts) coat protein mutants other temperature-sensitive mutants of TMV (Ni 2519) have been found [66, 132] which do not show amino acid replacements in the coat protein as compared with the parent TMV, nor could the defect be associated with the primary process of infection. It is likely that a function related with the spreading of infectivity in the tissue has been rendered temperature sensitive. If so, the corresponding mutation would then be located in the gene carrying the message for this still unknown function.

### 8.2. Potato Spindle Tuber Virus (PSTV)

In efforts to purify and characterize PSTV DIENER and RAYMER [32] found that the rather stable and easily extractable infectious agent of PSTV is not a conventional virus particle. Its properties suggest that the agent is rather a free nucleic acid which would make PSTV a defective virus.

PSTV exhibits the following RNA-like properties [32]: Although it is extractable with phosphate buffers, it cannot be concentrated by high-speed centrifugation but remains in the supernatant under conditions where normal virus is sedimented to give a pellet. In sucrose gradients the infectious agent sediments at approximately the same rate as particles with a sedimentation coefficient of 10 s. Like free RNA it is insensitive to treatment with organic solvents and can be concentrated by precipitation with ethanol. Treatment with phenol does not change its infectivity or sedimentation properties. Incubation with deoxyribonuclease has no effect on infectivity whereas incubation with ribonuclease at low ionic strength completely destroys infectivity. In media of high ionic strength, however, infectivity is only partially destroyed by ribonuclease. In equilibrium density-gradient centrifugation infectivity bands in a cesium sulfate gradient but in a cesium chloride gradient it forms a pellet like RNA.

PSTV appears to be able to replicate independently. If PSTV is a conventional nucleoprotein, it would have to be very small or contain a component of low density which would reduce its rate of sedimentation. The smallest plant virus known is the defective satellite virus (SV), which cannot replicate by itself because of the limited size and information of its RNA. Since the rate of sedimentation of SV is 50 s it is, by comparison, unlikely that PSTV sedimenting at 10 s could be a virus particle containing an RNA with the necessary information for independant replication. The stability of PSTV in organic solvent, on the other hand, rules out a possible association of this virus with a lipid which might reduce its sedimentation rate. Interestingly, however, a number of similarities could be demonstrated between the properties of PSTV and the properties of the double-stranded replicative form of RNA viruses which is known to occur in virus infected tissue [6, 35, 89, 96]. It is especially the sedimentation rate of PSTV (10 s) and its resistance to ribonuclease in media with low ionic strength which strongly suggest that PSTV is a double-stranded RNA.

Although the nature of its infectious agent is not yet fully revealed, it appears justified to assume that PSTV is a defective virus. It is interesting to note that BAGNALL [10] succeeded in preparing an antiserum specific for PSTV. The antigen particles, however, exhibited an unusually high diffusion rate in the gel diffusion test, which suggests that they might represent viral coat protein subunits. If correct, this observation would indicate that PSTV is able to induce the synthesis of a soluble virus-specific protein which might be related to the viral coat protein.

Assuming PSTV is defective, the defect may occur in one of the following ways:

a) PSTV is a virus with a double-stranded RNA like wound tumor virus [19] or rice dwarf virus [88]. But in contrast to these two "normal" plant viruses, PSTV would be defective. It would exist without a viral protein coat, because it produces a soluble but non-functional coat protein.

b) PSTV is a single-stranded RNA plant virus defective in a function(s) essential for the production or release of single stranded progeny RNA from the double stranded replicative intermediate form. In this case even a functional coat protein, if synthesized at all, would accumulate because it could not specifically aggregate with the unusual type of viral RNA present.

Additional investigations will clarify the exact nature of the defect in this interesting virus.

## 9. Defective Animal Viruses

Defective viruses, helper virus systems and complementing multicomponent virus systems have not only been found among plant viruses but also among viruses pathogenic to man and animal. Since tumorigenesis might be related to either a defective or synergistic interaction of defective viruses, investigations along these lines have been intensified recently. Some of the more general problems that might be encountered by defective viruses in mammalian hosts have been discussed by HERRIOTT [53], while the special problems of defectiveness in myxovirus infections have recently been reviewed by FRASER [37].

One of the first indications of a possible correlation between a conditionally defective cycle of viral replication and tumorigenesis was provided by ITO and EVANS [61] and ITO [59, 60]. They found that infectious papilloma DNA can be isolated from tumors free of detectable shope papilloma virus. This DNA was able to produce virus-containing tumors in cotton tail rabbits.

A well known example of a helper virus system is found among the RNA-containing Rous sarcoma viruses (RSV). In chicken cells the Bryan high titer strain of RSV is unable to produce virus infections for chicken cells unless related chicken leucosis viruses are superinfected. These superinfecting viruses provide the competent coat protein for RSV. It had been proposed that the RSV strain is defective in some late viral functions [47a, 48]. It is now evident that the Bryan high titer strain is able to produce infectious virus by itself. But, these particles remained undetected because their homologous coat protein restricts the host range. Although they are non-infectious for chicken cells they can infect cells of the japanese quail [126]. It appears that in the presence of the helper virus most of the RSV genomes are coated with the helper virus envelope, and it is this heterologous protein coat which enables "defective" RSV to infect chicken cells. Therefore, RSV is defective not in an absolute but only in a relative and quantitative sense. A comparable relationship seems to exist between the defective Maloney sarcoma virus (MSV) and certain murine leucosis viruses [52, 58]. The association of a viral genome with a non-homologous coat protein resembles "phenotypic mixing", a process which was first discovered with certain bacteriophage [30a, 90a].

A more complex interaction occurs between the DNA-containing adenoviruses and the PARA virus. These systems resemble the tobacco rattle virus system in function although some remarkable differences are found. Adenoviruses and PARA virus replicate in monkey kidney cells together but each one by itself is unable to replicate in these cells [97, 98, 104]. It was found that different adenoviruses are able to provide the coat protein for the PARA virus, a condition which has been called "transcapsidation" [98]. The PARA virus, on the other hand, provides a yet unknown function for the replication of the adenoviruses. However, the adenoviruses are defective only in monkey kidney cells, and they multiply normally in human kidney cells in which PARA virus is not replicated at all.

PARA virus is a rather unique virus itself because it is supposed to be a hybrid virus consisting of parts of the genome of an adenovirus and parts of the genome of simian virus 40 both wrapped together in a adenovirus protein coat [21, 98, 104]. PARA virus is unable to replicate in the absence of adeno viruses which is probably due to its hybrid and defective nature. From this unusual property the name of PARA virus is derived: Particles aiding and aided in replication of adeno viruses [97].

An other interesting phenomenon is the presence of virus-specific proteins in virus induced tumors. However, neither infectious virus nor virus coat protein are detectable in these tumors [105]. This has led to speculations that the tumor cells may contain only an incomplete or defective viral genome. This hypothesis is supported by the observation that defective PARA virus induces tumors which contain SV 40 specific proteins [97]. It has been shown that parts of viral genomes are present in the tumor cells even after numerous cell divisions and that in such cells a virus specific messenger-RNA is made [39]. There are indications that the genomes or parts of the genome of DNA containing viruses are associated with the genome of the cell. But the manner in which these (defective) viral genomes are present in the tumor cell is as yet unknown. They might be corporated into the chromosomes or be present as episomes. With RNA containing tumor viruses, however, the viral genome is present in such a form that it occasionally may separate from the host cell [87]. This results in a reversion of the tumor cell to a normal cell.

One has to bear in mind that the complete replication of tumorigenic viruses generally results in cell destruction whereas transformation of such cells to tumor cells implies the survival of the infected cells [17, 106]. It, therefore, appears plausible that only a part of the genetic information of tumorigenic virus is expressed in virus-transformed cells. This suppression of certain genetic informations of the virus, however, resembles in some way a functional defect in the expression of the viral genome.

Finally, an interesting parallelity in the structure of the viral genomes should be pointed out. Like in the case of the complementing multi-component plant virus systems the genomes of several animal viruses seem not to function as a continuous linear structure. They rather appear to be similarily multipartite and specialized in various segments with specific functions. This has been shown for the double-stranded RNA of reovirus, which produces three specific classes of messenger RNA in the infected cell [127]. A linear complex of subunits has also been suggested for the structure of influenza virus [13, 30, 54], Newcastle disease virus [54, 74] and vesicular stomatitis virus [25, 56, 57].

## 10. Concluding Remarks

Plant viruses in contrast to animal and bacterial viruses can readily be purified in large quantities which makes them excellent subjects for biochemical and structural investigations. Moreover, the availability of

defective viruses and segments of plant virus genomes found in complementing multi-component systems simplifies investigations of technically
complex problems. Questions concerning the correlation between molecular structure, genetic information and biological function are now open
to attack.

The importance of the results obtained with defective plant viruses
for general virology is obvious. The demonstration that persistent virus
infections are possible with "naked" viruses i.e. without the presence of
complete virus particles may stimulate investigations on certain obscure
viruses and virus-like diseases. Recent investigations demonstrate that
defectiveness seems to play an important role in certain viruses pathogenic
to man and animal. Thus, it has been postulated that tumorigenesis might
be related to defective viruses or the synergistic interaction of defective
viruses.

In the past it was generally assumed that the viral genome needs to
be a continuous linear structure in order to be able to infect and to direct
all the ensuing processes of viral replication. This was based on the observation that a single break of the nucleic acid destroys infectivity. This
concept needs revision since the genomes of several plant and animal
viruses, in analogy to the genomes of higher organisms, appear to be able
to exist and (or) to function in a multipartite state. This multipartite
genome provides additional changes for genetic interaction which increase
the survival potentials of the virus.

## Summary

Plant viruses have been described which exhibit certain defects in a biosynthetic
step of viral replication.

Defectiveness plays an important role in the complementing multi-component
system of tobacco rattle virus (TRV). This virus system is characterized by two
tubular RNA-containing particles of different lengths with only the long particle
being infectious. The "long" RNA of TRV appears to be specialized in early functions of viral replication only. It is able to initiate the synthesis of infectious "long"
progeny RNA but this RNA remains uncoated by a viral protein coat. The "short"
RNA of TRV, however, is specialized in late functions and carries the coat protein
gene of TRV. For its own replication it is dependent on the replicating "long" RNA
Therefore, a complete cycle of viral replication is based on the complementing
interaction of the RNA of an infectious and a non-infectious virus particle of TRV.
As the progeny of the "long" RNA of TRV certain defective coat protein mutants
of tobacco mosaic virus (TMV) and tobacco necrosis virus (TNV) exist in a "naked"
state in the infected cell. Defective TNV seems to produce no coat protein at all. But
in the case of TMV- non-functional coat proteins are synthetized which have lost
their ability to fold in the normal conformation due to certain amino acid replacements in the polypeptide chain.

Satellite virus (SV) is a stable virus defective in a function(s) related with the
synthesis of the viral RNA. Therefore, SV cannot induce its own replication but
requires tobacco necrosis virus (TNV) as a compatible helper virus for this function(s).

The dependency of SV is due to the small size of its RNA, which has only enough of genetic information stored sufficient to code for the SV coat protein.

Defective potato spindle tuber virus is described. Finally, the mechanisms of interaction in some multi-component plant virus systems and some of the condition of defectiveness in animal viruses are discussed.

## Acknowledgements

I wish to thank Miss KARLA RAMM for excellent technical assistance and Dr. GERTRUDE SCHLOER for critical suggestions and aid with the preparation of the manuscript. This work has been supported by the Deutsche Forschungsgemeinschaft.

## References

1. ABEL, P.: Multiplicity reactivation and marker rescue with vaccinia virus. Virology **17**, 511–519 (1962).

2. — Reactivation of heated vaccinia virus in vitro. Z. Vererbungsl. **94**, 249–252 (1963).

3. ALLEN, T. C., JR.: A strain of tobacco rattle virus from Oregon grown potatoes. Plant Dis. Rep. **47**, 920–923 (1963).

4. — Tobacco rattle virus from Oregon compared with pea early browning virus. Phytopathology **54**, 1431 (1964).

5. — Serological relationship between the Oregon strain of tobacco rattle virus and pea early browning virus. Phytopathology **57**, 97 (1967).

6. AMMANN, J., H. DELIUS, and P. H. HOFSCHNEIDER: Isolation and properties of an intact phage-specific replicative form of RNA phage M 12 f. J. molec. Biol. **10**, 557–561 (1964).

7. BABOS, P., and B. KASSANIS: Unstable variants of tobacco necrosis virus. Virology **18**, 206–211 (1962).

8. — — Serological relationships and some properties of tobacco necrosis virus strains. J. gen. Microbiol. **32**, 135–144 (1963).

9. BALD, J. G.: Symptoms and cytology of living cells infected with defective mutants of tobacco mosaic virus. Virology **22**, 388–396 (1964).

10. BAGNALL, R. H.: Serology of the potato spindle tuber virus. Phytopathology **57**, 533–534 (1967).

11. BANCROFT, J. B.: Association of infectivity with alfalfa mosaic virus bottom component virus only. Virology **14**, 296–297 (1961).

12. —, and P. KAESBERG: Size and shape of alfalfa mosaic virus. Nature (Lond.) **181**, 720–721 (1960).

13. BARRY, R. D.: The multiplication of influenza virus. Virology **14**, 398–405 (1961).

14. BAWDEN, F. C.: Speculations on the origins and nature of viruses. In: Plant Virology, p. 371. (Edit. M. K. CORBETT and H. D. SISLER). Gainsville: University of Florida Press 1964.

15. —, and N. W. PIRIE: A preliminary description of preparations of some of the viruses causing tobacco necrosis. Brit. J. exp. Path. **23**, 314 (1942).

16. — — Some factors affecting the activation of virus preparations made from tobacco leaves infected with a tobacco necrosis virus. J. gen. Microbiol. **4**, 464 (1950).

17. BENJAMIN, T. L.: Virus specific RNA in cells productively infected or transformed by polyoma virus. J. molec. Biol. **16**, 359–373 (1966).

18. Benzer, S., and S. P. Champe: Ambivalent rII mutants of phage T4. Proc. nat. Acad. Sci. (Wash.) **47**, 1025—1038 (1961).
19. Black, L. M., and R. Markham: Base-pairing in the ribonucleic acid of wound-tumor virus. Neth. J. Plant Path. **69**, 215 (1963).
20. Bockstahler, L. E., and P. Kaesberg: Isolation and properties of RNA from bromegrass mosaic virus. J. molec. Biol. **13**, 127—137 (1965).
21. Boeyé, A., J. L. Melnick, and F. Rapp: SV40-adenovirus hybrids: Presence of two genotypes and the requirement of their complementation for viral replication. Virology **28**, 56—70 (1966).
22. Bos, L., and J. P. H. van der Want: Early browning of pea, a disease caused by a soil- and seed-borne virus. T. Pl.-Ziekten **68**, 368—390 (1962).
23. Brandenburg, E., R. Eibner u. R. Tostmann: Untersuchungen über die Eisenfleckigkeit-Pfropfenbildung der Kartoffel als bodengebundene Virus-krankheit. Mittlg. BBA. **97**, 36—51 (1959).
24. Bratt, M. A., and W. S. Robinson: Ribonucleic acid synthesis in cells infected with Newcastle disease virus. J. molec. Biol. **23**, 1 (1967).
25. Brown, F., S. J. Martin, B. Cartwright, and J. Crick: The ribonucleic acids of the infective and interfering components of vesicular stomatitis virus. J. gen. Virol. **1**, 479—486 (1967).
26. Bruening, G., and H. O. Agrawal: Infectivity of a mixture of cowpea mosaic virus ribonucleoprotein components. Virology **32**, 306—320 (1967).
27. Cadman, C. H.: Evidence for association of tobacco rattle virus nucleic acid with a cell component. Nature (Lond.) **193**, 49—52 (1962).
28. —, and B. D. Harrison: Studies on the properties of soil-borne viruses of the tobacco rattle type occuring in Scotland. Ann. appl. Biol. **47**, 542—556 (1959).
29. Clark, J. M., jr., A. Y. Chang, S. Spiegelmann, and M. E. Reichmann: The in vitro translation of a monocistronic message. Proc. nat. Acad. Sci. (Wash.) **54**, 1193—1197 (1965).
30. Davies, P., and R. D. Barry: Nucleic acid of influenza virus. Nature (Lond.) **211**, 384 (1966).
30a. Delbrück, M., and W. T. Bailey, Jr.: Induced mutations in bacterial viruses. Cold. Spr. Harb. Symp. quant. Biol. **11**, 33—37 (1946).
31. Dezoeten, G. A.: California tobacco rattle virus, its intracellular appearance and the cytology of the infected cell. Phytopathology **56**, 744—754 (1966).
32. Diener, T. O., and W. B. Raymer: Potato spindle tuber virus: A plant virus with properties of a free nucleic acid. Science **158**, 378—381 (1967).
33. Eibner, R.: Untersuchungen über die „Eisenfleckigkeit" der Kartoffel. Unpublished dissertation, University Gießen 1959.
34. Epstein, R. E., A. Bollé, C. M. Steinberg, E. Kellenberger, E. Boy de la Tour, R. Chevalley, R. S. Edgar, M. Susman, G. Denhardt, and A. Lielausis: Physiological studies on conditional lethal mutations of bacteriophage T4D. Cold. Spr. Harb. Symp. quant. Biol. **28**, 375—394 (1963).
35. Fenwick, M. L., R. L. Erikson, and R. M. Franklin: Replication of the RNA of bacteriophage R17. Science **146**, 527—530 (1964).
36. Fraenkel-Conrat:, H. The role of the nucleic acid in the reconstitution of active tobacco mosaic virus. J. Amer. chem. Sci. **78**, 882—883 (1956).
37. Fraser, K. B.,: Defective and delayed myxovirus infections. Brit. med. Bull. **23**, 178—184 (1967).
38. Frost, R. R., B. D. Harrison, and R. D. Woods: Apparent symbiotic inter-action between particles of tobacco rattle virus. J. gen. Virol. **1**, 57—69 (1967).

39. Fujinago, K., and M. Green: The mechanism of viral carcinogenesis by DNA mammalian viruses, viral-specific RNA in polyribosomes of adenovirus tumor and transformed cells. Proc. nat. Acad. Sci. (Wash.) 55, 1567—1574 (1966).

40. Fulton, R. W.: Purification and some properties of tobacco streak and tulare apple mosaic viruses. Virology 32, 153—162 (1967).

41. — The effect of dilution on necrotic ringspot virus infectivity and the enhancement of infectivity by noninfective virus. Virology 18, 477—485 (1962).

42. Funatsu, G., and H. Fraenkel-Conrat: Location of amino acid exchanges in chemically evoked mutants of tobacco mosaic virus. Biochemistry 3, 1356—1362 (1964).

43. Gibbs, A. J., H. L. Nixon, and R. D. Woods: Properties of purified preparations of lucerne mosaic virus. Virology 19, 441—449 (1963).

44. —, and B D Harrison: A form pea-early-browning virus found in Britain. Ann. appl. Biol. 54, 1—11 (1964).

45. Gierer, A., and G. Schramm: Infectivity of ribonucleic acid from tobacco mosaic virus. Nature (Lond.) 177, 702—703 (1956).

46. —, and K. W. Mundry: Production of mutants of tobacco mosaic virus by chemical alteration of its ribonucleic acid in vitro. Nature (Lond.) 182, 1457—1458 (1958).

47. Gillaspie, A. G., and J. B. Bancroft: Properties of ribonucleic acid from alfalfa misaic virus and related components. Virology 27, 391—397 (1965).

47a. Hanafusa, H., T. Hanafusa, and H. Rubin: The defectiveness of rous sarcoma virus. Proc. nat. Acad. Sci. (Wash.) 49, 572—580 (1963).

48. Hanafusa, T., and H. Hanafusa: Interaction among avion tumor viruses giving enhanced infectivity. Proc. nat. Acad. Sci. (Wash.) 58, 818—825 (1967).

49. Harrison, B. D., and H. L. Nixon: Separation and properties of particles of tobacco rattle virus with different length. J. gen. Microbiol. 21, 569—581 (1959).

50. — — Purification and electron microscopy of three soil-borne plant viruses. Virology 12, 104—117 (1960).

51. —, and R. D. Woods: Serotypes and particle dimensions of tobacco rattle viruses from Europe and America. Virology 29, 610—620 (1966).

52. Hartley, J. W., and W. P. Rowe: Production of altered cell free in tissue culture by defective Moloney sarcoma virus particles. Proc. nat. Acad. Sci. (Wash.) 55, 780—786 (1966).

53. Herriott, R. M.: Infectious nucleic acids, a new dimension in virology. Science 134, 256—260 (1961).

54. Hirst, G. K.: Genetic recombination with Newcastle disease virus, polioviruses and influenca. Cold Spr. Harb. Symp. quant. Biol. 27, 303 (1962).

55. Hoggan, P., W. P. Rowe, P. H. Black, and R. J. Huebner: Production of "tumor-specific" antigens by oncogenic viruses during cytolytic infections. Proc. nat. Acad. Sci. U.S. (Wash.) 53, 12—19 (1965).

56. Huang, A. S., J. W. Greenawalt, and R. R. Wagner: Defective T-particles of vesicular stomatitis virus. I. Preparation, morphology and biologic properties. Virology 30, 161 (1966).

57. —, and R. R. Wagner: Defective T-particles of vesicular stomatitis virus. II. Biologic role in homologous interference. Virology 30, 173 (1966).

58. Huebner, R.: The murine leukemia-sarcoma virus complex. Proc. nat. Acad. Sci. (Wash.) 58, 835—842 (1967).

59. Ito, Y.: A tumor producing factor extracted by phenol from papillomatous tissue (Shope) of cottontail rabbits. Virology 12, 596 (1960).

60. — Relationship of components of papilloma virus to papilloma and carcinoma cells. Cold Spr. Harb. Symp. quant. Biol. 27, 387—394 (1962).

61. —, and C. A. Evans: Induction of tumors in domestic rabbits with nucleic acid preparations from partially purified Shope papilloma virus and from extracts of the papillomas of domestic and cottontail rabbits. J. exp. Med. 114, 485—500 (1961).

62. Jaspars, E. M. J., and J. B. Moed: The complexity of alfalfa mosaic virus. In: Viruses of plants, pp. 188—195. (Beemster, A. B. R., and J. Dijkstra, eds.) Amsterdam: North-Holland Publ. Comp. 1966.

63. Jokusch, H.: In Vivo- und In Vitro-Verhalten temperatursensitiver Mutanten des Tabakmosaikvirus. Z. Vererbungsl. 95, 379—382 (1964).

64. — Relations between temperature sensitivity, amino acid replacements, and quaternary structure of mutant proteins. Biochem. biophys. Res. Commun. 24, 577—583 (1966).

65. — Temperatursensitive Mutanten des Tabakmosaikvirus. I. In Vivo-Verhalten. Z. Vererbungsl. 98, 320—343 (1966).

66. — Temperatursensitive Mutanten des Tabakmosaikvirus. II. In Vitro-Verhalten. Z. Vererbungsl. 98, 344—362 (1966).

67. Kassanis, B.: Properties and behaviour of a virus depending for its multiplication on another. J. gen. Microbiol. 27, 477—488 (1962).

68. — Properties of tobacco necrosis virus and its association with satellite virus. Ann. Inst. Phytopath. Benaki 6, 7—26 (1965).

69. — Properties and behaviour of satellite virus. In: Viruses of plants, pp. 177 to 187. Amsterdam: North-Holland. Publ. Comp. 1966.

70. —, and H. L. Nixon: Activation of one plant virus by another. Nature (Lond.) 187, 713 (1960).

71. — — Activation of one tobacco necrosis virus by another. J. gen. Microbiol. 25, 459—471 (1961).

72. —, and G. W. Welkie: The nature and behaviour of unstable variants of tobacco necrosis virus. Virology 21, 540—550 (1963).

73. Kelly, J. J., and P. Kaesberg: Biophysical and biochemical properties of top component a and bottom component of alfalfa mosaic virus. Biochim. biophys. Acta (Amst.) 61, 865—871 (1962).

74. Kingsbury, D. W.: Newcastle disease virus RNA. I. Isolation and preliminary characterisation of RNA from virus particles. J. molec. Biol. 18, 195 (1966).

75. Knight, C. A.: Chemistry of viruses. In: Protoplasmatologia. Handbuch der Protoplasmaforschung. Wien: Springer 1963.

76. Köhler, E.: Über eine reversible, durch die Jahreszeit induzierte Virulenzänderung beim Tabak-Rattle-Virus. Nachr. Bl. dtsch. Pfl.Sch.D. 8, 93—94 (1956).

77. — Die Viren des Kartoffel-Stengelbunt (Tabak-Rattle) und der Pfropfenbildung (Spraing). Eine Stellungnahme zur Frage ihrer Verwandtschaft. Nachr.Bl. dtsch. Pfl.Sch.D. 12, 87—88 (1960).

78. Lister, R. M.: Possible relationships of virus-specific products of tobacco rattle virus infection. Virology 28, 350—353 (1966).

79. — A symptomatological difference between some unstable and stable variants of pea early browning virus. Virology 31, 739—742 (1967).

80. Luria, S. E.: Reactivation of irradiated bacteriophage by transfer of self-reproducing units. Proc. nat. Acad. Sci. (Wash.) 33, 253—264 (1947).

81. Luria, S. E., and R. Dulbecco: Genetic recombinations leading to the production of active bacteriophage from ultraviolet inactivated bacteriophage particles. Genetics 34, 93–125 (1949).
82. Maat, D. Z.: Pea early-browning virus and tobacco rattle virus — two different, but serologically related viruses. Neth. J. Plant Path. 69, 287–293 (1963).
83. Markham, R., and L. M. Smith: Studies on the virus of turnip yellow mosaic. Parasitology 39, 330–342 (1949).
84. Matthews, R. E. F.: Evidence for steps in the assembly of turnip yellow mosaic virus nucleoprotein. Biochem. biophys. Res. Commun. 1, 165–170 (1959).
85. — Turnip yellow mosaic virus nucleoprotein particles with differing biological and physical properties. Nature (Lond.) 184, 530–531 (1959).
86. —, and R. K. Ralph: Turnip yellow mosaic virus. Advanc. Vir. Res. 12, 273–328 (1966).
87. McPherson, J.: Malignant transformation and reversion in virus infected cells. In: Recent results in cancer research, pp. 1–11. Berlin-Heidelberg-New York 1966.:Springer 1966.
88. Miura, K., I. Kimura, and N. Suzuki: Double-stranded ribonucleic acid from rice dwarf virus. Virology 28, 571–579 (1966).
89. Montagnier, L., and F. K. Sanders: Replicative form of encephalomyocarditis virus ribonucleic acid. Nature (Lond.) 199, 664–667 (1963).
90. Mundry, K. W., u. A. Gierer: Die Erzeugung von Mutationen des Tabakmosaikvirus durch chemische Behandlung seiner Nucleinsäure in vitro. Z. Vererbungsl. 89, 614–630 (1958).
90a. Novick, A., and L. Szilard: Virus strains of identical phenotype but different genotype. Science 113, 34–35 (1951).
91. Nyland, G.: Possible virus-induced genetic abnormalities in tree fruits. Science 137, 598–599 (1962).
92. Offord, R. E.: Electron microscopic observation on the substructure of tobacco rattle virus. J. molec. Biol. 17, 370–375 (1966).
93. —, and J. I. Harris: The protein sub-unit of tobacco rattle virus. Fed.European Biochem. Soc. 2nd Meeting; Abstracts, p. 216–217. Wien: Verlag der Wiener Mediz. Akad. 1965.
94. Parish, C. L., and M. Zaitlin: Defective tobacco mosaic virus strains: Identification of the protein of strain PM1 in leaf homogenates. Virology 30, 297–302 (1966).
95. Paul, H. L., u. O. Bode: Elektronenmikroskopische Untersuchungen über Kartoffelviren. II. Vermessung der Teilchen von drei Stämmen des Rattle-Virus. Phytopath. Z. 24, 341–351 (1955).
96. Ralph, R. K., R. E. F. Matthews, A. I. Matus, and H. G. Mandel: Isolation and properties of double-stranded viral RNA from virus-infected plants. J. molec. Biol. 11, 202–212 (1965).
97. Rapp, F.: Complementation between defective oncogenic viruses. In: Recent results in Cancer Res. Vol. 6, pp. 77–94. Berlin-Heidelberg-New York: Springer 1966.
98. — J. S. Butel, and J. L. Melnik: SV40-adenovirus "hybrid" populations: Transfer of SV40 determinants from one type of adenovirus to another. Proc. nat. Acad. Sci. (Wash.) 54, 717–724 (1965).
99. Rappaport, I., and M. Zaitlin: Antigenic study of the protein from a defective strain of tobacco mosaic virus. Science 157, 107–108 (1967).
100. Ravenswaay Claasen, J. C., van: Synthesis of plant viral specific protein in the cell-free system of Escherichia coli. Thesis Univ. of Leiden 1967.

101. Ravenswaay Claasen, J. C., van, A. B. J. van Leeuwen, G. A. H. Duijts, and L. Bosch: In vitro translation of alfalfa mosaic virus RNA. J. molec. Biol. 23, 535—544 (1957).
102. Reichmann, M. E.: The satellite tobacco necrosis virus: A single protein and its genetic code. Proc. nat. Acad. Sci. (Wash.) 52, 1009—1017 (1964).
103. — M. W. Rees, R. H. Symons, and R. Markham: Experimental evidence for the degeneracy of the nucleotide triplet code. Nature (Lond.) 195, 999—1000 (1962).
104. Rowe, W. P.: Studies of adenovirus-SV40 hybrid viruses: III. Transfer of SV40 gene between adenovirus types. Proc. nat. Acad. Sci. (Wash.) 54, 711—717 (1965).
105. Sabin, A. B., and M. A. Koch: Source of genetic information for specific complement-fixing antigens in SV40 virus-induced tumors. Proc. nat. Acad. Sci. (Wash.) 52, 1131—1138 (1964).
106. Sachs, L.: A theory on the mechanism of carcinogenesis by small deoxyribonucleic acid tumor viruses. Nature (Lond.) 207, 1272—1274 (1965).
107. Sänger, H. L.: Untersuchungen über die beiden charakteristischen Partikel des Ratelvirus und das serologische Verhalten verschiedener Isolate. Unpublished dissertation, University Gießen 1960.
108. — Untersuchungen über schwer übertragbare Formen des Rattle-Virus. Proc. 4th Conf. Potato Virus Diseases, Braunschweig, pp. 22—29 (1960).
109. — Characteristics of tobacco rattle virus. I. Evidence that its two particles are functionally defective and mutually complementing. Molec. Gen. Genetics 101, 346—367 (1968).
110. — Characteristics of tobacco rattle virus. II. Nature and behaviour of defective tobacco rattle virus. In preparation (1968).
111. —, u. E. Brandenburg: Über die Gewinnung von infektiösem Preßsaft aus „Wintertyp"-Pflanzen des Tabak-Rattle-Virus durch Phenolextraktion. Naturwissenschaften 48, 391 (1961).
112. Semancik, J. S., and M. R. Kajiyama: Comparative studies on two strains of tobacco rattle virus. J. gen. Virol. 1, 153—162 (1967).
112a. Siegel, A.: Studies on the induction of tobacco virus mutants with nitrons acid. Virology 11, 156—167 (1960).
113. —, and M. Zaitlin: Defective plant viruses. In: Perspectives in virology, Vol. 4, pp. 113—125. New York: Harper & Row 1965.
114. — —, and I. P. Seghal: The isolation of defective tobacco mosaic virus strains. Proc. nat. Acad. Sci. (Wash.) 49, 1845—1851 (1962).
115. Schmidt, P., P. Kaesberg, and W. W. Beemann: Small-angle X-ray scattering from turnip yellow mosaic virus. Biochim. biophys. Acta (Amst.) 14, 1—11 (1954).
116. Scholtissek, C., R. Rott, u. W. Schäfer: Verhalten von Viren gegenüber dem Bayer Präparat A139. Z. Naturforsch. 17b, 222—227 (1962).
117. Takahashi, W. N.: The role of an anomalous noninfectious protein virus synthesis. Virology 9, 437—445 (1959).
118. —, and H. Gold: Serological studies with X protein, tobacco mosaic virus, polymerized X protein, and virus reconstituted from nucleic acid and X protein. Virology 10, 449—458 (1960).
119. —, and M. Ishi: A macromolecular protein associated with tobacco mosaic virus infection: Its isolation and properties. Amer. J. Bot. 40, 85—90 (1953).
120. Tomita, K., and A. Rich: X-ray diffraction investigations of complementary RNA. Nature (Lond.) 201, 1160—1163 (1964).

121. Tsugita, A., and H. Fraenkel-Conrat: The composition of proteins of chemically evoked mutants of TMV RNA. J. molec. Biol. **4**, 73—82 (1962).
122. — D. T. Gish, J. Young, H. Fraenkel-Conrat, C. A. Knight, and W. M. Stanley: The complete amino acid sequence of the protein of tobacco mosaic virus. Proc. nat. Acad. Sci. (Wash.) **46**, 1436—1469 (1960).
123. Vloten-Doting, L. van, and E. M. J. Jaspars: Enhancement of infectivity by combination of two ribonucleic acid components from alfalfa mosaic virus. Virology **33**, 684—693 (1967).
124. Vloten-Doting, L. van, and E. M. J. Jaspars: Personal communication.
125. — — On the infectivity of the RNA's from alfalfa mosaic virus. Fed. of European Biochem. Soc. 4th Meeting, Oslo. Universitätsvorlaget, Abstracts-Nr. 446, p. 112 (1967).
126. Vogt, P.: A virus released by "nonproducing" rous sarcoma cells. Proc. nat. Acad. Sci. (Wash.) **58**, 801—808 (1967).
127. Watanabe, Y., L. Prevec, and A. F. Graham: Specificity in transcription of the reovirus genome. Proc. nat. Acad. (Wash.) **58**, 1040—1046 (1967).
128. Wittmann, H. G.: Proteinuntersuchungen an Mutanten des Tabakmosaik-virus als Beitrag zum Problem des genetischen Codes. Z. Vererbungsl. **93**, 491—530 (1962).
129. — Proteinanalysen von chemisch induzierten Mutanten des Tabakmosaik-virus. Z. Vererbungsl. **95**, 333—344 (1964).
130. — Ein Stamm des Tabakmosaikvirus mit freier N-terminaler Aminosäure in der Hüllproteinkette. Z. Vererbungsl. **97**, 204—208 (1965).
131. — Die Proteinstruktur der Defektmutanten PM2 des Tabakmosaikvirus. Z. Vererbungsl. **97**, 297—304 (1965).
132. Wittmann-Liebold, B., J. Jauregui u. H. G. Wittmann: Die primäre Proteinstruktur temperatursensitiver Mutanten des Tabakmosaikvirus. II. Chemisch induzierte Mutanten. Z. Vererbungsl. **20b**, 1235—1249 (1965).
133. Wittmann, H. G., u. H. L. Sänger: Phytopathogene Viren. Fortschr. Bot. **27**, 250—290 (1965).
134. Wood, H. A., and J. B. Bancroft: Activation of a plant virus by related incomplete nucleoprotein particles. Virology **27**, 94—102 (1965).
135. Yamazaki, H., J. Bancroft, and P. Kaesberg: Biophysical studies of broad bean mottle virus. Proc. nat. Acad. Sci. (Wash.) **47**, 979—983 (1961).
136. —, and P. Kaesberg: Biophysical and biochemical properties of wild cucumber mosaic virus and two related virus-like particles. Biochem. biophys. Acta (Amst.) **51**, 9—18 (1961).
137. Zaitlin, M., and W. R. Ferris: Unusual aggregation of a nonfunctional tobacco mosaic virus protein. Science **143**, 1451—1452 (1964).
138. —, and C. L. Keswani: Relative infectivities of tissue containing a defective tobacco mosaic virus strain. Virology **24**, 495—498 (1964).
139. —, and W. F. McCaughey: Amino acid composition of a nonfunctional tobacco mosaic virus protein. Virology **26**, 500—503 (1965).

## Discussion

*Schuster:* Is it possible to degrade the small RNA particles by an exonuclease (for example snake venome diesterase) to a certain extent without loosing its activity?

*Sänger:* I have not tried it yet, so I cannot tell; UV-light destroys activity, but this is not relevant.

*Arber:* I guess you looked for very large particles, that could contain perhaps an RNA molecule formed by sealing the "long" and the "short" RNA and thus form an active virus.

*Sänger:* A distinct fraction of particles containing the long and short RNA in a joint linear structure has not been found in TRV preparations. It is puzzling that all TRV isolates found so far in different continents do contain at least two particles of different length. There is also no TRV-like "ancestor"-virus from which TRV could have originated by assuming a breakage of the genome.

*Bockstahler:* What happens of you add a mixture of short German and short USA particles to one of the large (German or USA) particles?

*Sänger:* A mixed TRV with 2 fractions of short particles and the fraction of the corresponding long particle is obtained.

# Gene Functions of Influenza Virus

Institut für Virologie, Justus Liebig-Universität, Gießen

Influenza viruses are RNA viruses with a rather complex structure [1]. According to the amount of RNA per virus particle its molecular weight is about 2,000,000; but at present this RNA has not been isolated in a single piece. All reports concerning a high molecular weight RNA from influenza virus seem to deal with aggregate formation of the RNA [2, 3, 4]. Because of the extremely high rate of recombination it is believed that the RNA of these viruses has either preferential breakage points or consists of loosely bound subunits [3, 5].

There are some properties, which are unique for influenza viruses and are not shared by e.g. parainfluenza viruses:

a) The high rate of multiplicity reactivation [6, 7, 8].

b) The formation of genetically incomplete virus after passage of the virus at high multiplicity [9].

Since these incomplete viruses do not show any sign of multiplicity reactivation as found after chemical or UV-inactivation, it has been suggested that all these incomplete viruses have lost at least one identical piece of this RNA [10, 11].

c) The third unusual property is the ability to inactivate the virus stepwise (Fig. 1) [12, 13].

A short treatment of fowl plague virus (which is an influenza A virus) with the ethylene iminoquinone Bayer A 139, which destroys nucleic acids without reacting significantly with proteins [8, 14], abolishes the infectivity. However, the capacity to induce the synthesis of the various viral subunits is not lost. Longer treatment destroys the capacity to produce hemagglutinin, a viral shell component. The next capacity lost is the ability to form viral neuraminidase, another virus-coded shell component. This is followed by the inability to synthesize a viral component which reacts positive in the complement fixation (CF) test with recon-valescent FM 1-antiserum. This test is believed to measure the viral inner component (RNP-antigen), but should be positive also with any antigen common between fowl plague virus and influenza FM 1-strain. Therefore this viral product will be called "common antigen". The capacity to

22 4. Symposium Naturforscher

synthesize viral RNA is lost in parallel with that of the "common antigen". With the latter test we measure the intactness of the gene which codes for the RNA-dependent RNA polymerase ("early protein").

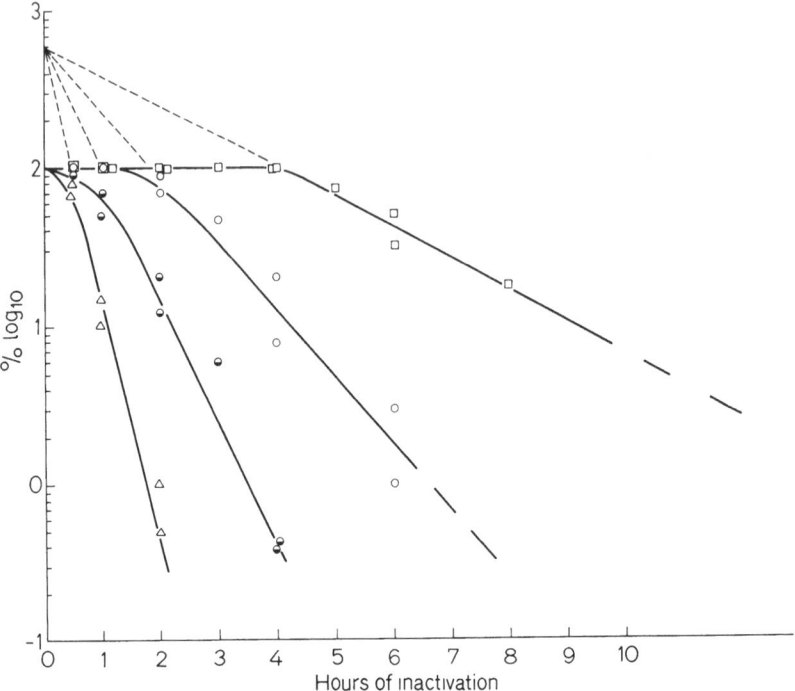

Fig. 1. Synthesis of viral components by cells infected with stepwise inactivated fowl plague virus [13]. △ = plaque forming units; ○ = neuraminidase units; ◕ = hemagglutinating units; □ = "common antigen" (complement fixation test). The curve representing the synthesis of viral RNA follows that of the "common antigen" [12]

The multiplicity of infection in Fig. 1 was 6 PFU per cell in order to have synchroneously infected cells and to avoid chance delay effects. From the data presented in Fig. 1 the target size of the various capacities can be determined. The target size for the hemagglutinin is about half of that for the infectivity etc. These targets can be arranged in two different ways. It is assumed that the total amount of the viral RNA is necessary for the production of infectious virus particles. Half of the RNA is necessary for the synthesis of the hemagglutinin etc. In scheme 1 the targets are arranged one after the other along the viral RNA. This means that the synthesis or biological activity of each viral subunit is independent of the presence of another subunit or intactness of the corresponding piece of RNA. Scheme 2 shows an overlapping arrangement of the various targets. This scheme would mean that the synthesis of the first viral subunit is indispensable for the synthesis or biological activity of the next subunit. For example, if that piece of RNA (gene) responsible for the synthesis of the RNA-dependent RNA-polymerase is destroyed, we do not expect the synthesis of viral RNA and therefore not of any other viral product in detectable amounts, independent of whether or not the residual RNA of the invading virus is still intact. At least for the target of the "early protein" the overlapping arrangement appears to be correct.

Since the molecular weight of the RNA is only about 2,000,000, then the targets are already in the order of magnitude of the size of genes. Taking advantage of scheme 2 for the "early protein" one comes to the reasonable figure of 220 amino acids per "functional unit", if one assumes that the size of the target is equal to the corresponding gene.

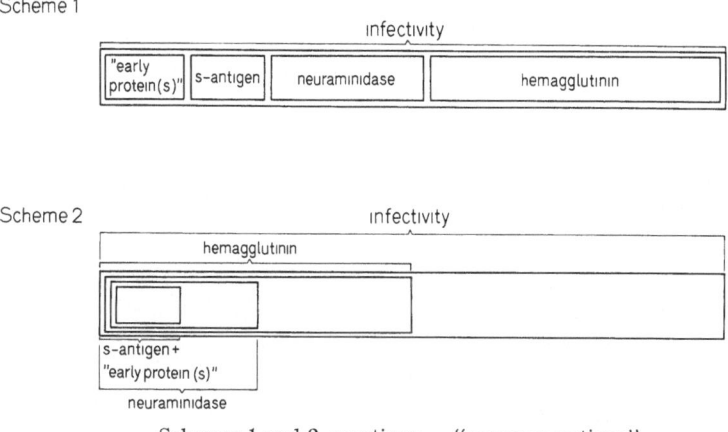

Scheme 1

Scheme 2

Schemes 1 and 2. s-antigen = "common antigen"

Using the method of stepwise inactivation it was found that a viral component with a target size of that of neuraminidase is responsible for the cytopathic effect [15]. This component is probably not the neuraminidase itself, since some myxoviruses, which induce the synthesis of neuraminidase, do not show any sign of cytopathogenicity. However, the important point is that it is possible to inactivate the virus so far that some of the infected cells survive but still produce one or a few viral products. The question is now, what happens to these cells? The following experiment was done [16]: Virus was inactivated for 3 h with Bayer A 139. This prolonged treatment destroys most of the capacity to kill cells. When HeLa cells, which were used for technical reasons in this experiment, were infected with this inactivated virus, then about 20 % of the infected

22*

HeLa cells survived and grew to again form a monolayer. These cells were subdivided into three new plates and thus were serially cultured. The results are presented in Fig. 2. After two passages the ability to synthesize

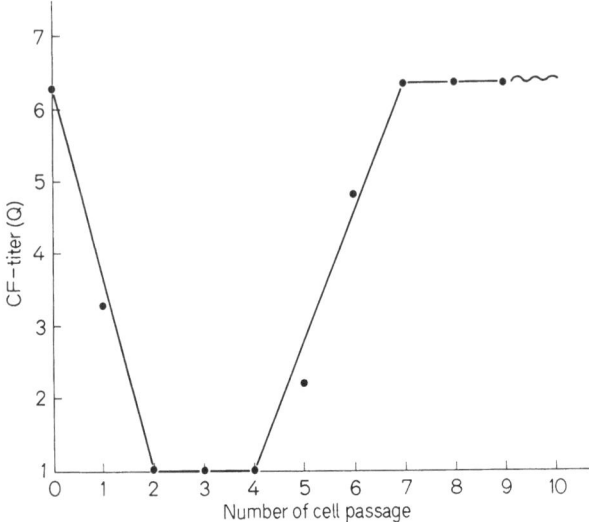

Fig. 2. Passages of HeLa cells infected with fowl plague virus which was inactivated by Bayer A 139 for 3 h

the "common antigen" (CF-test) was lost, but it reappeared after 5 passages. At this time the cells grew about twice as fast as non-infected HeLa cells. After the "common antigen" titer had reached its maximum, the cells stopped dividing. The reason for this is not yet known. Infectious particles, hemagglutinin, or neuraminidase were not found in these cells, which synthesized only a viral product positive in the CF-test (RNP-antigen and/or RNA-dependent RNA-polymerase?). The interpretation of this experiment is that only a few cells obtained the message to produce the "common antigen" and at the same time to grow faster than the other cells. The reason to stop growing might be that there is no good equilibrium between cell growth and synthesis of the "common antigen". The switch to produce the "common antigen" and to overgrow the other cells which is a rather rare event somehow resembles cell transformation and can be revealed only, if one avoids the cytopathic effect. With other words, there might exist several viruses with oncogenic capacities, which normally kill the cell and therefore do not exhibit this property.

In summary, influenza viruses exhibit some properties, which are in agreement with the idea, that their genetic material (RNA) exists in fixed subunits which can easily be exchanged and can function more or less independently. By this way it is possible to obtain infected cells, which carry the information of only one or a few viral products.

## References

1. ROTT, R., and C. SCHOLTISSEK: In: Mod. Trends Med. Virology 1, 25 (1967).
2. AGRAWAL, H. O., and G. BRUENING: Proc. nat. Acad. Sci. (Wash.) 55, 818 (1966).
3. PONS, M. W.: Virology 31, 523 (1967).
4. DUESBERG, P. H., and W. S. ROBINSON: J. mol. Biol. 25, 383 (1967).
5. HIRST, G. K.: Cold Spr. Harb. Symp. quant. Biol. 27, 303 (1962).
6. HENLE, W., and O. C. LIU: J. exp. Med. 94, 305 (1951).
7. BARRY, R. D.: Virology 14, 398 (1961).
8. SCHOLTISSEK, C., R. ROTT, and W. SCHÄFER: Z. Naturforsch. 17b, 222 (1962).
9. MAGNUS, P. VON: Advanc. Virus Res. 2, 59 (1954).
10. ROTT, R., and C. SCHOLTISSEK: J. gen. Microbiol. 33, 303 (1963).
11. SCHOLTISSEK, C., R. DRZENIEK, and R. ROTT: Virology 30, 313 (1966).
12. —, and R. ROTT: Nature (Lond.) 199, 200 (1963).
13. — — Virology 22, 169 (1964).
14. — Z. Krebsforsch. 62, 109 (1957).
15. — H. BECHT, and R. DRZENIEK: J. gen. Virol. 1, 219 (1967).
16. ROTT, R., and C. SCHOLTISSEK: unpublished.

# SPRINGER-VERLAG
## BERLIN · HEIDELBERG · NEW YORK

# Funktionelle und morphologische Organisation der Zelle

Wissenschaftliche Konferenz der Gesellschaft Deutscher Naturforscher und Ärzte in Rottach-Egern 1962. Herausgegeben von **P. Karlson**, Physiologisch-Chemisches Institut der Universität München.
Mit 91 Abbildungen. IV, 253 Seiten. (9 Beiträge in deutscher und 6 Beiträge in englischer Sprache). 1963.
Geheftet DM 36,—; US $ 9.00

# Sekretion und Exkretion

**Funktionelle und morphologische Organisation der Zelle.**
2. wissenschaftliche Konferenz der Gesellschaft Deutscher Naturforscher und Ärzte, Schloß Reinhardsbrunn bei Friedrichroda 1964.
Herausgegeben von **K. E. Wohlfahrt-Bottermann**, Bonn.
Mit 180 Abbildungen. XII, 404 Seiten (davon 100 Seiten in englischer Sprache). 1965.
Geheftet DM 58,—; US $ 14.50

# Probleme der biologischen Reduplikation

**Funktionelle und morphologische Organisation der Zelle.**
3. wissenschaftliche Konferenz der Gesellschaft Deutscher Naturforscher und Ärzte, Semmering bei Wien 1965.
Herausgegeben von **P. Sitte**, Heidelberg.
Mit 142 Abbildungen. VIII, 412 Seiten (davon 96 Seiten in englischer Sprache). 1966.
Geheftet DM 58,—; US $ 14.50